E-Learning and Virtual Science Centers

Leo Tan Wee Hin
Nanyang Technological University, Singapore

R. Subramaniam
Nanyang Technological University, Singapore

 Information Science Publishing

Hershey • London • Melbourne • Singapore

Acquisitions Editor:	Renée Davies
Development Editor:	Kristin Roth
Senior Managing Editor:	Amanda Appicello
Managing Editor:	Jennifer Neidig
Copy Editor:	Alana Bubnis
Typesetter:	Cindy Consonery
Cover Design:	Lisa Tosheff
Printed at:	Integrated Book Technology

Published in the United States of America by
Information Science Publishing (an imprint of Idea Group Inc.)
701 E. Chocolate Avenue
Hershey PA 17033
Tel: 717-533-8845
Fax: 717-533-8661
E-mail: cust@idea-group.com
Web site: http://www.idea-group.com

and in the United Kingdom by
Information Science Publishing (an imprint of Idea Group Inc.)
3 Henrietta Street
Covent Garden
London WC2E 8LU
Tel: 44 20 7240 0856
Fax: 44 20 7379 3313
Web site: http://www.eurospan.co.uk

Library of Congress Cataloging-in-Publication Data

E-learning and virtual science centers / Leo Tan and R. Subramaniam, editors.
    p. cm.
  Summary: "The book provides an overview of the state-of-the-art developments in the new and emerging field of science education, called virtual science centers"-- Provided by publisher.
  Includes bibliographical references and index.
  ISBN 1-59140-591-2 (hardcover) -- ISBN 1-59140-592-0 (softcover) -- ISBN 1-59140-593-9 (ebook)
  1. Science--Study and teaching--Computer network resources. 2. Internet in education. I. Leo Tan, 1944- II. R. Subramaniam, 1952-
  Q182.7.E18 2005
  025.06'5--dc22

                          2005004518

British Cataloguing in Publication Data
A Cataloguing in Publication record for this book is available from the British Library.

All work contributed to this book is new, previously-unpublished material. The views expressed in this book are those of the authors, but not necessarily of the publisher.

E-Learning and Virtual Science Centers

Table of Contents

Foreword vii
    Bill Peters, The Calgary Science Centre, Canada

Preface x
    Leo Tan Wee Hin, Nanyang Technological University, Singapore
    R. Subramaniam, Nanyang Technological University, Singapore

Section I: Theoretical Issues

Chapter I. Establishing Identification in Virtual Science Museums: Creating Connections and Community 1
    Billie J. Jones, Shippensburg University, USA

Chapter II. Free-Choice Learning Research and the Virtual Science Center: Establishing a Research Agenda 28
    Kathryn Haley Goldman, Institute for Learning Innovation, USA
    Lynn D. Dierking, Institute for Learning Innovation, USA

Chapter III. Contextualized Virtual Science Centers 51
    Andreas Zimmermann, Fraunhofer-Institute for Applied Information Technology, Germany
    Andreas Lorenz, Fraunhofer-Institute for Applied Information Technology, Germany
    Marcus Specht, Fraunhofer-Institute for Applied Information Technology, Germany

Chapter IV. Starting With What We Know: A CILS Framework for Moving from Physical to Virtual Science Learning Environments 68

    Bronwyn Bevan, Exploratorium, USA

Chapter V. Weaving Science Webs: E-Learning and Virtual Science Centers 93

    Susan Hazan, Israel Museum, Jerusalem

Chapter VI. Resource-Based Learning and Informal Learning Environments: Prospects and Challenges 110

    Janette R. Hill, University of Georgia, USA
    Michael J. Hannafin, University of Georgia, USA
    Denise P. Domizi, University of Georgia, USA

Section II: Design Considerations

Chapter VII. Interactivity Techniques: Practical Suggestions for Interactive Science Web Sites 127

    Michael Douma, Institute for Dynamic Educational Advancement, USA
    Horace Dediu, Handheld Media, USA

Chapter VIII. From the Physical to the Virtual: Bringing Free-Choice Science Education Online 163

    Steven Allison-Bunnell, Educational Web Adventures, LLP, USA
    David T. Schaller, Educational Web Adventures, LLP, USA

Chapter IX. Storytelling-Based Edutainment Applications 190

    Anja Hoffmann, ZGDV e.V. - Computer Graphics Center, Darmstadt, Germany
    Stefan Göbel, ZGDV e.V. - Computer Graphics Center, Darmstadt, Germany
    Oliver Schneider, ZGDV e.V. - Computer Graphics Center, Darmstadt, Germany
    Ido Iurgel, ZGDV e.V. - Computer Graphics Center, Darmstadt, Germany

Chapter X. Revolutionizing Information Architectures within Learning-Focused Web Sites 215

    Ramesh Srinivasan, Harvard University, USA

Chapter XI. From Information Dissemination to Information Gathering: Using Virtual Exhibits and Content Databases in E-Learning Centers 228

    Joan C. Nordbotten, University of Bergen, Norway

Chapter XII. Challenges in Virtual Environment Design: An Architectural Approach to Virtual Spaces 251

    Renata Piazzalunga, Information Technology Research Institute, Brazil
    Saulo Faria Almeida Barretto, Information Technology Research Institute, Brazil

Chapter XIII. Personalization Issues for Science Museum Web Sites and E-learning 272

    Silvia Filippini-Fantoni, The University of Paris I Sorbonne, France
    Jonathan P. Bowen, London South Bank University, UK
    Teresa Numerico, London South Bank University, UK

Chapter XIV. E-Learning and Virtual Science Centers: Designing Technology Supported Curriculum 292

    John Falco, Schenectady City School District, USA
    Patricia Barbanell, Schenectady City School District, USA
    Dianna Newman, State University of New York, USA
    Suzanne Dewald, Schenectady City School District, USA

Section III: Case Studies

Chapter XV. A Virtual Museum Where Students Can Learn 308

    Nicoletta Di Blas, Politecnico di Milano, Italy
    Paolo Paolini, Politecnico di Milano, Italy
    Caterina Poggi, Politecnico di Milano, Italy

Chapter XVI. Open Learning Environments: Combining Web-Based Virtual and Hands-On Science Centre Learning 327

    Hannu Salmi, Heureka, The Finnish Science Centre, Finland and
        University of Dalarna, Sweden

Chapter XVII. Use of Log Analysis and Text Mining for Simple Knowledge Extraction: Case Study of a Science Center on the Web 347

    Leo Tan Wee Hin, Nanyang Technological University, Singapore
    R. Subramaniam, Nanyang Technological University, Singapore
    Daniel Tan Teck Meng, Singapore Science Centre, Singapore

Chapter XVIII. The Development of Science Museum Web Sites: Case Studies 366

    Jonathan P. Bowen, London South Bank University, UK
    Jim Angus, National Institutes of Health, USA
    Jim Bennett, University of Oxford, UK
    Ann Borda, The Science Museum, UK
    Andrew Hodges, University of Oxford, UK
    Silvia Filippini-Fantoni, The University of Paris I Sorbonne, France
    Alpay Beler, The Science Museum, UK

Chapter XIX. The Educational Approach of Virtual Science Centers: Two Web Cast Studies (The Exploratorium and La Cité des Sciences et de l'Industrie) 393

    Roxane Bernier, Université de Montréal, Canada

Chapter XX. Real Science: Making Connections to Research and Scientific Data **423**
   Jim Spadaccini, Ideum, USA

About the Authors **442**

Index **453**

Foreword

Four decades ago, a few visionaries — Frank Oppenheimer of the Exploratorium a leader among them — set out to transform how people learn about science and technology. They were convinced of the merit of transforming individual lives and thus transforming society by engaging a broad public with the ideas, methods and fruits of science and technology. With interactive exhibits as their most engaging tool, Oppenheimer and his colleagues ignited a dynamic, influential learning revolution.

Today the science center revolution is converging with the Internet revolution and virtual science centers are an increasing presence in cyberspace. *E-Learning and Virtual Science Centers* is the first book to document and explore this new phenomenon. It accesses the collective talent of leading scholars to take a snapshot of these converging revolutions, to probe the trends, assumptions and review the work to date. I'm excited about this book because its authors provide a structured platform for thinking as those of us in the profession take the next steps in evolving the virtual science center.

What is a virtual science center? In this rapidly evolving field we can best understand this by exploring the different examples cited throughout this book. At the most basic level the virtual science center is an Internet representation or reflection of an existing, physical institution. Over the Internet we are able to access some of the qualities and experiences offered by the physical institution. Billie Jones provides a number of good examples of this kind of virtual science center in "Establishing Identification in Virtual Science Museums: Creating Connections and Community." There are also excellent examples noted in other chapters.

A growing number of science centers exist only in cyberspace, a trend we can expect to continue. Jones draws our attention to the Leonardo: Interactive Virtual Science Museum, where the exhibits consist of a set of well-organized links to the best interactive science applets on the Internet. A very different, fully virtual example is wonderville.ca, a highly animated site crafted by the Science Alberta Foundation to enhance science learning and career interest among students in grades 4-6.

The science center revolution has inspired vigorous research interest in learning. Because of science centers, the way we understand both formal and "free choice" learning is profoundly more sophisticated — a sophistication reflected in the chapters of this book. I am encouraged to see chapters like Goldman and Dierking's "Free-Choice Learning

Research and the Virtual Science Center: Establishing a Research Agenda", which looks toward extending our quest for knowledge about learning to the virtual science center realm at this early stage of the field's development.

There are other dimensions of knowledge to be brought to the creation of virtual science centers. In the chapter "From the Physical to the Virtual: Bringing Free-Choice Science Education Online", Allison-Bunnell and Schaller point out that virtual science center work to date has sometimes been haphazard in that it is unguided by formal knowledge of information architecture and virtual reality design. I am pleased to note that *E-Learning and Virtual Science Centers* makes a tool kit of ideas and methods available to science center professionals to enable them to address the next stages in the development of the virtual science center in a more intentional manner.

There is also a convergence taking place of a variety of science publishers, filmmakers and media producers with the Internet. Filmmakers and media producers bring skill sets that will be of value to the development of virtual science centers. An example of a virtual science center that intensively uses these skills is becominghuman.org, a site produced by the Institute for Human Origins that describes itself as an "interactive documentary experience". The trend to bring other media and other professional skills, from outside the conventional science center professions, to creating virtual science centers is one to be monitored as we move into the future.

An important distinguishing characteristic of physical science centers is the significant trust the public places in them in two domains: (1) as tellers of scientific truth and (2) as creators of learning experiences that are understood to be important in people's lives. The Internet environment, which is rife with inaccurate, misleading and biased information, provides virtual science centres with the opportunity to be valued and popular for their ethical stance and integrity, similar to the value placed in physical science centres. The same Internet provides an opportunity for education and information exchange in this vital area.

In *E-Learning and Virtual Science Centers*, I am appreciative that a number of the authors express how the ethical character and integrity of a virtual science center's offerings is important. Billie Jones captures this especially well in drawing our attention to sites that demonstrate "good character, good sense, and goodwill" toward the user. I am pleased to see Spadaccini's chapter, "Real Science: Making Connections to Research and Scientific Data", because connecting our audiences to real science and scientists is fundamental to the perceived integrity of science centers.

Research about science centre learning, a topic touched on in many chapters, is a key underpinning of the ethics of science centres. This research is essential in keeping us true to our missions and true to our audience's expectations. It distinguishes science centres from commercial attractions or commercial web sites. Although commercial attractions may do market research, which assists them in a business sense, they typically do not document and learn from the value they produce in people's lives. Documenting our value in people's lives gives science centers a significant strategic advantage.

What does the future hold for virtual science centers? Two major trends are clear. First an increasing fraction of humanity will continue to come online. With 13% of the world's population already using the Internet and use exceeding 50% of the population in 22 countries, science center professionals must appreciate that a far larger fraction of humanity already has access to virtual science center experiences than has access to the physical ones. Thus we can imagine an increasing emphasis on the development of virtual science centers.

The second trend is that broadband and wireless Internet connectivity will become almost universally available in many areas of Asia, Europe and North America over the next few years. Simultaneously, the speed and capacity of home and school computers will continue to increase. The obvious result of this is the potential to create increasingly sophisticated virtual experiences. The less obvious trend will be the parallel demand on virtual science center workers to comprehend the tools of the trade and fully master the methodologies of Internet learning and virtual interaction at a high level of professionalism.

What of real science centers? One way to think about what is evolving is to compare real science centers to live theatre and virtual science centers to film. Film and theatre are both strong and important attributes of modern civilization. Film and theatre have become distinct professions, though they share some common history and many common professional attributes. Similarly, it is not hard to imagine the real and virtual science center professions diverging, as have the theatre and film professions.

Likewise I have no doubt that both real and virtual science centers will exist in the same mutually supportive way that film and theatre exist together. Both real and virtual science centers have roles to play in enhancing how people learn, continuing the important work of transforming individual lives and thus strengthening society.

In concluding this foreword, I congratulate the editors, Leo Tan Wee Hin and R. Subramaniam, and the authors on a ground breaking book — the first on virtual science centers, a work that I'm confident will be a most important resource in this new field.

Bill Peters, Chief Executive Officer
The Calgary Science Centre, Canada

Preface

Science centers play a key role in many countries in popularizing science and technology to the public. By sensitizing the public through the provision of a context that allows them to realize how science and technology impact on their everyday lives and society, science centers provide an upstream initiative for the fostering of literacy in informal science. Their impact is to a significant extent confined to people who enter their premises to savor their attractions, though many science centers are also involved in outreach initiatives, which help to expand their influence further. Avenues that help science centers to reach out to more people will allow them to fulfill their mission objectives even more purposefully.

In recent times, the Internet has made a profound impact on various aspects of society, including education. The availability of a PC and a network point is opening up new possibilities for learning and other fruitful endeavors. It is testament to the vision of science centers that they have not overlooked the potential of the Internet to aid them further in their popularization efforts. This has led to the birth of the virtual science center, the Web-based equivalent of a traditional science center. Whilst pioneering efforts were limited to the offering of simple information on their Web sites, advances in technology have allowed science centers to unleash a slew of resources and other attractions on their Web sites, so much so that the virtual science center now constitutes a new genre of learning in informal science education.

This book aims to provide an overview of the state-of-the-art developments in the field of virtual science centers and address the needs of practitioners in this fast developing field. It is the first book on Web-based science centers to appear in the market. Addressed at science and technology centers, science museums, and researchers and practitioners in Web-based education, it offers an overview on developments in a new and emerging field of science education. Other target audiences for the book include teachers, multimedia developers, educational administrators, developers and managers of Web technologies, and Web content developers.

The 20 chapters accepted for publication in this book span a diverse spectrum of topics. It represents contributions from science centers/science museums, academia and other organizations. A total of 44 authors from 27 institutions are represented in this effort. The international flavor of the book can be seen from the fact that the authors come from eleven countries: Brazil, Canada, Finland, France, Germany, Israel, Italy, Norway, Singapore, UK, and the US.

The 20 chapters are grouped into three sections: Theoretical Issues, Design Considerations and Case Studies. The choice of chapter assigned to a particular section is primarily determined by content coverage and, secondarily, by convenience for readers. Inevitably, there is bound to be some overlapping coverage in the three sections. A brief commentary on the various chapters follows.

Section I features six chapters on theoretical issues. The platform afforded by theoretical issues presents an opportunity to analyze issues from fundamental considerations. In the chapter titled *Establishing Identification in Virtual Science Museums: Creating Connections and Community*, Billie Jones draws on the identification metaphor from the communication domain to advance the view that science museums on the Web need to connect with their audience through suitable site design and content. This is indispensable for attracting site traffic as well as drawing repeat visitations from the Web-surfing public. Kathyryn Haley Goldman and Lynn Dierking, in their chapter *Free-Choice Learning Research and the Virtual Science Center: Establishing a Research Agenda,* suggest that the contextual model of learning can be a useful tool to understand the virtual science center and help frame a research agenda for free choice learning — learning that is self-directed and voluntary. They suggest that better understanding of the nature of such virtual experiences and the factors that contribute to online learning will enable science centers to better design their virtual domains as well as build a knowledge base on how people engage in free choice learning online. In their chapter *Contextualized Virtual Science Centers*, Andreas Zimmermann, Andreas Lorenz and Marcus Specht argue the need to consider issues such as modality, reception style, technical limitations, location and time when contextualizing information delivery on the Web-based science center. They draw on information brokering techniques to advance an approach for the development and maintenance of context-sensitive systems and techniques suitable for virtual science centers. Bronwyn Bevan, in her chapter *Starting With What We Know: A CILS Framework for Moving from Physical to Virtual Science Learning Environments*, considers how the essential characteristics of learning within science centers can be translated and applied to learning in Web-based science centers. She advances the view that science centers need to leverage on their intrinsic strengths and unique pedagogy to fill an educational niche in the Web landscape rather than compete with commercial and other educational agencies engaged in the development of online learning environments. In the chapter on *Weaving Science Webs: E-Learning and Virtual Science Centers*, Susan Hazan suggests the need for science and technology to be placed in a social context that solubilizes the perceived boundaries between art, culture and science that are inscribed in institutional activities. Using examples from museums and other online architectures, she explores innovative systems that harvest data across electronic highways, online collaborations between museums and their public, and other narratives that invigorate community knowledge and stimulate science discourse. In the chapter *Resource-Based Learning and Informal Learning Environments: Prospects and Challenges*, Janette Hill, Michael Hannafin and Denise Domizi explore how a resource-based learning approach can be implemented on the Web sites of science centers. They describe opportunities and challenges associated with such an endeavor.

Section II features eight chapters that focus on design considerations. The allure of a virtual science center hinges significantly on its design as much as its content. Design elements that are compelling have the advantage of adding value to the content hosted on the Web sites of science centers. Michael Douma and Horace Dediu, in their chapter *Interactivity Techniques: Practical Suggestions for Interactive Web Sites* draw on their experience and expertise in creating interactive Web sites to offer comprehensive advice to science centers on making online exhibits interactive. Indeed, interactive exhibits are a major consideration in drawing online visitors to the Web sites of science centers as well as helping to extend their dwell times there. In the chapter *From the Physical to the Virtual: Bringing Free-Choice Science Education Online*, Steven Allison-Bunnell and David Schaller draw on their experience and expertise to propose strategies for recreating science center exhibits online. They argue that while physical and virtual exhibits share certain common features, interactive virtual exhibits need to be configured in terms of the strengths and limitations of the Web. Anja Hoffmann, Stefan Göbel, Oliver Schneider and Ido Iurgel, in their chapter *Storytelling-Based Edutainment Applications*, underscore the need for science centers not to overlook the potential of storytelling contexts when interpreting science content online. Storytelling has the advantage of fostering easy connection with the audience, and interactives incorporating such narratives can confer yet another dimension to the learning process. While existing technological contexts have served Web-based science centers well, Ramesh Srinivasan argues on the need for science centers not to overlook the potential of new information architectures in his chapter *Revolutionizing Information Architectures Within Learning-Focused Web Sites*. He presents two approaches for information design: community driven ontologies and social information filtering agents, and addresses the question of how to share knowledge across a community of visitors without physical co-assembly. In her chapter titled *From Information Dissemination to Information Gathering: Using Virtual Exhibits and Content Databases in E-Learning Centers*, Joan Nordbotten draws on some of the experiences of traditional museums to provide pointers for science centers to use virtual exhibits and content databases in the e-learning process. She discusses how different e-learning architectures can support different forms of learning in Web-based science centers and science museums. Renata Piazzalunga and Saulo Barretto discuss some fundamental questions concerning creation and development of interfaces in their chapter *Challenges in Virtual Environment Design: An Architectural Approach to Virtual Spaces*. They address three levels of complexity and offer useful tips for science centers to re(design) their virtual space to promote better interaction with cyberspace visitors. The subject of personalization — how to tailor an e-learning experience for an individual visiting the Web site of a science center, is the focus of the chapter *Personalization Issues for Science Museum Web Sites and E-Learning*, by Silvia Filippini- Fantoni, Jonathan Bowen and Teresa Numerico. Science centers and science museums have yet to tap into this tool to a significant extent but it has the potential to draw repeat visitations and enhance user experience. John Falco, Patricia Barbanell, Dianna Newman and Suzanne Dewald, in their chapter *E-Learning and Virtual Science Centers: Designing Technology-Supported Curriculum*, propose a partnership model involving virtual science content providers that creates technology-infused science curriculum using interactive videoconferencing technologies and supporting Web resources. They argue that such enriched content can promote new structures of pedagogy that can motivate students to enhance their cognitive development.

Section 3 on case studies features six chapters. Case studies help to bridge the gap between theory and practice, and thus offer useful insights for science centers to learn from the experiences of others. In the chapter *A Virtual Museum Where Students Can Learn*, Nicoletta Di Blas, Paolo Paolini and Caterina Poggi use the example of the collaboration between the Israel Museum and the Politecnico di Milano to show how Internet and multimedia technologies can be effectively exploited to deliver complex scientific and cultural concepts to middle and high school students. It is based on a shared 3-D online environment where students from five different countries meet together to learn, discuss and play, thus dissolving the boundaries of space and time when reaching out to new audiences and promoting outreach. In his chapter *Open Learning Environments*: *Combining Web-Based Virtual and Hands-On Science Centre Learning*, Hannu Salmi recounts the experience of the Finnish Science Centre and other European science centers to show that the virtual science center need not be a disparate endeavor. By judiciously integrating elements of the gallery experience with the Web experience, he adds that new avenues for promoting effective learning can emerge. In their chapter *Use of Log Analysis and Text Mining for Simple Knowledge Extraction: Case Study of a Science Center on the Web*, Leo Tan Wee Hin, R. Subramaniam and Daniel Tan Teck Meng use log analysis of server data to study the Web site of the Singapore Science Centre. They present a commentary on the use of log analysis, an overlooked tool, for studying the effectiveness of content hosted on the Web sites of science centers. In the chapter *The Development of Science Museum Web Sites: Case Studies*, Jonathan Bowen, Jim Angus, Jim Bennett, Ann Borda, Andrew Hodges, Silvia Filippini-Fantoni and Alpay Beler trace the historical development and features of a number of pioneering science museum Web sites. These historical developments present useful insights for science museums and science centers, including those that have yet to transplant themselves onto the Web. Roxane Bernier, in her chapter *The Educational Approach of Virtual Science Centers: Two Web Case Studies*, uses the examples of The Exploratorium and the La Cité des Sciences et de l'Industrie to cement her argument that innovative pedagogy leveraging on real-time Web casting can act as a focus to enrich people's interest in the notion of public understanding of research. She suggests that a reflective perspective drawn from a multidisciplinary approach can aid science centers to form their own viewpoints on contemporary issues ranging from genetic engineering to space exploration. Jim Spadaccini, in his chapter *Real Science: Making Connections to Research and Scientific Data*, shows how compelling scientific imagery in a variety of fields of study that are publicly available can help science centers open up another niche area to help the public better understand scientific research and the scientific process. He draws on his experience and expertise with four key educational Web sites to provide useful perspectives.

Leo Tan Wee Hin and R. Subramaniam

National Institute of Education

Nanyang Technological University

Singapore

Acknowledgments

We thank all authors for their assistance and cooperation in bringing about this book. Most of the authors also served as referees for the chapters. Their valuable feedback has helped to improve all chapters, and this is gratefully appreciated.

We thank the National Institute of Education in Nanyang Technological University for their assistance and support in the course of working on this book project for a year.

We also thank Dr. Bill Peters, Chief Executive Officer of The Calgary Science Center, Canada, for graciously consenting to write the Foreword for this book.

The staff at Idea Group has been extremely supportive in the process of bringing out this book. Our grateful thanks go to Ms. Michele Rossi, Ms. Jan Travers and Ms. Amanda Appicello.

We would also like to place on record our special thanks as well as gratitude and appreciation to Dr. Mehdi Khosrow-Pour of Idea Group for giving us this great opportunity to edit this book.

Leo Tan Wee Hin and R. Subramaniam
National Institute of Education
Nanyang Technological University
Singapore

Section I

Theoretical Issues

Chapter I

Establishing Identification in Virtual Science Museums:
Creating Connections and Community

Billie J. Jones, Shippensburg University, USA

Abstract

The concept of identification from the field of communication studies is used in this chapter to discuss the ways to, and the benefits of, develop(ing) identification with an audience through the design and arrangement of virtual science sites (i.e., aid in the comprehension of the new, and potentially difficult, content of the site; and help in the creation of regularly returning site visitors/users). The chapter concludes with checklists to aid Web site designers/managers and educators analyzing/using virtual science sites to maximize the benefits of identification.

Introduction

Museums are a culture industry that has evolved over time. Historically, museums have been the place of rare and often valuable collections, preserved and displayed predominantly for their aesthetic value, rather than for an educational goal. However, instead of a collection as a museum's commodity, Eilean Hooper-Greenhill (1995) concludes that "[k]nowledge is now well understood as the commodity that museums offer" (p. 2).

It was modern science museums that took the lead in creating museums that were designed around the message they wished to espouse rather than a collection they wished to exhibit (Weinberg & Elieli, 1995, p. 50). As Bonnie Pitman (1999) chronicles, "The Museum of Science and Industry in Chicago, established in 1926 . . . [and] the New York Museum of Science and Industry (now defunct) represents (*sic*) the arrival of science technology centers, founded without collections, that focus on their role as educational institutions to promote an understanding of scientific principles" (p. 7). About this evolution of science and technology centers (as well as children's museums), which he calls "great pioneers in improving the process of learning by the young" (Skramstad, 1999, p. 118), president emeritus of the Henry Ford Museum and Greenfield Village, Harold Skramstad (1999) writes: "These new types of museums developed out of community concerns that more traditional, collection-focused museums were not meeting the learning needs of their audiences" (p. 118). Kenneth Hudson claims that "' . . . the most fundamental change that has affected museums during the [past] half century . . . is the now almost universal conviction that they exist in order to serve the public'" (as cited in Weil, 1999, p. 232). Not only did science museums take the lead in this change to serve the public, but according to many their change was also the most profound. Jorge Wagensberg (2000) writes in his essay, *In Favor of Scientific Knowledge: The New Museums*, "[i]t is the science centers and museums that have most changed their content, their methods, their role in society, and their attitude toward their public" (p. 129).

This evolution of modern museums, particularly science ones, as educational outlets to serve the public has continued to evolve with the advent of computer technology and the World Wide Web. As Pitman (1999) notes, "Museums have developed marketing policies to attract new audiences, to increase the access to their educational resources both at the museums themselves and through the World Wide Web" (pp. 13-14). Furthermore, Pitman (1999) continues, "The explosion of the World Wide Web has added yet another dimension to the role of museums as forums. Museums are becoming 'virtual museums' with beautifully produced pages that summarize their offerings, take you on a virtual tour of their galleries, and provide access to the collections and exhibitions with images and audio" (p. 23). Considering the change that the World Wide Web has brought to the relationship of communities and their museums, Pitman (1999) continues, ". . . we are not required to go to museums to see certain objects, hear lectures, conduct research, or participate in discussion" (p. 26); users can experience much from the comforts of their own home or classroom through the use of virtual science centers. San Francisco's *Exploratorium* (<http://www.exploratorium.edu/>), Philadelphia's *The Franklin Institute Online* (<http://sln.fi.edu/tfi/welcome.html>), and London's *Science Museum* (<http://www.sciencemuseum.org.uk/>), which will be discussed later in this chapter, are three of many science museums that have become impressive virtual centers, as complementary outreach centers on the World Wide Web; however, other "centers" are entirely virtual, residing wholly in cyberspace with the intention of educating their audience about scientific concepts and thinking. The final virtual center that will be discussed here, *Leonardo: Interactive Virtual Science Museum* (<http://www.ba.infn.it/~zito/museo/leonardoen.html>), is one such virtual-only center. Common to all four of these sites are the centers' attempts to meet public need toward educational goals in science.

The marriage of science education and the Internet is a natural one, but like any relationship, some virtual science sites are better than others — better at making its contents easily accessible, seemingly credible, and comprehensible. The use of state-of-the-art Web technologies is important; however, those should not be employed at the exclusion of long-recognized principles of communication. By borrowing the concept of identification from the field of communication studies, this chapter will discuss the ways, and the benefits, of developing identification with an audience through the design and arrangement of the virtual sites of science centers (e.g., aid in the comprehension of the new, and potentially difficult, content of the site; and help in the creation of regularly returning site visitors/users). The chapter will conclude with checklists to aid Web site designers/managers of virtual science sites and educators who use such sites to maximize the benefits of identification.

The Concept of Identification

Twentieth-century literary and social critic, Kenneth Burke (1967, 1950) claims the term *identification* from the field of psychology as the defining term for modern rhetoric in his germinal article, *Rhetoric — Old and New*, in which he writes:

> If I had to sum up in one word the differences between the "old" rhetoric and a "new" (a rhetoric reinvigorated by fresh insights which the "new sciences" contributed to the subject), I would reduce it to this: The key term for the old rhetoric was "persuasion" and its stress was upon deliberate design. The key term for the "new" rhetoric would be "identification," which can include partially "unconscious" factor of appeal. (p. 63)

Burke (1950) deals with the concept of identification as a rhetorical strategy, linking the concepts of *identification* and *persuasion* when he claims that "*Wherever there is persuasion, there is rhetoric. And wherever there is 'meaning,' there is 'persuasion'*" (p. 172). In short, rhetoric is any use of meaning-laden communication; not only textual communication, but non-textual as well — images, architecture, spatial arrangement, even body language all fall under the defining umbrella of rhetoric.

Burke (1950) further writes that, "You persuade a man only insofar as you can talk his language by speech, gesture, tonality, order, image, attitude, idea, *identifying your ways with his*" (p. 55). By demonstrating commonalities between the speaker/writer/designer and his/her audience, the speaker/writer/designer increases the likelihood that the audience will be persuaded. Consequently, in order for a science center, physical or virtual, to educate its audience, it must establish identification with that audience.

Issues of Audience and *Ethos*

Audience-awareness, the act of recognizing and consciously adapting information to "his (*sic*) ways," is integral to establishing identification, and museums have recognized that. In its content and presentation science centers must address a diverse audience. About the "museums of the future," Maxwell Anderson (1999) wrote, "The nature of a museum visit will be palpably different, since the visitor will have at his or her command a massive database delivered at levels appropriate to schoolchildren or to scholars" (p. 135). This needed breadth of content, and presentation and production values may be difficult to achieve, but others note its importance as well. Although Harold Skramstad (1999) was quoted earlier as recognizing the pioneering success of science museums in "improving the process of learning by the young" (p. 118), he adds that "the museum model of education should not be limited to the younger years. Museums can and should provide educational experiences for adult learning that are just as powerful. In museums adults can learn at their own speed and in their own way in a setting that is multisensory and engages the emotions as well as the intellect. With no mandated curriculum, learners can organize themselves by almost any criteria of interests" (p. 119). Rather than seeing this diverse audience with equally diverse needs and interests as a hindrance, Skramstad continues, "The mixing of education, age, gender, and race can become a strong asset in a shared learning experience. The museum can provide a place that encourages and enables intergenerational learning" (p. 119). And Susanna Sirefman (1999) agrees, adding that "Very few cultural institutions are so effective in dissolving generational gaps" (p. 317).

While Anderson, Skramstad, and Sirefman are referring to physical-space museums, the same should be said for virtual museums and centers as well. The potentially metaphoric "database" of Anderson's "museum of the future," can become quite literal in a virtual science center. And Skramstad's call for "multisensory" settings and multiplicitous interests aptly describe Web sites' potential and its audience. Because the audience of the World Wide Web is so diverse, and so amorphous, provisions must be made for users from a variety of interest and intellectual levels.

Science centers have recognized this complex audience and have responded to their needs. In writing about a *New York Times* section on museums, headlined "Culture's Power Houses," Emlyn H. Koster (1999) notes that "The essential claim [of the *Times* section] is that science centers are a relatively flexible type of museum, having shown repeatedly over a relatively short history an ability to be highly responsive to community needs" (p. 278). That "responsiveness" is an effort to identify the museum's ways with those of its audience. Related to the action of "responsiveness" is the demonstration of "relevancy." Pitman (1999) describes "an audience-driven, educationally-active museum that positions itself as a relevant community resource" (p. 287), and Koster (1999) speaks of science centers' "mission of relevancy to the community" (p. 293). In Burke's terms that "relevancy" would be the museum's demonstration of goodwill — caring concern and likeminded-ness — for the audience, which should lead toward identification. Goodwill is also demonstrated by museums' shift from appreciation to activism, not only in shifting from offering aesthetic collections of objects to offering opportunities for learning, but also an activism toward responsibility and conservation. Pitman (1999) writes of the "profound shift from . . . objects of natural history [displayed] in order to excite wonder and curiosity to exhibits that focus on science and are concerned with the preservation of species and life on earth" (p. 19). By caring about the world, museums not only demonstrate their concern for their community, but they also entice and prepare their visitors to become activists themselves.

That responsiveness and relevancy to community demonstrates not only the museum creators'/designers' goodwill toward its audience, but also its credibility — its *ethos*. Embedded in the process of establishing identification is the concept of *ethos*. Part of Aristotle's three-part strategy for creating a convincing argument, *ethos* deals with the character of the speaker/writer/designer (Cherry, 1994, p. 86). In short, *ethos* is the character of the speaker/writer/designer, which is ever-present in a text. This *ethos* can be invented, constructed by the speaker/writer/designer through the discourse, or *ethos* can be situated, inherent to the speaker/writer/designer based on her/his reputation or the particular rhetorical occasion (Crowley, 1994). Museums, by general reputation, are perceived as possessing good character, so they are in an enviable situation to be seen as credible by that audience; however, science centers and museums can enhance, or "invent" their *ethos*, by making prominent their associations with other museums, educational institutions, other socially-concerned and ethically-aware non-profit organizations, and even some corporations.

"Inventing" one's *ethos* is frequently a reciprocal act, with each bolstering the credibility of the other. For example, demonstrating (or enhancing) one's ethical credibility is one of the reasons for corporate sponsorship of museum exhibitions, in the same way that corporate support for a museum exhibition lends credibility to the museum, as visitors/user subconsciously believe that the supported venue must be "good" if it is being supported by corporate funds. On the other hand, corporate sponsorship can be seen as sullying a museum's good name or co-opting a museum's content, and a corporation's presumably well-meaning intention for sponsorship may be interpreted as insincere or self-serving. While one's *ethos* may need bolstering through invention, care must be taken to demonstrate earnestness and sincerity.

The concept of *ethos* can be divided into three sub-categories: *phronesis*, *arete*, and *eunoia*, which help speakers/writers/designers to develop their character in order to establish goodwill and thereby be rhetorically persuasive. Although there is a great deal of overlap among these three characteristics, examining their separate identities will help to more clearly understand their combined *ethos*. The first of the three, *phronesis*, uses "good sense" to prompt persuasion. Clearly, throughout their exhibits — physical-space or virtual — science centers and museums demonstrate "good sense" through a myriad of facts presented in an authoritative, objective air, in order to establish their goodwill toward an audience. As Koster (1999) notes, "Science centers have a particular opportunity to capitalize on the public's need to function in a technologically advanced world in which science-technology-society issues are interwoven" (p. 289). By offering their audience the "good sense" to function in this technologically advanced world, science centers and museums enhance their own *ethos*. *Arete* is the moral virtue, the "good character," the "accepted and embraceable moral sensibility," that the speaker/writer/designer demonstrates to her/his audience (Sipiora, 1994, p. 268). As has already been said, museums enjoy, for the most part, the perception of being ethically upright. And the third, *eunoia*, is the mutuality of goodwill — goodwill demonstrated by the speaker/writer/designer that elicits feelings of identification from the audience, and goodwill on the part of the audience toward the speaker/writer/designer as a result of that identification. Because the museum cares about its community's needs, its audience will care about the museum — using and supporting its resources on a recurring basis. Basically, people believe the sources they trust because of the source's *ethos*, and they consult repeatedly sources in which they believe because they identify with those sources.

Identification in Physical-Space Science Centers

Physical-space science centers are rife with attempts to establish identification. Skramstad (1999) writes, "The goals are always to make a connection with an audience, to establish a relationship of trust, and to cause some specific outcome, whether it be knowledge, fun, insight, or the purchase of a product or service" (p. 121). In the rhetorical terms discussed here, the goals of a museum are to demonstrate its good sense — its *phronesis* — and its good character — its *arête* — to create a mutuality of goodwill — *eunoia* — for/with its audience through caring consideration and likeminded-ness, thereby showing the museum's credible *ethos*, leading then to audience identification, that "connection," and ultimately persuasion, that "specific outcome."

Skramstad (1999) elaborates:

> More and more Americans expect that their social, economic, and cultural activities, though shaped by a variety of sources, will engage them in a way that is vivid, distinctive, and out of the ordinary. This is even more the case for children who are being brought up in a world of interactive media, which sets up new expectations of active participation. They expect to be treated as individuals who have a significant capacity to influence as well as be influenced by any experience in their lives. This means that the world of the next century [writing of the 21st century], these experiences will have to be developed with close interaction between their producers and their consumers. This interaction and the resulting relationship of trust is what increasingly will give authority to any experience. (p. 121)

That trust can be fostered by demonstrating the museum's relationships with other institutions, organizations, and even corporate sponsorship, which can help to show a museum's connectedness with its community and its citizenry.

Even paying attention to visitor comforts demonstrates a museum's goodwill toward its audience. Paving the way for easy visitor access to centers by providing informational materials, such as maps and exhibit guides, and as well having staffed information kiosks and exhibit areas on site with which visitors can interact all help to establish identification as they demonstrate goodwill for the audience on the museum's part. About the expectations of today's museum-goers, Anderson (1999) writes:

They have certain expectations of comfort and convenience . . . [and he cautions that] museums often fall below their expectations . . . [with] . . . poor directional signage, infinitesimal and inscrutable labels, minimal restroom facilities, a shortage of elevators, inadequate seating, hard stone floors, and overpriced eateries. (p. 141)

Such failure, although likely not intentional, can be interpreted as a lack of goodwill toward the museum's audience. Instead Anderson (1999) insists that: "[m]useums should be working now to make the museum visit the easiest and most attractive option possible, through networked collaboration, the use of Web casting, the availability of up-to-date information on activities in museums, and the provision of services including on-line ticketing, reservations, and interactivity, in preparation for and in the aftermath of a visit" (p. 145). The efforts that Anderson calls for are ones that would demonstrate a museum's goodwill toward its audience.

Clearly, audience trust and identification is a critical goal, but so too is the integral component of interactivity. Speaking of the changes that the "wired world" will precipitate, Anderson (1999) writes: "The re-villaging of our sensibilities, independent of physical location, will draw us toward an experience with which we are inextricably involved: no longer as passive viewers, but as, at the very least, 'viewsers' — viewers and users of information" (p. 134). It should be noted, however, that the sort of interactivity that computer technology allows for in both physical and virtual science museums and centers is not necessarily a new concept. As Anderson (1999) notes:

The temptation to animate the motionless object is not new. It dates at least as far back as the Capitoline Museums, which were founded in 1471 but extensively renovated in the nineteenth century. Numerous classical statues there and in the Vatican museums to this day have brass spindles that allow the visitor to turn the work on a cylindrical pedestal so as to see it in the best possible light. (p. 141)

This sort of early concern for museum visitors' experience is mimicked in advanced technological ways through the manipulation of digitized objects.

It should be noted too that in the physical-space museum field, digitizing objects has sparked considerable controversy over the possible loss of an original object's "sacred aura," and subsequent loss of physical-space museum visitors. Experts such as Anderson (1999) claim:

The situation need not be as bleak as we might fear. Just as slides, postcards, and posters of famous works of art have encouraged generations of college students and members of the public to visit art museums, the promulgation of digital images and information about works of art should encourage future audiences to visit art museums. No less importantly, the growing surfeit of "virtual" experiences in daily life is likely to result in a growing appetite for the authentic, and especially for encounters with those priceless touchstones of human creativity that cannot be adequately experienced in a virtual medium, provided that museums do not accede to the impulse to satisfy the lowest common denominator. (p. 144)

Certainly, the appropriateness of digitizing objects is moot in virtual science center where that very same mimesis is integral to the virtual exhibit; however, it should be added that in writing about the effect of museum Web sites on the increased draw that museums are currently experiencing, Susanna Sirefman (1999) writes that ". . . museum web sites [of which digitized objects are a component] are included as a positive influence on this draw" (p. 311). Rather than satiating a potential physical-space museum or center visitor's desire for scientific knowledge, virtual sites seem to whet their appetites.

About the interaction to which Skramstad and Anderson refer above, here, too, science centers and museum seem to be responsive. Of science museums Wagensberg (2000) writes:

Do not touch has become *Please touch.* The concept of the display case has changed into the idea of the experiment; the idea of the academic label has turned into the text with literary quality; using the sense of sight as the only method of perceiving has given way to using all the senses (or almost all); the emphasis has shifted from the preparation of answers to the preparation of questions; and love of the past alone has given way to a desire to use the museum as a tool for change. (pp. 129-130)

Questioning is central to the scientific method, so it is only fitting that science museums and centers valorize questioning as a potential educational end in and of itself. Furthermore, the opportunity to interact with exhibits, either physical-space or as will soon be discussed, virtual centers, helps visitors to feel a part of — to identify — with the center.

Identification in Virtual Science Centers

Turning to virtual science centers, many of which are related to physical-space science centers, designers need to consider ways to convey a sense of goodwill to their visitors, and the ways in which identification is established in virtual science centers are both the same and different to those of physical-space centers. Traditional learning strategies and visitor accommodations incorporated in physical-space science centers must be *remediated* to take best advantage of the virtual environment.

Remediation, as defined by Jay David Bolter & Richard Grusin in *Remediation: Understanding New Media*, is a concept by which newer media "define themselves by borrowing from, paying homage to, critiquing, and refashioning their predecessors [. . .]" (cited in Bolter, 2001, 24). In that text, Bolter & Grusin offer examples from new media like computer graphics and the World Wide Web in which those media have borrowed from older media like film and photography to create themselves. In *Writing Space:*

Computers, Hypertext, and the Remediation of Print (2nd ed.), Bolter (2001) extends the concept of remediation to include the transition from the print age to an electronic, hypertextual age, saying that: "Hypertext in all its electronic forms — the World Wide Web as well as the many stand-alone systems — is the remediation of print" (p. 42). While the medium of the WWW allows for hypertextuality in a way that print text never could, many of the same rules of print-based texts: reading from left to right, from top to bottom are usually observed in hypertexts as well. This mixing of new and old offers users a certain familiarity in an unfamiliar medium.

More than simply borrowing, remediation "involves both homage and rivalry, for the new medium imitates some features of the older medium, but also makes an implicit or explicit claim to improve on the older one" (Bolter, 2001, p. 23). Remediation is at the heart of the evolution in science centers from "cemeteries of knowledge" (Galluzzi, 2000, p. 107) to places of interactive engagement, and that evolution to interactive engagement has evolved or been remediated to include multimedia materials. As Koster (1999) writes, "The science center's methodology — one initially described as hands-on and now increasingly multi-sensory and multimedia in nature — has been an important innovation in the museum field" (p. 291). Clearly the move to "multi-sensory and multimedia" is seen as an improvement over the past.

And this remediation has continued as physical-space science centers have moved into cyberspace. For example, arranging a virtual site to mimic the arrangement of a physical space science museum remediates the traditional expectations of a physical space museum into its new virtual environment. The Franklin Institute's virtual science center, which will be discussed shortly, quite literally does this by using their building's image as a central graphic in its site, from which the three major components of the site are linked. This connection with a physical space building taps what is familiar about a science center to ease visitors into a potentially unfamiliar location in cyberspace.

The act of remediation is frequently reciprocal, because as a newer medium borrows from the older, so too may the older medium borrow from the new in an effort to regenerate itself. Bolter (2001) offers the national newspaper, *USA Today* as an example ". . . of the refashioning of the newspaper itself" (p. 51). He continues:

> In graphic form and function, the newspaper is coming to resemble a computer screen, as the combination of text, images, and icons turns the newspaper page into a static snapshot of a World Wide Web page. In many newspapers the index now consists of summaries gathered in a column running down the left-hand side of the page, and a small picture is often included with the summary. Anyone familiar with multimedia presentations can easily read such a picture as an iconic button, which the user would press in order to receive the rest of the story. *USA Today* in fact makes considerable use of "hypertextual" links back and forth throughout its pages, and these links are sometimes cued by small graphics. The purpose of these icons together with the other pictures and graphics is not merely decorative. Together they help to redefine the function of the newspaper, which is no longer only to transmit verbal information, but also to provide an appropriate visual experience and through that experience to dictate an appropriate reaction to the stories being told. (Bolter, 2001, p. 51)

This "refashioning" occurs as a result of remediation, which is a fluid process. The use of images borrowed from the WWW to refashion newspapers had originally been borrowed from film and television, and today the computer screen-like environment which Bolter describes above is now common on many television networks, such as *ESPN*, where the television screen is frequently a composite of multiple "windows:" with news bytes and sports scores running in ticker fashion across the top and or bottom of the main "window" — the current broadcast.

This complex give-and-take of remediation can be seen in science centers as well. As physical-space science centers moved to remediate their hands-on interactive environments to include computer-based multimedia technology, virtual science centers, wholly dependent on computer-based technology, borrowed from the "original" physical-space as well. As in the example of *The Franklin Institute Online* discussed above, sometimes the literal physicality of the physical-space center is borrowed, as is also material from the physical-space exhibits. In some cases, technologies have advanced so quickly that it is difficult to ascertain which medium is borrowing from which.

Many of the means of establishing identification in the audience of a physical-space science center, or at least remediations of the same means of establishing identification, can be used in virtual science centers; however, demonstrating goodwill toward the audience of a virtual center, a largely anonymous and widely diverse audience, is more difficult than establishing identification with visitors to a physical space. First of all, the demographics of the location of a physical-space center make it somewhat easier to conceptualize its potential audience; however, such conceptions of audience are much more difficult to ascertain in cyberspace (technology does allow for some virtual demographics based on interest, of course). Furthermore, there are no physical beings with whom users can identify in a virtual space; cyberspace breeds an impersonal environment; thereby making identification more difficult. Moreover, designers of online museum exhibits do not have the luxury of spinning knowledge with a narrative thread; cyberspace valorizes hypertextuality — not linearity. Nevertheless, virtual center designers can still work to establish identification with their cyber visitors.

Virtual science centers are able to establish identification with their audiences primarily through demonstrating their goodwill for audiences and creating a familiar "architectural" environment. Goodwill for an audience can be demonstrated in a variety of ways. First of all, making materials available on the WWW in and of itself is a demonstration of goodwill, not only for potential physical-space visitors but also for the many virtual visitors who will never set foot in the physical-space center. Beyond simply making materials available, organizing them in ways that are easily accessible through user-defined hierarchies (sections/links for different audience types: e.g., educators, physical-space visitors, students, families) and through search engines is another way that virtual science centers can demonstrate good sense and goodwill. Because of the hypertextual context of online museum exhibits, designers have a responsibility and actually an opportunity to guide site visitors through the inclusion of navigational cues and devices, and ease of access and navigation are two ways that online exhibition designs can demonstrate goodwill for their audiences. Certainly, too, the overall look of the site — its thoughtful use of colors, graphics, multimedia, and interactive components can all work to create Burke's "unconscious factor of appeal" — to foster the idea that a virtual science center believes in its mission and cares about its audience.

Discussing the role of "real-time architecture" of "a physical destination place" (Sirefman, 1999, p. 311), Sirefman writes: "Connection to a local neighborhood and relevance to a specific community begin with a building attentive to its surroundings...[, with] an architecture that is *substantive* and *welcoming* [italics added]" (p. 317). This mandate for "substantive and welcoming" "real-time architecture" holds true for a virtual site's architecture as well.

Creating a familiar virtual space with common format schemes can also demonstrate that good sense and goodwill, but when the virtual space takes advantage of its relationships with existing science organizations and facilities, sometimes going so far as to borrow from the physical architecture of its buildings, the virtual center seeks to locate visitors in a safe space from which to navigate the center's new and exciting terrain. Exploring and learning new concepts and ideas can be exciting for some and intimidating for others — familiar touchstones in site organization can provide users with the familiar foundation from which to safely explore the unknown.

To demonstrate these principles of demonstrating good sense, good character, and goodwill toward an audience in order to establish identification, the sites of four virtual science centers will be discussed: *The Franklin Institute Science Museum Online* (http://sln.fi.edu/tfi/welcome.html); the *Science Museum* (http://www.sciencemuseum.org.uk/); the *Exploratorium: The museum of science, art and human perception* (http://www.exploratorium.cdu/); and *Leonardo: Interactive Virtual Science Museum* (http://www.ba.infn.it/~zito/museo/leonardoen.html). By examining the ways that each of these centers demonstrates goodwill toward their audiences, one can extrapolate principles for center design and usage to best establish identification.

Figure 1. Franklin Institute Online welcome page (http://www.fi.edu/tfi/)

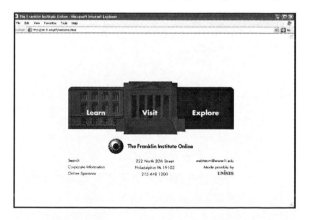

The Franklin Institute Online
(http://sln.fi.edu/tfi/)

The Franklin Institute Online, the online arm of The Franklin Institute in Philadelphia, borrows heavily from its physical-space counterpart for its virtual architecture and design; thereby, blending the physical space with the virtual site. As has already been mentioned, a tri-colored graphic based on its building façade serves as the central point of its Welcome page (see Figure 1) (http://sln.fi.edu/tfi/welcome.html). It should be noted that a smaller version of this building façade image map is located on the upper left-hand corner of each of the site's pages, giving users an easy way to reconnect to the central tripartite organization of the site.

From there, users may click on one of three wings of the building to "Learn," "Visit," or "Explore" the site.

As one might expect, the "Visit" wing is full of information about the physical-space Institute: hours of operation, ticketing, special events, and information on the specific exhibits. Additionally, there are links for visitors traveling in groups or from outside the area, which will take them to tourist information beyond the Institute itself. All of this information demonstrates the Institute's goodwill toward its audience. Here, too, invitations to join the museum through membership are frequent, but virtual site users are also invited to share their memories of "the Heart," a fifty-year old exhibit that is currently closed for renovation (until October 2004). Using a simple form, users can share their memories of visiting this exhibit over the years (http://sln.fi.edu/biosci/memory/; links to this "Memory Box" are also available through the Heart's online exhibit as well). Users can also read remembrances of other contributors. Such invitations allow virtual center users to feel a part of the center, and that sense of belonging, of ownership, leads to identification.

Figure 2. Franklin Institute "Explore" page (http://sln.fi.edu/exploreF.html)

The "Explore" page, too, (see Figure 2) features recurring and frequent connections with the Institute's physical space, with links to five-day weather forecast for the area, details about the Exhibit Halls, and even a Web cam showing live shots of the surrounding Philadelphia skyline. The "Explore" page also invites visitors to the "'inQuiry Attic' [which] offers an online exploration of . . . a century's worth of scientific instruments" collected by The Franklin Institute. There, users are invited to "Step into the Attic and explore our treasures" (http://sln.fi.edu/qa98/atticindex.html). This use of the term "attic" quite palpably situates virtual site users in a common physical-space area to rummage through previously exhibited artifacts and information.

The borrowing from building architecture to organize the center is carried throughout the site. Each of the three main pages: "Learn," "Visit," and "Explore" uses the same arrangement with a larger center panel surrounded on three sides by eight smaller rectangular panels, each with a linked graphic to another part of the site. This arrangement could loosely be seen as a blueprint drawing of rooms in a building, thereby continuing the architectural theme of the site, and its repetition throughout the site demonstrates good sense. Central to each of these pages is a search opportunity, and the smaller version of the building façade image map, which allows users ready access to the central organization of the site. Unfortunately, below this arrangement of rectangles are a series of boxes with links to additional information. Falling below the primary grouping of rectangles, and below or nearly below the original screen view (that which is visible without scrolling downward) in many browsers, these boxes are seen more as an afterthought to the page's primary arrangement, and materials there may frequently go unseen.

Common to all three divisions of the site, in addition to the architectural features discussed above, are the links to information for teachers and to interactive materials. While teachers seem to be the primary audience for the "Learn" page, there is information throughout the entire site that would be useful for teachers. Links are available to field trip information, teachers' guides to exhibits and other activities, correlative materials to state standards in science, as well as professional development opportunities. In fact, teachers seem to be the primary audience throughout *The Franklin Institute Online's* organizational architecture. While site redundancy, making frequently sought after information readily available at various portals of entry, is appropriate, overall, *The Franklin Institute Online* seems to be designed primarily for educator access, to prepare for physical-space visits or to share online exhibits with students, as opposed to engaging students working on their own. While the site itself is attractive if not a bit boxish, its colorful, sometimes cartoonish graphics might fail to engage adolescents seeking to explore a virtual science center. Furthermore, while there are seven online exhibits, links to which are scattered about the site, or available centrally on a page (http://sln.fi.edu/tfi/virtual/vir-summ.html) that can be found only by using the search function for "online exhibits," the overall site of *The Franklin Institute Online* seems primarily intended to enhance physical-space visits, demonstrating goodwill to those users, and particularly toward teachers of science.

Figure 3. Science Museum welcome page (http://www.sciencemuseum.org.uk/)

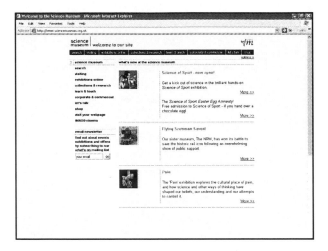

Science Museum (UK)
(http://www.sciencemuseum.org.uk)

Although not in its site design, the virtual site of London's *Science Mus*eum (see Figure 3) makes strong connections with its physical counterpart as well. Rather than relying on the metaphor of building architecture, geometric arrangement, or graphics like *The Franklin Institute Online*, the virtual architecture of the London's *Science Museum* Welcome page relies much more on text, giving the Welcome page a sparser and potentially *less* engaging appearance. In fact, most of the textual links in the main content area of the *Science Museum's* Welcome pages seem primarily to be an advertisement for the physical-space center, featuring only brief online components to entice users to visit the physical-space center. For example, the site's Welcome page is primarily a World Wide Web outreach site for the physical-space center's exhibit, complete with floor plan, ticket information, calendar of upcoming events, merchandise, timetable for an IMAX movie, *Top Speed*, and photographs of some of the exhibits. It should be added that there are, however, several online activities including a clip from a movie shown at the museum, and a game, again intended to connect to the physical-space exhibit.

Another example of their use of virtual center as advertisement for their physical-space center is related to the "Science of Sport" exhibit (see Figure 4). The descriptive text reads as follows: "Science of Sport explains the science behind sport and encourages children (and adults!) to get themselves interacting with our state-of-the-art exhibits.' Instead of looking at displays, you'll actually be in them, taking part in races, games and quizzes to test your skills and knowledge of sport! And while you're playing, you're learning too!" (http://www.sciencemuseum.org.uk/exhibitions/sport/site/index.asp). Obvious efforts to bridge the generations and to ensure visitors of the opportunities for interactivity are clear, thereby engaging its physical-space audience.

Figure 4. Science Museum "Science of Sport" site (http://www.sciencemuseum.org.uk/ exhibitions/sport/site/index.asp)

As another way to engage its audience and demonstrate its goodwill toward that audience, the virtual site advertises "free" admission to the "Science of Sport" exhibit for the cost of one chocolate Easter egg the day after Easter weekend. As the site explains:

> The amnesty - which is supported by the British Dietetic Association and Sport England - is designed to offer all young people the chance to take part in the hands on exhibition and learn about the importance of balancing diet and exercise. Children will be able to try a number of specially designed activities, from running on a treadmill while wearing a specially designed "fat pack," to measuring their fat levels using a body mass indicator.

Clearly, caring enough about children's physical health and fitness demonstrates good character, good sense, and goodwill, but the benefits are to be reaped by visitors to the physical-space center. This initial seeming imbalance of physical-space activities over virtual ones leaves the Web user feeling a bit overlooked in favor of the physical-space visitor.

However, by exploring the site menu on the left hand side of the Welcome page, and also across the top, users find a wealth of virtual materials rife with good sense and goodwill. At the "Exhibitions online" page, users will find links to virtual "exhibitions and interactives" as diverse as "Exploring Leonardo," which looks at Leonardo da Vinci as a "scientist, inventor, and artist" (http://www.sciencemuseum.org.uk/on-line/leonardo/ LeoHomePage.asp), to "Apollo 10" (http://www.sciencemuseum.org.uk/on-line/apollo10/ index.asp), which, in addition to telling about the Apollo space program, offers users an opportunity to design a rocket, which they will later launch to an authentic sounding audio countdown. "Interactives" such as these are extremely sophisticated, and do an excellent job of engaging the user in a rich multimedia learning experience. Furthermore, the breadth, depth, and sophistication of the materials available demonstrate the Science

Figure 5. Science Museum "Launchpad" comments (http://www.sciencemuseum.org.uk/ on-line/launchpad/3XX.asp)

Museum's good sense and goodwill toward the virtual center user. Most importantly, these opportunities for interaction allow users to identify with the museum by becoming "owners" in their own exploration and knowledge making.

This demonstration of goodwill is continued on the "Learn & Teach" page (http://www.sciencemuseum.org.uk/education/index.asp), where links are available for different audience types, such as "teachers," "adults," "students," and "families." At the "teacher's" link, there are, in addition to information and ancillary materials to support a physical-space museum field trip, curricular guides and activity sheets, searchable by age or type of material. The "students" and "families" sites are designed with virtual users in mind with links to fun activities as well as material to be used in school projects. While all of these sites mention activities available at the physical-space Science Museum, there is much for the virtual center user here. One other group of learners who appears to be the subject of greater goodwill here are adult learners, rating their own site and activities. Clearly, the Science Museum does an excellent job of demonstrating its goodwill toward the virtual center users who may never enter its physical-space walls.

One other way that the virtual site of the Science Museum establishes identification with its virtual users is through its frequent use of phrases like "let's talk," "what do you think?" or "tell us what you think." Through email links and invitations to participate in online debates (http://www.sciencemuseum.org.uk/lets_talk/debate.asp) about questions such as "Telepathy: Fact or Fraud" and "Should we be collecting objects from the wreck of the Titanic, or should we leave the dead to rest?," not only are virtual users invited to participate in these moderated online discussions, but also each discussion is accompanied with some background material to contextualize the debate. One more accommodation that the virtual site of the Science Center makes for its virtual users is the use of scalable fonts on its pages, allowing users to change the size of the page's font within their Web browser. This demonstrates goodwill toward those with visual impairments, but also for users with a variety of browsers. Again, users have a choice

— and an opportunity to interact with the site to make it theirs. As was mentioned previously — making a site theirs allows users to identify with that site.

In addition to requesting participation from virtual users, the site includes audio feedback from museum visitors, accompanied by the speakers' photographs (Figure 5) (http://www.sciencemuseum.org.uk/on-line/launchpad/3XX.asp). The use of these faces and voices makes human other center visitors, thereby helping to establish identification between the virtual users and the physical-space visitors.

While at first glance the *Science Museum* would seem to suffer from the same imbalance favoring physical-space visitors over virtual users noted about *The Franklin Institute Online*, a closer examination reveals a wealth of opportunities and interactivity for virtual users of many ages and interest levels. The designers of the virtual site for the Science Museum have combined the goals of using the WWW to demonstrate their good sense and goodwill for both their potential physical-space visitors, and their virtual users quite well.

Exploratorium (http://www.exploratorium.edu)

While both *The Franklin Institute Online* and the main page of the virtual site of the *Science Museum* (UK) utilize the Web to offer enticement and easy access to their physical-space visitors, the *Exploratorium*'s virtual center has much more to offer the Web user from the outset (see Figure 6).

Figure 6. Exploratorium Welcome Page (http://www.exploratorium.edu)

While there are links to information that would attract and inform visitors to/of the physical-space, including Web cams showing exhibits and the surrounding skyline through its "Roof Cam" (in real time during museum/daylight hours), similar to the one at *The Franklin Institute Online*, these physical-space reminders are not the primary focus of the *Exploratorium*'s virtual science center.

At first glance, the user is met with a riot of colors, images, and texts, and the hypertextuality of the center is visually conveyed, particularly as compared to the linearity of the Science Museum's opening page. The Exploratorium's site is divided into five major categories: "Explore," "Educate," "Visit," "Partner," and "Shop." Common to all of the pages are their "screen size" format, search function, site map link, and an invitation to "Get Involved" in some way, whether through becoming a member, a volunteer, a donor, or simply registering for a newsletter.

As would be expected, the "Visit" link takes potential physical-space visitors to not only the necessary visitor information, such as upcoming events, hours, admission fees, maps, and etcetera.; it also provides would-be visitors a sampling of exhibits available in the physical-space center. As the name would imply here too, the "Shop" link takes users to a store, but this time a fairly sophisticated online store, where a wide variety of educational toys and gifts, as well as publications and other educational materials can be purchased. While the opportunity to purchase items can be seen as a for-profit motive for the center, most users realize that non-profit organizations must support their operations. Furthermore, making quality science-related items available to virtual center users demonstrates goodwill toward that audience. (E-commerce plays an interesting role in virtual science centers, particularly as it relates to museum shops. All of the sites related to physical-space museums and centers offer virtual users an opportunity to shop online. *The Franklin Institute Online's* has a limited store with most of its products connected closely with the exhibits of the physical-space center, little more than advertisements for the physical-space center, but the *Exploratorium* and *Science*

Figure 7. Exploratorium "Educate" (http://www.exploratorium.edu/educate/index.html)

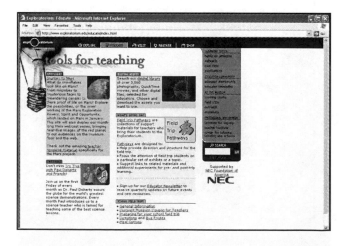

Museum stores are much larger, offering a wide variety of science-related toys, clothing, publications, etc. On one hand this availability may seem at cross-purposes to a non-profit, caring, like-minded institution; however, if the centers' mission is to engage users in science education, then the stores, which make one-stop shopping for ancillary materials available do seem like an act of goodwill.)

As was noted previously about opportunities to become a member of The Franklin Institute Online, evidence of and opportunities to "partner" with the Exploratorium demonstrate goodwill toward its audience by offering users "ownership" of the center — and in turn their learning there. Furthermore, by sharing information about other partnering institutions, agencies, and organizations, the *ethos* of the center is bolstered by its associations there.

The two most notable sub-sites in Exploratorium are "Explore" and "Educate" (see Figures 7 and 8). The "Explore" link takes users to a variety of highly interactive mini-sites on topics as diverse as earthquakes to frogs to the origin of languages. A separate menu at the "Explore" site organizes the material differently, taking users to online activities, exhibitions, Web casts, and other hands-on activities for the virtual user. Clearly, the Exploratorium's virtual science center demonstrates its goodwill toward its virtual audience by offering so many accommodations for those who may never set foot inside their physical-space.

Exploratorium's goodwill toward its educator audience is clear in the amount and breadth of materials intended for teachers under the "Educate" category: materials to enhance class visits to the physical site, but also a variety of learning tools, including directions for hands-on activities, links to Web casts, and etcetera. In addition to the information they can receive there, teachers have the opportunity to participate in the "Educate" site by subscribing to an educator's newsletter and to contribute by sharing worthwhile educational Web sites (outside of the Exploratorium's site) to become part of a monthly

Figure 8. Exploratorium "Explore" (http://www.exploratorium.edu/explore/index.html)

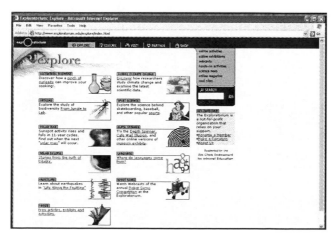

feature, "Ten Cool Sites" (http://www.exploratorium.edu/learning_studio/sciencesites.html). There, too, former "cool sites" are archived, accessible by topic.

Melding the definitions of museum as collection and museum as learning environment, one collection that the *Exploratorium* makes available under its "Educate" category is their "Digital Assets Collection:" over 3,000 "digitized museum materials related to interactive exhibits and scientific phenomena, including images, educational activities, PDFs, QuickTime movies, and audio files . . . for educational use" (http://www.exploratorium.edu/educate/edam/index.html). The availability of this digital collection with free access for educational uses, as with the many other materials available for educators, creates strong identification between educator-users and the virtual center.

That the *Exploratorium*'s virtual science center is a step above other such centers in establishing identification with its virtual audience is not surprising. Writing about the founding of the physical-space Exploratorium, Kathleen McLean (1999) describes it as:

> a new kind of museum altogether, born from the philosophies of self-directed learning, interactivity, and individual discovery that were growing out of a burgeoning educational reform movement. At the heart of the new Exploratorium — " A Museum of Science, Art, and Human Perception" — was a fundamental mission to empower the public and "bridge the gap between the experts and the laymen" with exhibits and experiments that visitors could activate on their own. (p. 90)

Figure 9. Leonardo: Interactive Virtual Science Museum (http://www.ba.infn.it/~zito/museo/leonardoen.html)

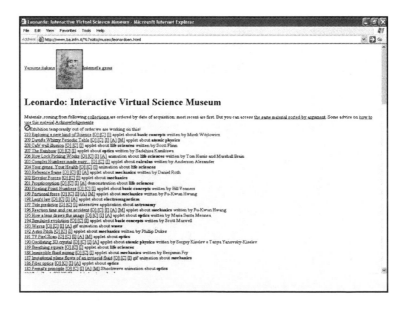

The virtual site of the *Exploratorium* is one more way in which they seek to "empower the public . . . with exhibits and experiments that visitors could activate on their own," and in empowering the public they are also successful in establishing identification with its users, thereby creating regular users.

Leonardo: Interactive Virtual Science Museum (http://www.ba.infn.it/~zito/museo/leonardoen.html)

Leonardo: Interactive Virtual Science Museum is, to use Maria de Lourdes Horta's phrase, "A museum without walls and without objects, a *true* virtual museum" (cited in Weil, 1999, p. 236). Unlike the three virtual centers previously examined, *Leonardo* has no physical-space museum related to it. In truth, *Leonardo* is little more than a series of over 200 textual links to sites related to scientific concepts; primarily physics (see Figure 8). There are no visually engaging images (save one of da Vinci) or even a spatial arrangement on the page that draws readers into its textual density. However, despite its sparseness, this site *is* a virtual science center representative of substantial Web searching, and through its good sense it makes some surprising efforts to demonstrate its goodwill toward its audience.

First of all, the site is offered in both English and Italian, and accommodations are made in different ways throughout the site for translation. In addition, the designer has created a series of alphabetic codes that give users additional information about each linked site before going to the site itself. This coding allows users to preview each site without taking time to open each, many of which might take a long time to do so because of their interactive nature (frequently Java applets). Care is taken to note any broken links as well, further demonstrating the site creator's goodwill for its audience. Furthermore, this site actually features two different pages — one with the links arranged by date of acquisition (to the site), and the other arranged by topics to better enable users to access the appropriate linked material. It is in the complexity of the topics covered at the site that also demonstrates goodwill, but endeavoring to make available more comprehensible presentations of particularly dense material. By focusing on potentially less engaging topics and offering the materials is a less than engaging "package," than the other three virtual science centers discussed here, *Leonardo*'s designer(s) lose points in appeal, but win points in the efforts made to help its users better understand difficult concepts.

One other benefit of looking at the *Leonardo* site is that, because of its organizational simplicity, relying on the overall difficulty of its content and the possible sophistication of its constituent sites from which to draw its own brand of sophistication, one is reminded how easily such a site can be created. Certainly, that the site is multilingual is no easy feat, neither is its availability in two different organizational hierarchies (order of acquisition and subject matter) and coding system; however, its overall concept of providing a list of links to other sites is something that educators can do to create their own "virtual science center," by organizing the virtual exhibits of others (to whom credit would be given of course).

Several Final Caveats

Before concluding that virtual science center designers should employ all of the methods to establish identification with their audiences, and that educators using those sites should use only those that do, several caveats need to be stated. Balance is an important concept to remember — balance in catering to one's audience, avoiding unduly focusing on one constituency to the exclusion of another, while at the same time not trying to be everything to every user. Balance, too, in content choice: offering what is scientifically relevant, what users need, what will attract users, and what users are interested in.

Care must be taken by museum site designers to achieve a balance between sophisticated content and possibly less than sophisticated site visitors and their systems. Site sophistication through the use of graphics and images, or audio, video, and Flash files that are large or that may require special plug-ins is another aspect of virtual science center design that can contribute in both positive and negative ways to the establishment of identification. Within the architecture of these sites, photographs are a common method by which a virtual center can exhibit its holdings, both photographs and artifacts. The inclusion of photographs is an obvious way to establish identification between a center's audience and its subject, and photographs are one of a virtual science center's best ways of conveying a sense of their presence to site visitors, and thereby establishing identification with their audience. However, in an online exhibition, photographs can have both positive and negative effects on building identification. While photographs can provide an important bridge across the impersonality of cyberspace, the additional drag that multiple image files place on pages' loading speed can sometimes irritate or confuse site visitors, or even overtax certain older systems' capacities. Care must be taken to reduce, as much as possible, the size of image files, without unduly sacrificing visual quality. In such cases where the designer determines that the contribution that an image makes to the page is worth the extra loading time, then the designer would be wise to add a notice to advise site visitors about possibly slow loading times.

The same consideration would apply for files — video, audio, Shockwave, and applets — that require special plug-ins or programs that are not widely available on most, particularly older computer systems. Movie clips like the one available at the Science Museum's virtual site, and Flash movies that launch automatically at *The Franklin Institute Online* are certainly engaging, but encountered on older or ill-equipped hardware, they can result in frustration rather than engagement. Any obstacles that stand between the Web exhibit and its audience can hinder identification, prompting them to "question how welcome he or she is" (Anderson, 1999, p. 142), so designers must consciously consider whether such large, slow loading files, some which may even necessitate the addition of less common system features, add sufficiently to the site to justify their inclusion. Ironically, the very means by which interactivity can be infused into a site can also hinder users from developing identification. One way to handle this possible obstacle is to make mention of special system requirements at the beginning of the Web exhibit, or to use software that searches users' systems, notifying them of needed downloads, or even automatically launching those downloads. On one hand, in order to support access, obtaining the appropriate application must be simple and free; but on the other hand, monitoring and particularly auto-launching can be seen as intrusive, thereby destroying any goodwill established by easy and ready access.

Finally, while the concept of identification has been tested in disciplines such as communications and English, and even in our own interpersonal relationships (e.g., we are best persuaded by those who are most like us or who seem to have our best interests at heart), studies of visitors' perception/interpretation of "goodwill" through user evaluation and monitoring would be an important verifying step in the formal study of virtual science center design. Such studies would help determine that all important balance point mentioned at the outset of these caveats.

Conclusion

Here are a few final words of advice about utilizing/testing identification to develop user-loyalty and as a tool of persuasion. As a final act of goodwill, checklists follow to help each group follow that advice:

- To designers, know your audience — their interests and needs — and meet those while demonstrating your center's credibility. By doing so, users will return to the center over and over, identifying with this site, trusting it to be accessible, credible, and comprehensible.

- To educators using virtual science centers in their classrooms — make your and your students' interests and needs known — become a part of the centers you use. Carefully evaluate the sites you use with your students, as you do with all other educational materials, so that students are exposed to the most engaging resources available, so that they can identify with those sites, becoming independent, questioning users themselves.

Checklist for Virtual Science Center Web Designers

*Considerations that Demonstrate Center's Credibility (*Ethos*)*

- Make clear any relationship(s) with physical-space institutions (e.g., museums, educational institutions)

- Note partnerships with other institutions, agencies, and/or organizations [e.g., educational, AAM (American Association of Museums), ASTC (American Science-Technology Centers)]

- Be *visible*, proactive partners with educational institutions

- Explain collection and research processes

- Identify curators, researchers, and etcetera and their associations

- Document sources, links, and etcetera

- Cover "hot" or controversial topics, without compromising research standards

- Employ as sophisticated Web technologies as can be well-maintained

Considerations that Demonstrate Goodwill Toward One's Audience

- Offer a balance of resources for physical-space and virtual visitors

- Accommodate audience diversity with a variety of topics, each of which is likely to interest particular contingent(s) in that audience

- Create a visually organized and appealing site

- Provide access to resources in a variety of hierarchies (e.g., "user-type," subject, resource type)

- Provide search capabilities

Establishing Identification with Users Will

- Create supportive learning environment

- Enhance understanding of difficult or unfamiliar concepts

- Create loyal users

Checklist for Educators Teaching with Virtual Science Centers

Evaluate the Center's Credibility (Ethos)

- Who created and maintains the site? What is their reputation and training?

- Are the references, sources, and researchers made known?

- Is the site exemplary of the scientific process?
- Are the external links to credible sites as well?
- Is contact information made clear?

Evaluate the Center's Goodwill Toward You and Your Students as an Audience

- Does the site try to address the interests of you and your students?
  - Is the overall site or sub-site age appropriate for your students?
  - Are the topics and/or approaches to those topics of interest to your students?
- Does the site try to serve the needs of you and your students?
  - Are there curricular materials made available on the site?
  - Are these materials connected to local/state curricula/standards, and etcetera?
- What efforts are there made to help you use the site?
  - "Searchability"
  - Variety of menus based on interest groups, topics, and etcetera
  - Mailings sent through schools, educational organizations, or center newsletters (virtual or "snail mail")
- Does the center offer professional development opportunities in connection with science education?

Ways to Use Virtual Science Centers to Best Fit Your Situation

- Access curricular materials
- Prepare for field trip
- Lead students on "virtual field trip" from center to center
- Create your own virtual science center through links on a simple Web page
  - Done for your students, you have a resource that is ultimately more tailored to their needs.

Done with or by your students, you have provided a learning opportunity that is unsurpassed in fostering knowledge-production.

References

Anderson, M.L. (1999). Museums of the future: The impact of technology on museum practices. *Daedalus, 128*(3), 129-162.

Bolter, J.D. (2001). *Writing space: Computers, hypertext, and the remediation of print* (2nd ed.). Mahwah, NJ: Erlbaum.

Bolter, J.D., & Grusin, R. (2000). *Remediation: Understanding new media.* Cambridge, MA: MIT.

Burke, K. (1950). *A rhetoric of motives.* New York: Prentice-Hall.

Burke, K. (1967). Rhetoric—Old and new. In M. Steinmann, Jr. (Ed.), *New rhetorics* (pp. 69-76). New York: Scribners (originally published in 1950/1957).

Cherry, R.D. (1994). Ethos versus persona: Self-representation in written discourse. In P. Elbow (Ed.), *Landmark essays on voice and writing* (pp. 85-106). Davis: Hermagoras (original work published in 1988).

Crowley, S. (1994). *Ancient rhetoric for contemporary students.* New York: Macmillan.

Exploratorium: The Museum. (n.d.). Retrieved March 15, 2004, from *http://www.exploratorium.edu*

The Franklin Institute Online. (n.d.). Retrieved March 15, 2004, from *http://www.fi.edu/tfi/*

Galluzzi, P. (2000). New technologies and the objects of science: Reflections on the use of multimedia. In S. Lindquist, M. Hedin, & U. Larsson (Eds.), *Museums of modern science* (pp. 107-116). USA: Science History Publications.

Hooper-Greenhill, E. (1995). *Museum, message, media.* London: Routledge.

Koster, E.H. (1999). In search of relevance: Science centers as innovators in the evolution of museums. *Daedalus, 128*(3), 277-296.

Leonardo: Interactive Virtual Science Center. (n.d.). Retrieved March 15, 2004, from *http://www.ba.infn.it/~zito/museo/leonardoen.html*

McLean, K. (1999). Museum exhibitions and the dynamics of dialogue. *Daedalus, 128* (3), 83-107.

Pitman, B. (1999). Muses, museums, and memories. *Daedalus, 128*(3), 1-31.

Science Museum. (n.d.). Retrieved March 15, 2004, from *http://www.science museum.org.uk/*

Sipiora, P. (1994). Ethical argumentation in Darwin's origin of species. In J.S. Baumlin & T. F. Baumlin (Eds.), *Ethos: New essays in rhetorical and critical theory* (pp. 265-292). Dallas: Southern Methodist UP.

Sirefman, S. (1999). Formed and forming: Contemporary museum architecture. *Daedalus, 128*(3), 297-320.

Skramstad, H. (1999). An agenda for American museums in the twenty-first century. *Daedalus, 128*(3), 109-128.

Wagensberg, J. (2000). In favor of scientific knowledge: The new museums. In S. Lindquist, M. Hedin, & U. Larsson (Eds.), *Museums of modern science* (pp. 129-183). USA: Science History Publications.

Weil, S.E. (1999). From being about something to being for somebody: The ongoing transformation of the American museum. *Daedalus, 128*(3), 229-258.

Weinberg, J., & Rina, E. (1995). *The Holocaust Museum in Washington.* New York: Rizzoli.

Chapter II

Free-Choice Learning Research and the Virtual Science Center:
Establishing a Research Agenda

Kathryn Haley Goldman, Institute for Learning Innovation, USA

Lynn D. Dierking, Institute for Learning Innovation, USA

Abstract

Societies are in the midst of change, witnessing an explosion in out-of-school learning. From the proliferation of educational programming through film, television, museums and science centers, there are more opportunities for free-choice learning, self-directed and voluntary, than ever before. However, most virtual learning research is focused on classroom-based practices with little research on how learning occurs virtually. This chapter describes an appropriate research agenda, suggesting some of the research questions of highest priority. Authors suggest that models such as the Contextual Model of Learning are useful tools to understand the virtual science center experience and frame a research agenda for the future. Better understanding the nature of such virtual experiences and the factors that contribute to learning online will enable the field to better design such science centers, as well as begin to build a body of knowledge about how people in the 21st century engage in free-choice learning online.

Introduction

Globally, societies are witnessing a virtual explosion in out-of-school learning. From the proliferation of educational programming through film and television, museums and science centers, there are more opportunities for self-directed learning than ever before. In a typical day, an individual might surf the Internet to track down a book in a local library, attend a play or a book discussion group, watch a nature documentary on television or interact with exhibitions at the local science center. All of these events are free-choice learning[1] experiences — self-directed, voluntary, and rather than following a set curriculum, are guided by the individual learner's needs and interests.

The rise of the Internet in particular is fanning the flames of free-choice learning. Individuals all over the world are increasingly seeing learning in a broader sense — not simply something that occurs in the classroom or even in a place per se, but an activity that one can engage in virtually as well. Increasingly, these learners understand learning is the way individuals make meaning of, and survive in their world and they are empowered to pursue learning in their own ways. Traditionally, museums and science centers have been great places for free-choice learning, however, with the virtual explosion of new media, these institutions are poised to become an even greater resource.

Currently, the majority of virtual learning research is focused on classroom-based practices, not free-choice learning situations. Consequently, there is very little research on how learning occurs at places such as virtual science centers and therefore our ability to know how to best design such centers is limited. Research in this area is critical, since the learning potential of virtual science centers is enormous. The virtual nature of these science centers enables centers to break free of geographic, physical and time constraints, reaching an even greater share of the public, thus engaging them in science, and hopefully increasing science and technology awareness and literacy.

This lack of research exists for several reasons, including most significantly the methodological obstacles in conducting research on "non-captive audiences" in virtual environments. The evaluation and research that does exist tends to focus disproportionately on usability issues, such as reduction of system-critical errors and ease of navigation. This focus on usability is important and contributes to the ease with which users can access the resources of virtual centers, but unfortunately it also obscures larger, more critical issues at stake. For instance understanding how, why and to what end people use virtual science centers would help designers and "curators" of such sites better select, organize and present e-learning resources and activities. Understanding the impact of such experiences can also provide insights into how best to position this resource in relationship to the physical science center, other museum Web sites and other free-choice learning resources (books, magazines, television). It is akin to concentrating solely on whether the door of your science center is unlocked, without paying attention to how visitors are using the resource, whether it could be improved, and whether they are having enjoyable and meaningful experiences.

There is an even more important need for research in this area though. The existence of learning potential in no way suggests that this potential *has* been realized. Thus, on a more practical level, better understanding of the nature of the virtual science center experience and the factors that contribute to learning online will enable the field to use virtual museums to engage and inform the public in more meaningful and effective ways. In this chapter we will describe what we feel would be an appropriate research agenda in this area, suggesting some of the research interests and questions that we feel are of highest priority. It is critical that practitioners in this area be aware of both the issues at stake and the research conducted thus far, in order to enhance the impact of virtual science centers and ensuring that they do not become flat de-contextualized versions of the physical science center.

Initiating a Research Agenda

It is one thing to say that research in this area is critical, quite another to decide what models or frameworks could best guide such research and what research interests and questions we feel are of the highest priority and will result in the most usefulness for designers and curators. We propose that one place to start would be to build on established research in the free-choice learning arena. Thirteen years ago John Falk and Lynn Dierking formulated a framework for thinking about learning that tried to accommodate much of the diversity and complexity surrounding learning, a framework that at the time was called the Interactive Experience Model. More recently they built upon and refined this model, recasting it as the Contextual Model of Learning. This model posits that there are hundreds of factors fundamental to science center learning experiences. Among those hundreds, 12 critical suites of factors, clustered into the three contexts of importance in these settings (Personal, Physical & Sociocultural), have emerged through research (cf., Falk & Dierking, 2000), as those which individually and collectively influence the meaning-making process of visitors to free-choice learning settings like museums and science centers. No one factor is dominant; indeed the interaction of these factors is unique to every individual. This model has proven to be a convenient way to think about learning: as a process/product being constructed over time as people move through their sociocultural and physical worlds. Over time meaning is built-up, layer upon layer. However, this model does not quite capture the true dynamism of the process, since even the layers themselves are not static or necessarily even permanent. All the layers, particularly those laid down earliest, interact and directly influence the shape and form of future layers; the learners both form and are formed by their environments. For convenience, we have distinguished three separate contexts, but it is important to keep in mind that these contexts are not really separate, or even separable.

The twelve suites of factors identified in the Model are:

Personal Context

1. Motivation and expectations
2. Prior knowledge and experience
3. Prior interests and beliefs
4. Choice and control

Sociocultural Context

5. Within group social mediation

6. Facilitated mediation by others

7. Cultural background and upbringing

Physical Context

8. Advance organizers

9. Orientation to the physical space

10. Architecture and large-scale environment

11. Design of exhibits and content of labels

12. Subsequent reinforcing events and experiences outside the museum

The Institute for Learning Innovation, established in 1986 as a not-for-profit learning research and development organization, has been an international leader in investigating free-choice learning, developing and applying innovative models for learning such as the Contextual Model of Learning, and understanding new media and its potential use in these innovative approaches to learning. Institute researchers, and other researchers as well, have used this model to frame recent research on learning in and from museums (Falk & Storksdieck, in revision). It has also served as a useful tool for practitioners as they attempt to design free-choice learning experiences that take into account these characteristics. For instance, in the design and redesign of exhibitions and programs, the model has been a useful prism with which to examine these offerings from multiple viewpoints, in order to address the visitor experience as completely as possible.

These factors have been shown to be robust and to validly describe the nature of learning within the physical science center. For instance, factors related to choice and control have emerged as critical variables, with research demonstrating that learning is maximized when visitors have choices about what, when, how and with whom to learn — in other words, learning is enhanced when the learner feels in control of his/her learning. This research has been used to effectively influence practice with successful science centers working to incorporate aspects of visitor choice and control into their exhibitions and programming.

Just as these factors contribute to and influence the visitor experience in physical science centers, it is logical to think that they, and perhaps other factors, play a role in virtual science centers as well, albeit likely in different ways, and therefore that this model could form an effective foundation for a research agenda in this area. For example, choice (visitors making choices about what they see and do based on their interests, attitudes and prior experiences) and control (visitors' actual and perceived control over the experience) are factors that greatly influence a physical visit to a physical science center, and are likely to influence virtual visitors. Yet the factors of choice and control in virtual settings may also encompass issues of access to software, plug-ins and download times. In addition, it is highly possible that some of the twelve suites of factors are not as critical to a virtual environment or that there are other suites of variables that emerge as important specifically to understanding learning in virtual environments.

It would be impossible to comprehensively describe all of the contextual factors related to on-line learning that might be important to investigate in this one chapter. Therefore we have chosen to focus on specific factors within the Contextual Model of Learning that may have different implications in the virtual world and thus, make sense to investigate and incorporate into an online learning research agenda. These issues include items such as motivation for visiting, choice and control within the visit, the social interaction that surrounds the visit, wayfinding, and concurrent and subsequent events that reinforce the virtual visit.

Personal Context

Motivation and Expectations

Museum learning researchers have documented a variety of motivations that visitors have described for visiting physical science centers (Rosenfeld, 1980; Moussouri, 1997; Falk, Moussouri, & Coulson, 1998; Ellenbogen, 2003). This research suggests that the motivation and expectations of any group of visitors is likely to include one of these factors or to be composed of some combination of these motivations:

1.   Fun/entertainment/recreation
2.   As a social activity
3.   For general educational purposes
4.   As a life cycle event, whether through personal connection or as from a sense of duty (my mother always took me here, so I will take my children)
5.   As a "site" to be seen while on vacation or having visitors
6.   For specific content related to the Science Center or a particular exhibition
7.   For practical reasons (too cold to take the children to the park)

Preliminary research indicates that the motivations and expectations for virtual visits are quite different than they are for physical science center visits (Haley Goldman, & Schaller, 2004). For the most part, visitors do not visit virtual science museums as a life cycle event, certainly not as a place to take out-of-town visitors or while on vacation — virtual museums lack a sense of place. And although a virtual visit might be undertaken for practical reasons, such as it being too cold outside to take the children to the park, this motivation is rarely reported. In addition, visiting a virtual institution for social reasons is rarely reported.

The most popular motivations for virtual visitors have been characterized as:

- Gathering information for an upcoming visit to the physical site

- Engaging in very casual browsing

- In search of specific content information (self-motivated research)

- In search of specific content information (assigned research such as a school or job assignment)

(Bowen, 1999; Sarraf, 1999; Chadwick et al., 2000; Kravchyna & Hastings, 2002; Schaller et al., 2002; Ockuly, 2003; Haley Goldman, & Schaller, 2004).

The multiple studies confirm the findings, each study slightly differing due to the phrasing of the questions. Kravchyna and Hastings (2002) for instance asked, "What is the purpose of your visit?" but framed each possible answer in terms of information-seeking: "To find information about recent exhibits, to find information about special events in the museum, to find additional materials for the research needs, to find information on how to contact museum staff." Chadwick et al. (2000), Schaller et al. (2002), and Ockuly (2003), developed a wider variety of motivations, including "to learn about art for personal enrichment" and "personal growth." Bowen (1999) includes "it's fun and interesting" as one of the three main motivators.

These motivations are logical when situated in the larger context of how people nationwide are using the Internet today. The latest report from the Pew Internet and American Life Project (Madden, 2003) details that 26 million Americans surf the Internet just for fun or to pass the time per day, 22 million do research for their job, 22 million look for information on a hobby or interest, 22 million to "answer a question" and 12 million do research for school or training. The above uses of the Internet rank as some of the most popular activities to do online — only getting news and using e-mail rank higher. These numbers emphasize that the motivations of the general Internet visitor are the same as the virtual museum-goers — fun and information seeking.

We do not currently know in what combinations these motivations are "bundled" and in what ways these motivations, either individually or collectively, influence the expectations and outcomes of these experiences. In research using this motivational framework in physical visits, Falk, Moussouri, and Coulson (1998) found that these motivations did bundle together, that fun and learning were not necessarily independent variables (someone could be visiting both for educational *and* entertainment reasons) and that these motivations strongly influenced the learning that resulted from physical visits. There is no reason to believe that similar processes are not in action in virtual visits but currently we have no idea whether and in what ways motivations and expectations influence online learning. This certainly is an area worthy of investigation.

One other important variable to add to the mix is the interaction of motivation and expectation with a virtual visitor's prior knowledge and understanding of the information they seek. Currently most virtual science centers contain content designed as "one size fits all" and there has been little experimentation with offering information and Web-based interactives at different levels, other than at the most gross level (e.g., a section for adults and one for children). This is also an area worth investigating.

As revealed by the number of research and information-based motivations above, content is still king in the world of choices on the Internet. Quite a few history and natural history museums design portions of their site to interest the amateur or professional researcher. This emphasis on collections information can be a disadvantage for the science center, which traditionally does not have extensive collections and instead tends to emphasize content facilitated through interactives. Some content matter can be adapted for the Web and some are even improved via that medium, for instance, flat panel displays of cross-sections are more intuitive and dynamic when shown on an animated Web site. However, interactives that the visitor uses their whole body to engage with might suffer by being translated to the flat screen. We know little about the effects of the loss of physicality, also an area that could benefit from a research focus. However, we do know that video games have found creative ways to incorporate elements of physicality into the experience of playing. For years, game players have been able to use alternatives to joysticks, such as elements that point at the screen. More recent applications include items such as Dance Revolution, a game where visitors mimic the dance moves shown onscreen on a specialized mat, with the moves becoming more and more complex. This type of game has proved to be a creative way of incorporating physicality into the experience and has been very popular with a wide range of ages.

Another potentially rich research focus could be investigating the novelty and "fun" elements of the experience, although now that the sense of the Web as a new frontier has faded and commercialism has grown, some Web surfers are becoming disenchanted with Web sites as a place to have fun. The exuberance is gone, claims *The New York Times*, stating "the toy box has turned into a tool box" (Guernsey, 2002). Following up on this assertion to see whether this is indeed the case with virtual science centers and other free-choice learning Web sites would be very useful for designers and educators alike. Perhaps virtual science centers could position themselves as places where fun can still be found online.

Choice and Control

Issues of choice and control are complex in both physical and virtual sites. As Falk & Dierking state in their book, *Learning from Museums*, "Learning is at its peak when the individual can exercise choice over what and when they learn and feel that they control their own learning." There has been much rhetoric that the Internet can provide boundless choices and absolute user control, however, the actual choice and control factors are significantly more nuanced. Even mundane things, such as dial-up speed and the availability of plug-ins, may influence the virtual visitor's perception of their own choice and control. If the visitor has a slow connection speed and the content takes a long time to download, they may feel frustrated, and not in control of the experience at all.

Since their development, plug-ins have been a "hot button" issue for teachers and schools. Although the issue has not been at the forefront of discussions in the Internet museum community recently, teachers using interactive Web-based museum resources still report difficulties in using even basic plug-ins, such as Flash, in classroom settings, due to school restrictions on software (Haley Goldman, 2003). Such technical difficulties can also lead to frustration and alter the perception of control during a virtual visit. Although museum Web designers frequently tout the possibilities of using Flash, new research has indicated that visitors have no strong preference for flash-based museum sites over the same site in HTML (Schaller et al., 2004).

Some of the most intriguing issues of choice and control online can be found grouped under the umbrella issue of easier opt-out or lack of opportunity costs in virtual visits. There is considerable investment of effort, time and money ("opportunity costs") in the traditional physical museum visit. Visitors must know the hours of operation, find the science center and parking, generally pay to enter, and then figure out how to accommodate their other needs, such as food and restroom breaks. Typical science center visits average approximately two hours, not including the time spent planning the visit and traveling to the institution. These opportunity costs influence the visit in a variety of ways; much of the research in physical sites has actually not taken into account the influence these costs are likely have on the typical visit. For instance, the time spent in an institution is partially a function of an informal cost-benefit analysis carried out by one or more of the group members, calculating "how much time should we be spending here at this center, based on the trouble and money it took to get us here?"

By contrast, a virtual visitor is unlikely to have invested anywhere near the same amount of time and effort in their visit. For instance, unlike the visitor who is planning their outfit for the new exhibition opening, the virtual visitor can view the Science Center in their pajamas, at any time of day or night. The opportunity costs for a virtual visit are much less — one only needs to have a computer and an Internet connection.

The trade-off in comparative lack opportunity costs in virtual visiting result in substantial impacts. Among these are:

- Shorter visit time;
- Lack of "wait time";
- Ability to visit more frequently;
- Ability to visit at the time of the visitor's choice;
- Ability to visit science centers that are not geographically convenient; and,
- Ability to visit related/unrelated sites within the same session.

Shorter visit time: Virtual visits are significantly shorter than physical science center visits. The average physical visit to a science center is two hours, whereas the average virtual visit is generally under 12 minutes (Jensen, 1999; Semper et al., 2000). This difference has potentially serious impacts on meaning-making, because visitors who stay for such a short period of time are unable to significantly interact with the content or interactive experiences provided. Therefore, unless the visitor was visiting to gather very specific information that was easily found on the site such as institution hours or directions or a specific fact, it is less likely that the virtual visit itself will be memorable and result in meaningful impact. However, this is only a supposition which further research could affirm or negate.

Lack of "wait time": Early museum Web site research suggested that half of all visitors who arrive at a home page leave after viewing only the home page. This can be interpreted in two ways. Either as we suggested earlier, the visitor was able to get exactly what they were interested in, perhaps the institution's opening hours, and therefore was satisfied, or for whatever reason they were unsatisfied and left immediately (Jensen, 1999). Although we have seen exceptions, generally once someone has made a commitment to visit a physical institution and enters the front door, they continue with their visit. Physical elements encourage the visitor to continue, such as glimpses of possibly interesting items just around the corner. This is not true of virtual institutions. Virtual science center designers must address home page design with an eye to inviting visitors to continue to click further into the Web site. Applied research that tests how best to do this would be extremely useful. This poses an interesting question for virtual science museums — can we get visitors to increase their visit time by increasing their investment in the visit?

Ability to visit more frequently: Due to the lack of opportunity costs, virtual visitors are able to make more frequent visits than they would to a physical science center location. This advantage has made the Internet invaluable for free-choice learning researchers, whether they are schoolchildren working on a report who need to make repeated visits for their research or hobbyists who bookmark a site so they may continue to access an institution's information or collections in an on-going manner, or professional researchers who are able to electronically access data and other information once only accessed by searching archives or behind-the-scenes collections. Anecdotal evidence shows the development of the Internet has increased hobbyist use of institutions. We might then speculate that this ability increases the user's connection or attachment to the institution. There is little research that has been conducted yet to understand how repeat visitation changes the way visitors use virtual sites, or the impact that has on both institutions and visitors.

Ability to visit at the time of the visitor's choice: As virtual science centers are available 24 hours a day, seven days a week, any visitor with Internet access has the choice of exactly when to visit. One would hypothesize that it is this instant access at inconvenient hours that subtly defines the motivations of the visitors that attend. Free-choice learners, whether they are schoolchildren, teachers, or hobbyists with day jobs, have limited time to actually visit the physical museum in order to satisfy their needs for particular content or interactive experiences. However, if they are able to fulfill those needs on their own schedule, regardless of time of day, then increasing numbers of them will take advantage of the resources within virtual science center. Does the ever-present ability to visit increase the visitor's sense of control? Or does the lack of constraints in visiting increase the visitor's expectations of all possible content available at any moment? One possible simple study could involve surveying visitors about their expectations of the breadth and depth of the online offerings. This is an area of great research potential; we know little about how free-choice learners perceive their online experiences.

Ability to visit science centers that are not geographically convenient: This is another of the much–heralded advantages of the Internet; it has mixed implications for the virtual science center. Virtual visitors are frequently from distant and even international sites. Some of these visitors are undoubtedly canvassing the site for information about an upcoming visit, but many may be visitors that would have no way of attending the institution other than virtually. However, those in search of particular content and/or interactive experiences have choices they may not have in the physical world. If the opportunity costs for visiting any institution in the world are equal, then science centers are no longer competing with other science learning resources or leisure-time choices in their own town, but instead are competing with all other such institutions worldwide. A discussion of the laws of gravity, for instance, might be found on thousands of Web sites worldwide. What makes the design, presentation or approach of the laws of gravity at one site different from any other? Does the use of search engines as one of the primary methods of finding virtual museums negate any "home field" advantages local virtual science centers might have? What are the configurations of use by visitors? Are they connecting a variety of experiences from sites or do they form an attachment to a particular virtual institution, which becomes their own "local" site? These are all unanswered questions. These questions are sometimes addressed via analysis of log files. It might prove interesting to conduct a study in the reverse fashion, such as where one would track all of the Internet resources used by an 8th grade during a science project and question the students about their Internet resource choices. A study of this type would help practitioners understand how to differentiate their virtual science museum from others.

Sociocultural Context

Within Group Social Mediation

Investigators over the years have documented the highly social nature of visits to physical science centers (Falk & Dierking, 2000). Visitors view their visits as not only a chance to learn, but a chance to interact with family members and friends; in fact this may be their primary motivation in visiting and important social learning outcomes, such as learning about family members and how to learn collaboratively, are important outcomes of physical visits. This within-group social mediation is not only key for visit satisfaction, but plays a strong role in meaning-making for individuals (Dierking, 1987; Borun, Chambers, & Cleghorn, 1996; Borun & Dritsas 1997; Borun, Dritsas, Johnson, Peter, Wagner, & Fadigan, 1998; Crowley & Callanan, 1998; Crowley, Galco, Jacobs, & Russo, 2000; Schauble & Gleason, 2000; Luke, Coles, & Falk, 1998; Dierking & Anderson, 1998).

Currently it appears that virtual visits are far more likely to be made alone than visits to physical science centers (Haley Goldman & Schaller, 2004; Semper et al., 2000). There are notable exceptions to this, such as the growing use of the Internet in classrooms, whether for Web casts and virtual field trips, or as an electronic content resource. Family groups use the Internet to conduct research for homework and reports. Yet for the most part, the Internet remains primarily a solitary experience.

When a pair or group of people, either schoolchildren or co-workers, make a virtual visit together, it differs from a group making a physical visit in other key ways. In a visit to a physical science center, group members gather and split away, look at two different displays simultaneously and generally interact as if they were molecules, joining together, moving in one direction, then splitting and going elsewhere, then rejoining, in a frequent pattern. A group can walk through the same exhibition hall, more or less together, and yet see entirely different pieces of the hall based on their particular interests and pace. And they will continue to talk about their experience with each other as they explore the center. This allows individuals, even children, to control aspects of the visit by choosing what to focus their attention on.

By contrast, the virtual visit takes place in a flat, non-surrounding medium — the screen. Groups of virtual visitors generally choose one person, the person in control of the mouse, as the driver in the driver's seat. The visitors may discuss the content and make choices together. They may even choose to look at different aspects of the same screen. (Research into gaze-tracking suggests that individuals viewing the same object, such as a painting, see completely different paintings, as their attention focuses on different aspects of the same object.) Gaze-tracking not withstanding, it is still fundamentally only one person who is controlling what the others view. This means the details of the visit and the pacing is regulated for everyone by one person. Knowing how this impacts the visit might allow designers to change the experience so that it is engaging for those who are not in the driver's seat — or possibly develop sites that allow multiple person interaction, much like a multi-player game does.

Considerable research has been conducted into how students interact and collaborate while using school-assigned hypermedia or through distance learning. Unfortunately, this research is not easily applicable to free-choice learning situations, and little has been conducted solely within the virtual free-choice learning realm. There is tremendous potential for research in this area. For example, due to the distributed nature of the medium, whole new methods of interacting socially have developed within virtual visits. Take, for instance, the possibility of the "telephone visit," which currently we have little evidence either of its frequency or what the implications of such interactions might be. As an example, Harry calls his sister Eve, to tell her about this great science Web site he is viewing online. Eve pulls up the site on her computer and the two of them continue to explore the site, separately yet together. Each is in control of what they are viewing at the moment, yet they also are able to converse and "see" what the other individual is seeing, very different than if Harry merely sent Eve a link to the site, and they both viewed it separately. The telephone visit is actually in many ways more similar to a visit in a physical science center than if Eve and Harry were sitting side-by-side, yet only one of them in control. The social interaction between the two likely increases the length of time of the visit, as well as enriches the meaning-making possible via conversation.

More commonly, users refer their friends and colleague to virtual science centers. Word of mouth is a critical element in drawing visitors, whether to a physical or a virtual site. Several museums and science centers have incorporated electronic postcards into their sites. Other sites, such as the Lemelson Center's Invention at Play site, have developed small Web-based interactives, which one can then e-mail to friends. Although anecdotal information suggests this "referral" form of interaction is quite popular, there is little usage or impact data available to suggest the scope or nature of the interaction. Efforts to investigate such social configurations as "telephone visits" and referrals and their impacts on online learning are wide open to research possibilities.

Despite the lack of research, it is not unreasonable to hypothesize that the opportunity for changing social configurations online can provide opportunities for virtual science centers to take on new roles. For example, increased contact between visitors with other visitors, as well as visitors with staff, not only from e-mail, but from newsletters, forums, guest books, discussion mailing lists, bulletin boards and online surveys, pushes such virtual institutions into the role of gathering place and community center. This is also an area rich with research possibilities.

Facilitated Mediation by Others

The other type of social interaction that occurs during physical science center visits is visitor interaction with staff and volunteers. This can take the form of a guided tour, an on-the-spot lab demonstration, a facilitator demonstrating an interactive, a live performance or theatre piece, to name a few methods. On first glance, it would appear that the Internet would be a poor place to integrate facilitated mediation by staff or volunteers. Yet the research and developments in this field, both for the Internet in general and for museums and science centers specifically, indicates a wide variety of opportunities in use and others in development. One hopes that as the Internet medium continues to develop, designers and educators will remember one of the greatest strengths of the Internet, as well as its original application — the ability to communicate.

Developments to increase facilitated communication include methods such as Web casts and virtual field trips that are "broadcast" online, robots and museum wearables that allow co-visiting, and virtual avatars that can explore physical or virtual space and add personalized interpretation. Science centers and other free-choice learning institutions have used these communication formats alone and in concert. Each type has benefits and drawbacks.

One could even argue that communication between institution staff and the public has actually increased since the development of the Internet. For example, the ability to send e-mail queries and correspondence has allowed visitors access to staff that previously remained behind the scenes and more difficult to contact (Gaia, 2001). Building on the ability for increased communication, some institutions have incorporated methods of both asynchronous and real-time communication between staff and visitors.

Real-time contact, of course, must be scheduled in advance. Such real-time contact may include purely "chat events" or something closer to Web casts where institutions are generally able to "broadcast" live over the Internet, and users may communicate via telephone (Jacobson & Swiader, 2003) or via e-mail (Steinbach, 2001). Like a television show with some surrounding interactive features, a Web cast is generally a live "broadcast" over the Internet, allowing a tour or demonstration for all those interested and online at that time. The Exploratorium offers the ability to communicate with staff via e-mail both before and during the actual Web cast, and then archives past Web casts for virtual visitors to view later. On an experimental basis, they even have explored physical visitors interacting with virtual visitors in real time. One might hypothesize that the impact of this type of programming could be very similar to attending programming in person at the institution. As of yet we have little research to compare the audience distribution, usage, or learning generated from such programs, nor do we understand the economic implications of staffing and other costs. These efforts represent fruitful areas for research that have very direct impact for professionals making cost-benefit program decisions.

Virtual field trips represent another way that institutions are trying to provide facilitated experiences for virtual visitors. In these cases, a physically distant school group might have a virtual tour at a pre-arranged time with science center staff members as facilitators. Virtual field trips also have the benefit of the possibility of two-way communication, allowing students to ask questions or educators to tailor the material to the group's needs. Students are able to virtually visit places they would not be able to otherwise, such as the Liberty Science Centers' Live From: Cardiac Classroom (Recipient of the Association of Science-Technology Center's 2002 Innovation Award) and the Museum of Science and Industry's Live at the Heart project. In these visits, students are able to view open heart surgery, while asking questions and talking to the surgeons after they perform the surgery. In the Jason Project, students are able to work with scientists in diverse environments such as rainforests and wetlands. Although there are some opportunities for home-schooling individuals, currently both of these projects are run through schools. They highlight the possibilities of the Internet as a science learning resource, but are unavailable to the average Internet browser. Participants in these virtual field trips find them highly memorable — how can this type of experience be shared with the general public? New projects are in development that may provide the opportunity to research such experiences. Indiana State Museum is currently planning a program of virtual field trips aimed at community groups and families, hosted at libraries and other like organizations. The Institute will be involved with assessing this effort.

Some co-visiting experiments have dealt directly with the issue of social interaction by focusing specifically on the relationship between the physical and virtual site and encouraging the distant visitor to make a journey to the location some time after his virtual visit, in order to continue a relationship with the physical museum. These institutions have pioneered different methods of "co-visiting," primarily through physical robots (Goebel & Hoffmann, 2003; Giannoulis et al., 2001) or museum wearables (Chalmers & Galani, 2002; Sparacino, 2002). In the case of the robots, they are equipped with a camera, microphone and speakers and then are connected to the Internet. Virtual visitors connect with the robot over the Internet and then are able to literally visit the physical location, not only viewing the exhibition or event, but also participating and influencing the outcomes at hand. These robots do allow mediated communication between those present in the institution and those who are connected online. Little research has been conducted to understand the impacts of such visits. For instance, does the novelty of the co-visiting experience increase or detract from the content matter of the exhibition or event? Do visitors spend more time due to the novelty of the equipment? Does the initial effort to become oriented to the equipment interfere with the learning process? One pilot project, The Virtual Gorilla Project , discovered that its proposed virtual reality headsets involved too much time and material costs to overcome the novelty factor, causing radical changes in the project's hardware approach. Research such as this can allow professionals to make more informed choices about designing experiences.

Museum wearables are similarly a camera, microphone and speakers, connected to the Internet, but worn by an actual individual visiting the physical site. This allows a geographically distant group to visit an institution jointly, provided one of the members of the group can be at the physical site with the appropriate equipment. The drawbacks to this approach are that the virtual experience is dictated by the constraints of the physical site — the exhibition available, the equipment available, the institution being open, and etcetera. Researchers at the Institute (Kate Haley Goldman, Martin Storksdieck, & Lynn Dierking) have just begun working on a European Commission-funded project called CONNECT [Designing the Laboratory of Tomorrow by using Advanced Technologies to Connect Formal and Informal (Free-Choice) Learning Environments]. This project aims to integrate virtual field trips, museum wearables and online content from science centers and museums across the continent within a "virtual science theme park" that would provide access to real-time and archived events. One of the aspects of the project we intend to investigate is whether it is possible to integrate social aspects into such virtual trips and the role that they may play in student meaning-making.

Other institutions have experimented with creating virtual personas to facilitate the visitor's experience. The best of these software agents can accommodate user interests and preferences and enhance the user interface. [The worse of them, such as the Microsoft paperclip, merely annoy and interfere (Economist, March 24, 2001).] The agents may take the form of digital animations with dialogue-based interactions built in, equipped with a knowledge base that allows them to greet users, answer questions appropriately and provide customized information (Almeida & Yokoi, 2003). It is still not clear how this form of social interaction influences learning, but it is clear from evaluation studies that users prefer animations which have as many human-like characteristics as possible. As argued by Bertoletti et al. (2001), a primary benefit to these agents is their ability to compensate for user unfamiliarity with the site, thus enhancing navigation and operation. In addition, the personalization features of these agents allow them rich records of user actions, providing useful evaluation and research data on the system and great potential to be "intelligent systems" that would remember the visitor on a return trip and help him or her extend the previous visit. One of the more unique uses of virtual personas was developed to help provide historical, cultural and social context for works of art (Iurgel, 2003). Project Art-E-Fact is a collaboration of several institutions in four European countries. The user interacts with multiple personality-laden personas, who help provide context by their interaction with each other and the user. Using software agents in this format is much more akin to museum theatre than to a guided tour.

This interaction with virtual characters taken to the natural extreme has led to the development of role-playing activities, which educate virtual visitors through a game-like situation. Examples include the Brookfield Zoo's *In Search of the Ways of Knowing Trail*, where virtual visitors learn about an ecosystem by interacting with animated characters as they journey through an African forest. Other developers have experimented directly with adapting established gaming technology to provide virtual immersive educational environments (Calef et al., 2002). In the Virtual Leonardo project, visitors are able to direct their software agents to interact with other visitor's agents in the multi-user virtual museum. This virtual social interaction suggests many questions — how does the virtual nature of the experience change the interaction? Are visitors more likely to interact with strangers in a virtual environment than they would be in at a physical science center? This could be an effective method to engage visitors who might not typically make use of explainer staff at a physical science center. Clearly virtual facilitation would be an interesting area of research in this domain.

As institutions struggle to provide more content and more interactive experiences for their virtual visitors, they must make choices on how to deal with several issues, such as real-time or asynchronous communication. Is this process dependent on someone in the physical science center, so that timing may be limited? Does this process provide communication with actual humans or is the connection with a virtual human? Are the resources available to the general public, or are they primarily targeted for schools? Each of these decisions on how to provide a communication-rich experience has benefits and limitations in terms of the strength of connection, dependency on the physical site and the restriction of the audience.

Due to the advantages and limitations of each of these types of interaction, several institutions, such as the Exploratorium, are making use of more than one format of communication. There are several projects that have taken the next step, creating a multi-purpose platform, allowing a user to access content, interactives and other features of multiple institutions. One of the larger of these collaborations, TryScience.org, has contributions from over 622 science centers and museums. TryScience incorporates Web cams, interactive activities, and online experiments. Although full visitor evaluation on the site remains to be done, visitor stay time is approximately five minutes per session, longer than it would be to the average individual site (Friedman & Marshall, 2004). Aspects of the CONNECT project described earlier will be similar to TryScience and one aspect of our research will be to investigate how students (and their families) make use of the resources within the virtual science theme park and to what end.

Physical Context

The physical context of virtual and physical visiting is entirely different. Although each set of visitors must find their way to the site and navigate within it, currently virtual visitors are less active physically during the visit, unable to move their bodies through the actual space, and unable to use their full range of sensory functions. Though the page may be interactive, so that the virtual visitor is providing input and interacting in meaningful ways, it is not a whole-body experience as interaction is within a science center. In addition, the presentation of the environment is restricted to the screen, not an immersive experience one would have within a physical science center. There can be no sense of impressive areas' monumental space, such as entrance to some science centers, or the conception of scale of the exhibitions.

Site Usability

Usability of a site is clearly a key element of any virtual visit. Due to the low opportunity costs incurred when making a virtual visit, any sort of difficulty in interacting with the site motivates the visitor to fulfill their needs and interests elsewhere. There is no doubt that it is critical to conduct usability research for free-choice learning sites. However, as we suggested in the opening of this chapter, the focus on usability obscures larger, more critical issues at stake, such as whether visitors are interacting with the content in effective and meaningful ways and what meaning they are making of these interactions. Usability issues, primarily the roles of orientation and design, clearly are one of the physical context factors of the Contextual Model of Learning, however, since this is an area in which much of the current research and evaluation has been conducted we feel there are more profitable areas of research to investigate.

Orientation and Wayfinding in the Space

When visiting a physical science center, the visitor is surrounded by signs, displays and other information, from which they take cues and make sense of the environment. Each center tends to be unique in the ways in which they provide information to their visitors. Web sites, for better or worse, even science center Web sites, are far more standardized. One can expect to find similar organizing principles, not only in other science center sites, but also in commercial sites. This may actually make someone who is an Internet user but a novice science center visitor more comfortable, as they can become quickly familiar with science center Web sites. Though there is a potential downside; if virtual science centers look so similar to other sites this can downplay their individuality and the resources they have to offer. Research into the ways that people perceive their use of science center Web sites as compared to commercial sites would be interesting. For instance, we could hypothesize that individuals who do not generally make visits to physical science centers but are proficient on the Internet would feel more comfortable making a virtual visit than a physical visit. In addition, we might hypothesize that they would have a better chance at a meaningful interaction with the site, since they were not as concerned with issues of orientation and familiarity. But as there is little research into this field, these remain largely unsubstantiated hypotheses.

One interesting aspect of online visits is the sheer number of visitors that do not walk in the "front door," even after a short visit to check the site out. The fear of the museum professional is that a virtual visitor could get lost and lose interest more easily and leave, without ever discovering the richness of the information and experience provided (Nordbotten, 2000). In looking at log-files over a three-year period, Nordbotten (2000) found that 85% of the virtual visits were started from a search engine request and of those 76% started at a detail page, rather than the home page. This study found that visits starting at this level were significantly shorter than sessions begun at an exhibition home page. Semper et al. (2000) found that at least 40% of the visitors to the Science Learning Network arrived to the virtual site via a search engine. In addition to the use of a search engine, visitors can access detail pages directly (without visiting the institution's home page) via referral links from other sites or persons, or as a repeat visitor using a bookmark to get back to a specific page. This ability for fluid connections among sites is another ideal research opportunity. For instance, do virtual visitors who enter via a "side-door" distinguish where the visit begins and ends?

Subsequent Reinforcing Events and Experiences Outside the Museum

Learning does not have an on-off switch. People constantly take in new information and experiences and integrate those experiences into their lives, whether they are at a grocery store or at a science center. The experience a visitor has at a science center, whether virtual or physical, continues after the visit in a variety of ways. Virtual visiting means that one can visit multiple museums and science centers, or even one science center and multiple other sites, within the same session. For the most part, physical science center visitors do not have that opportunity (there are a few cities, such as San Francisco, Washington, D.C., St. Louis and Los Angeles, in which science museums, science centers and aquariums are grouped within close proximity, allowing for joint visitation, but these are rare). This ability to make connections among sites, comparing and contrasting a variety of visits, is an ideal research opportunity in which to investigate how visitors connect their free-choice science learning experiences.

Take for example a hypothetical Internet session by Steven. Steven is going online to plan his trip to London, and is curious about what he might see there. He checks out possible timing and fares of flights, and then moves on to find out what the hours at the Buckingham Palace are, if there are any special programs at the Natural History Museum, where he can get theater tickets, the latest gossip on the royal family and what's on exhibit at the Tate Modern. Visiting the Tate reminds him of other museums that have more than one location, so he decides to check out the Guggenheim site. After about 27 minutes online, Steven shuts the computer down. During those 27 minutes, Steven has not only visited three museums, but numerous other sites. These visits are all connected by Steven's agenda at this moment; he has immediately placed them within a particular context, though it may be a shifting context, within his life. The fact that these three visits are together within a single session suggests that they have a relationship within a contextual framework. What research on virtual visits has been conducted has only focused on the visit to a single institution. There is a distinct need for research that takes into account the relationships between sites visited and the impact of visiting multiple sites.

Conclusion

As suggested throughout this chapter, the majority of virtual learning research is focused on classroom-based practices, rather than free-choice learning situations. There is very little research on how learning occurs at places such as virtual science centers and what learning results from these visitor interactions. Yet, research in this area is critical — clearly the learning potential of virtual science centers is enormous and to date has not been realized. Better understanding the nature of the virtual science center experience and the factors that contribute to learning online will enable the field to better design virtual science centers that effectively communicate to the public, as well as begin to build a body of knowledge about how people in the 21st century engage in free-choice learning online.

Models such as the Contextual Model of Learning are useful tools, because it is through comparison to the wealth of material on learning in science centers that we will be able to identify gaps in understanding about the virtual science center and thus frame a research agenda for the future. Practitioners can use this model to examine virtual experiences and think more deeply about how these experiences can be designed to enhance learning. Research progress has been made in some arenas. For instance, we know that virtual visitors are generally in search of content and enjoyable experiences, rather than using virtual museums for the purpose of social engagement. However, as we pointed out in the sociocultural section of this chapter, that could all change as virtual science centers and museums increasingly assume the role of gathering place and community center. We know that the lack of "opportunity costs" for virtual visits dramatically changes the nature of the virtual visit; allowing more frequent access but generally resulting in shorter visits. Other factors, even ones heavily investigated in the distance learning field like choice and control, have been neglected in the free-choice learning arena. Choice and control are factors that greatly influence a physical visit to a physical science center, and are likely to influence virtual visitors. Yet the factors of choice and control in virtual settings may also encompass issues of access to software, plug-ins and download times. And there are fascinating developments going on in the area of visitor interaction with facilitators, either real or virtual. As these technologies are more effectively and seamlessly integrated into the activities of institutions, it will be important to have a research agenda in place to investigate these efforts and a plan for how these findings can influence practice. Only then will we be able to both improve practice and inform the field of the value of such virtual experiences.

References

Adelman, L., & Dierking, L.D. (2000). *Girls at the center*. Technical report. Annapolis, MD: Institute for Learning Innovation.

Adelman, L., Dierking, L.D., & Adams, M. (1999). *Girls at the center, 1999*. Technical report. Annapolis, MD: Institute for Learning Innovation.

Almeida, P., & Yokoi, S. (2003). Interactive character as a virtual tour guide to an online museum exhibition. In *Museums and the Web, Selected Papers from Museums and the Web 2003*. Pittsburgh, PA: Archives & Museum Informatics.

Argyros, A. et al. (2001). Enhancing museum visitor access through robotic avatars connected to the Web. In *Museums and the Web, Selected Papers from Museums and the Web 2001*. Pittsburgh, PA: Archives & Museum Informatics.

Artz, K., Bartow-Melia, A., Glassman, B., & McCray, K. (1999). *Intergenerational experiences and the OurStory Program: History through children's literature program*. Evaluative survey report. Washington, D.C.: National Museum of American History.

Bernier, R. (2002). *The uses of virtual museums: The French viewpoint*. Museums and the Web Annual Conference, Boston. Archives & Museum Informatics.

Bertoletti, A. C. et al. (2001). Providing personal assistance in the SAGRES Virtual Museum. In *Museums and the Web, Selected Papers from Museums and the Web 2001*. Pittsburgh, PA: Archives & Museum Informatics.

Borun, M., & Dritsas, J. (1997). Developing family-friendly exhibits. *Curator, 40*(3), 178-96.

Borun, M., Chambers, M. & Cleghorn, A. (1996). Families are learning in science museums. *Curator, 39*(2), 123-38.

Borun, M., Dritsas, J., Johnson, J.I., Peter, N., Wagner, K., Fadigan, K., Jangaard, A., Stroup, E., & Wenger, A. (1998). *Family learning in museums: The PISEC perspective*. Philadelphia: The Franklin Institute.

Bowen, J. (Ed.). (1999). Time for renovations: A survey of museum Web sites. In *Museums and the Web, Selected Papers from Museums and the Web 1999*. Pittsburgh, PA: Archives & Museum Informatics.

Bruner, J. (1996). *The culture of education*. Cambridge, MA: Harvard University Press.

Buchner, K., Dierking, L.D., & Soren, B. (1999). *The story in history summative evaluation*. National Museum of American History, Washington, DC. Technical report. Annapolis, MD: Institute for Learning Innovation.

Calef, C. et al. (2002). Making it realtime: Exploring the use of optimized realtime environments for historical simulation and education. In *Museums and the Web, Selected Papers from Museums and the Web 2002*. Pittsburgh, PA: Archives & Museum Informatics.

Calvin, W.H. (1997). *How brains think*. New York: BasicBooks.

Chadwick, J. (1998). Public utilization of museum-based World Wide Web Sites. (unpublished Ph.D. Dissertation). University of New Mexico.

Chadwick, J., Falk, J.H., & O'Ryan, B. (2000). Assessing institutional Web sites. Council on Library and Information Resources. Retrieved on February 6, 2004, from *http://www.clir.org/pubs/reports/pub88/appendix2.html*

Chalmers, M., & Galani, A. (2002). Can you see me? Exploring co-visiting between physical and virtual visitors. In *Museums and the Web, Selected Papers from Museums and the Web 2002*. Pittsburgh, PA: Archives & Museum Informatics.

Crowley, K., & Callanan, M. (1998). Describing and supporting collaborative scientific thinking in parent-child interactions. *Journal of Museum Education, 23*(1), 12-17.

Crowley, K., Callanan, M., Lipson, J.L., Galco, J., Topping, K., & Shrager, J. (n.d.). *Shared scientific thinking in everyday parent-child activity* (in review).

Crowley, K., Galco, J., Jacobs, M., & Russo, S.R. (2000). *Explanatoids, fossils and family conversations.* (paper presented as part of a set, Museum Learning Collaborative: Studies of Learning from Museums, at the annual meeting of the American Educational Research Association, New Orleans).

Dierking, L.D. (1987). *Parent-child interactions in a free choice learning setting: An examination of attention-directing behaviors.* Unpublished PhD dissertation. University of Florida.

Dierking, L.D., & Anderson, D. (1998). *Summative evaluation of World We Create, Louisville Science Center,* Louisville, KY. Technical report. Annapolis, MD: Institute for Learning Innovation.

Ellenbogen, K.M. (2003). From dioramas to the dinner table: An ethnographic case study of the role of science museums in family life. *Dissertation Abstracts International, 64*(03), 846A (University Microfilms No. AAT30-85758).

Falk, J.H. (2001). Free-choice science learning: Framing the issues. In J.H. Falk (ed.), *Free-Choice Science Education: How We Learn Science Outside of School.* New York: Teachers College Press.

Falk, J.H., & Dierking, L.D. (1992). *The museum experience.* Washington, DC: Whalesback Books.

Falk, J.H., & Dierking, L.D. (1998, May/June). Free-Choice learning: An alternative term to informal learning? *Informal Learning Environments Research Newsletter.* Washington, DC: American Educational Research Association.

Falk, J.H., & Dierking, L.D. (2000). *Learning from museums.* Walnut Creek, CA: AltaMira Press.

Falk, J.H., & Dierking, L.D. (2002). *Lessons without limit: How free-choice learning is transforming education.* Walnut Creek, CA: AltaMira Press.

Falk, J.H., & Storksdieck, M. (in revision). A multi-factor investigation of a science center exhibition. *Science Education.*

Falk, J.H., Moussouri, T., & Coulson, D. (1998). The effect of visitors' agendas on museum learning. *Curator, 41*(2), 106-120.

Friedman, A., & Marshall, E. (2004). TryScience: Virtual synergy. *Museum News, 83*(2), 9-10

Gaia, G. (2001). Towards a virtual community. In *Museums and the Web, Selected Papers from Museums and the Web 2001.* Pittsburgh, PA: Archives & Museum Informatics.

Giannoulis, G. et al. (2001). Enhancing museum visitor access through robotic avatars connected to the Web. In *Museums on the Web.* Pittsburgh, PA: Archives & Museum Informatics.

Goebel, S., & Hoffmann, A. (2003). Designing collaborative group experience for museums with Telebuddy. In *Museums and the Web, Selected Papers from Museums and the Web 2003.* Pittsburgh, Archives & Museum Informatics.

Guernsey, L. (2002, March 28). As the Web matures, fun is hard to find. *The New York Times*, pg. E1-E5

Haley Goldman, K. (2003). *Atlanta History Center distance learning front-end report.* Technical report. Annapolis, MD: Institute for Learning Innovation.

Haley Goldman, K., & Schaller, D. (2004). Exploring motivational factors and visitor satisfaction in on-line museum visits. In *Museums and the Web, Selected Papers from Museums and the Web 2004*. Pittsburgh, PA: Archives & Museum Informatics.

Hay, K. E., Crozier, J., & Barnett, M. (2000). The Virtual Gorilla Project. Paper presented at the *Annual Meeting of the American Educational Research Association*, New Orleans, LA.

Hudson, J., & Nelson, K. (1983). Effects of script structure on children's story recall. *Developmental Psychology*, *19*, 525-635.

Iurgel, I. (2003). Experiencing art on the Web with virtual companions. In *Museums and the Web, Selected Papers from Museums and the Web 2003*. Pittsburgh, PA: Archives & Museum Informatics.

Jacobson, G., & Swiader, L. (2003). Integrating real time communications applications in a museum's Website. In *Museums and the Web, Selected Papers from Museums and the Web 2003*. Pittsburgh, PA: Archives & Museum Informatics.

Jensen, J. (1999). *Goals, server logs and other arcane lore: An evaluation of 2 Canadian Websites*. Presentation at the 1999 Annual Visitor Studies Conference, Chicago.

Johnston, D.J. (1999). *Assessing the visiting public's perceptions of the outcomes of their visit to interactive science and technology centres.* PhD dissertation. Curtin University of Technology

Koster, E.H. (1999). In search of relevance: Science centers as innovators in the evolution of museums. *Daedalus*, *128*(3), 277-296.

Kravchyna, V., & Hastings, S.K. (2002, February). Informational value of museum Web sites. *First Monday*, *7*(2). Retrieved February 9, 2004, from *http://www.firstmonday. org/issues/issue7_2/kravchyna/index.html*

Luke, J., & Dierking, L.D. (1999). *Tripod*. Technical report. Annapolis, MD: Institute for Learning Innovation.

Luke, J., Coles, U., & Falk, J.H. (1998). *Summative evaluation of DNA Zone*, St. Louis Science Center, St. Louis, MO. Technical report. Annapolis, MD: Institute for Learning Innovation.

Madden, M. (2003). *America's online pursuits: The changing picture of who's online and what they do*. Washington, DC: Pew Internet & American Life Project.

Milekic, S. (2003). The more you look the more you get: Intention-based interface using gaze-tracking. In *Museums and the Web, Selected Papers from Museums and the Web 2003*. Pittsburgh, PA: Archives & Museum Informatics..

Miltiadou, M., & Savenye, W.C. (2003). Applying social cognitive constructs of motivation to enhance student success in online distance education. *Educational Technology Review*, *11*(1).

Moussouri, T. (1997). *Family agendas and family learning in hands-on science museums.* Unpublished dissertation. University of Leicester, UK.

Nelson, K., & Brown, A.L. (1978). The semantic-episodic distinction in memory development. In P.A. Ornstein (Ed.), *Memory development in children* (pp. 233-41). Hillsdale, NJ: Erlbaum.

Nordbotten, J. (Ed.). (2000). Entering through the side door: A usage analysis of a Web presentation. In *Museums and the Web, Selected Papers from Museums and the Web 2000.* Pittsburgh, PA: Archives & Museum Informatics.

Ockuly, J. (2003). What clicks? An interim report on audience research. In *Museums and the Web, Selected Papers from Museums and the Web 2003.* Pittsburgh, PA: Archives & Museum Informatics.

Pintrich, P., & Schunk, D. (1996). *Motivation in education: Theory, research & applications.* Englewood Cliffs, NJ: Prentice-Hall.

Rosenfeld, S.B. (1980). Informal learning in zoos: Naturalistic studies of family groups. *Dissertation Abstracts International, 41*(07). (University Microfilms No. AAT80-29566).

Rountree, J. et al. (2002). Learning to look: Real and virtual artifacts. *Educational Technology and Society, 5*(1), 129-134.

Ruohotie, P., & Nokelainen, P. (2003). Practical considerations of motivation and computer-supported collaborative learning. In T. Varis, T. Utsumi, & W. R. Klemm (Eds.), *Global peace through the global university system.* Hameenlinna, Finland: University of Tampere.

Sarraf, S. (1999). A survey of museums on the Web: Who uses museum websites? *Curator, 42*(3).

Schaller, D.T., Allison-Bunnell, S., Borun, M., & Chambers, M.B. (2002). How do you like to learn? Comparing user preferences and visit length of educational Web sites. In *Museums and the Web, Selected Papers from Museums and the Web 2002.* Pittsburgh, PA: Archives & Museum Informatics.

Schauble, L., & Gleason, M. (2000). *What do adults need to effectively assist children's learning?* Paper presented as part of a set, Museum Learning Collaborative: Studies of Learning from Museums, at the Annual Meeting of the American Educational Research Association, New Orleans.

Semper, R. J. et al. (Eds.). (2000). Who's out there? A pilot user study of educational Web resources by the Science Learning Network. In *Museums and the Web, Selected Papers from Museums and the Web 2000.* Pittsburgh, PA: Archives & Museum Informatics.

Son of paperclip. (2001, March 24). *Economist.*

Sparacino, F. (2002). The Museum Wearable: Real-Time Sensor-Driven Understanding of Visitors' Interests for Personalized Visually-Augmented Museum Experiences. In *Museums and the Web, Selected Papers from Museums and the Web 2002.* Pittsburgh, PA: Archives & Museum Informatics.

Steinbach, L. (2001). *Using interactive broadband multicasting in a museum lifelong learning program*. Paper presented at the Museums and the Web Annual Conference, Pittsburgh.

Sylwester, R. (1995). *In celebration of neurons*. Alexandria, VA: Association for Supervision and Curriculum Development.

Endnote

[1]   The Institute for Learning Innovation has been advocating *free-choice* learning as a preferred alternative to *informal* learning to describe the learning that occurs from experiences in science centers. Free-choice learning is self-directed, voluntary, and rather than following a set curriculum, is guided by the individual learner's needs and interests. This term focuses on the characteristics of the learning rather than defining it by where it occurs. This would seem essential when discussing learning from virtual centers, free of geographic, physical and time constraints [See, for instance, Falk & Dierking (1998). Free-Choice learning: An alternative term to informal learning? *Informal Learning Environments Research Newsletter.* Washngton, D.C.: American Educational Research Association; Falk, J.H. (2001). Free-choice science learning: Framing the issues. In J.H. Falk (ed.), *Free-Choice Science Education: How We Learn Science Outside of School.* New York: Teachers College Press; Johnston, D.J. (1999). Assessing the Visiting Public's Perceptions of the Outcomes of their Visit to Interactive *Science and Technology Centres.* (Ph.D. Dissertation). Curtin University of Technology; Falk & Dierking (2000); Falk (1999)].

Chapter III

Contextualized Virtual Science Centers

Andreas Zimmermann,
Fraunhofer-Institute for Applied Information Technology, Germany

Andreas Lorenz,
Fraunhofer-Institute for Applied Information Technology, Germany

Marcus Specht,
Fraunhofer-Institute for Applied Information Technology, Germany

Abstract

Today it is not enough just to supply content without the consideration of the recipient, his/her current task and situation. Therefore the time, the location, the particular technical limitations, and the modality and style of reception are important parameters for contextualized interactions and information delivery. Context-sensitive content and information processing are especially assets for the generation of added value in information delivery. This chapter describes how contextualization can be performed in virtual science centers. The demand for context-sensitive functionalities constitutes a crucial challenge for application developers, system integrators and product designers. This chapter furthermore offers a potentially substantive approach for development and maintenance of context-sensitive systems and services by adapting information brokering techniques.

Introduction

Given the affordances of the World Wide Web, one of its true instructional applications of high potential are online simulations of science experiments and phenomena. Thus far in the evolution of the Web this has surfaced in the form of *virtual science museums*: Web sites that offer vicarious science experiences using Web technologies. Most of these sites are the online presence for actual museums located around the world (such as http://www.science.edu.sg/ssc/index.jsp, http://www.deutsches-museum.de/e_index.htm), but it is their specific orientation to science that can make them especially invaluable for, for example, a classroom.

Virtual science centers encompass hands-on exhibits, lab spaces, science fairs, and communication spaces in a multi-user setting. These will include interactive interfaces to simulations and visualizations created to communicate key concepts in science and technology to an increasingly diverse audience. For example, the Chabot Virtual Space & Science Centre (http://www.chabotspace.org/vsc/) explains the physical constraints for a Lunar Lander and lets the visitor experience the behavior of the Lander in an online simulation. Virtual science centers make science information readily available to anyone with access to the Internet, like a family exploring exhibits from their home computer, or a group of students operating telescopes from their classroom. Integrated portals (like the European Collaborative for Science, Industry and Technology homepage: http://ecsite.ballou.be/new/index.asp) provide access to innumerable virtual exhibitions for multiple topics.

For the flexible use of information about exhibits of a virtual science center we propose a centralized model around a domain ontology that is described in a software tool for information brokering. We think that the proposed model is a very flexible way to reuse existing information and support curators and exhibition experts in the design of a variety of personalized experiences ranging from virtual exhibitions, virtual science museums to virtual science centers. In such scenarios, the authoring, meta-tagging and distribution of this valuable information on exhibits is a non-trivial task.

A variety of guidance or information systems have been developed in the last few years for the support of a museum visit or for the preparation of such a visit. In most cases the authenticity and the possibility to contextualize the information presentation to the current position or situation of a user were seen as a central issue (Shilit et al., 1994; Dey & Abowd, 2000). Classic information systems in electronic space (like e-learning applications) and physical space (such as museum guides, city guides, navigation systems) have a common information filtering process (cf., Figure 1). The user, while interacting with his/her device, provides information (implicitly or explicitly expressed) to the application, which recognizes the changed parameters due to the interaction. The incoming information is used for maintaining the personal profile of the user, which determines the filtering process for relevant information. Furthermore, the filtering process is influenced by domain-dependent knowledge inside the application. Only the information provided by the Content-Management System that passed the filters is finally handed over to the internal presentation engine for being delivered to the (information-) consumer.

Figure 1. Classic approach covering information filtering inside the application

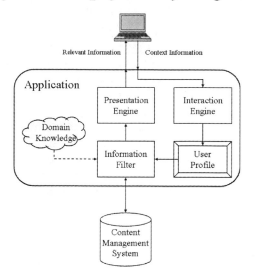

Personalization Techniques for Virtual Exhibitions

For museums and exhibitions, generally the individual visitor's level of factual, general and domain-specific *knowledge* is a valuable source for adaptive operations. The visitor's *experience* describes how familiar s/he is with the structure of the information space and how easy s/he can navigate in it. Possible indicators for knowledge and expertise are the type of information the visitor requests or the frequency of usage. The interaction engine records all interactions of the visitor with the system in an individual user model. For each page of a virtual exhibition, the current information unit is assigned with the corresponding knowledge and expertise state of the visitor (e.g., "learned," "not completed," "unknown," etc.). During the course, the corresponding unit is marked as visited in the user model. An assessment of these indicators and the conclusion, that one specific visitor might not need further support, still remains a non-trivial task.

Being aware of the visitor's *interests* and *disinterests* is another important prerequisite for the content adaptation. The interests configure filters for the information presented to the visitor. One opportunity to identify interests relies on explicit feedback: After closing one information unit, visitors can either rate the unit as "interesting" or "not interesting," request additional information about the current story, or let the system know that they already knew the content. Additionally, the interaction engine covers the observation of the online visitor's browsing behavior (implicit feedback). By positive feedback (e.g., following a recommended link) and analyzing the time (e.g., time the user spent while reading the page), interest in the associated information is expressed; whereas concluding disinterest from browsing behavior remains an unsolved problem.

Another prominent and often implemented opportunity is the adaptation to user *preferences*, like language, colors, modality of interaction, and etcetera. If the user does not inform the online application directly about such parameters, individual preferences cannot be deduced automatically by the system. On the other hand, usage *habits* own a longer-term sense and can be obtained by an analysis of the interaction history.

The identified features related to the current individual visitor influence the information selection process, for example, a qualified visitor can be provided with more detailed and deep information while a novice can receive additional explanations. Beside *adaptive presentation* of content regarding knowledge, interest and preferences, Brusilovsky (1996) defines *adaptive navigation support* to help visitors to find their paths through the information space. With the simplest technology, direct guidance recommends the next "best place" to visit according the visitor's interests or level of knowledge — in turn, *adaptive hiding* restrict the navigation space, for example, by hiding links to "not interesting" pages. To avoid "follow me or no help" problems, *adaptive ordering* can be used to sort all links of a particular page according to the user profile. To support the visitor's decision of what link to follow, *adaptive annotation* augments links with some form of comments giving a look forward to what is behind the annotated link.

Structuring the Information Domain

A central issue in context-aware information selection and presentation is to structure an information domain not only appropriate from an information engineer's point of view but also from the user's viewpoint. Especially for designing personalized information services the intuitiveness of the information structure is essential for the successful application of user modeling and personalization methods. Usually, applications for non-virtual information systems have several models in common, such as Location Model, Augmentation Model, Domain Model and User Model. For illustration, we will briefly describe the models we have developed for our applications for personalized presentation of multimedia information at the Fraunhofer Institute for Applied Information Technologies (Fraunhofer FIT).

The *World Model* (*Space Model*, *Location Model*) describes the physical environment the user moves through while interacting with the system (Goßmann & Specht, 2001). For this environment, the space model contains the geometric information of the exhibition space and its objects. The geometric information can be formulated as a floor plan illustrating positions, size and shape of objects, doors and windows, or physical barriers the visitor cannot get over. More conceivable, the world model of the LISTEN-project (a system for an immersive audio augmented environment applied in the art exhibition domain) is a detailed VR-based geometric model. It is described as a geometric scene graph and can be tested and prototyped in a CAVE system (Eckel, 2001). Through a wireless motion-tracked headphone the virtual space can be explored in real space with virtual audio content displayed.

The *Augmentation Layer* on top of the World Model defines virtual areas like zones, segments, and triggers within the world model. Additionally, virtual objects (e.g., sound sources or switches) are placed in the environment. By defining zones and segments, the visitor's location and focus obtain valuable meanings in correlation to physical objects. The augmentation layer filters the position and motion of the user by dividing the dimensions the user moves through into meaningful constraints and deriving continuous parameters. Switches and triggers contain active information or sound objects the users of the system interact with.

The *Domain Model* holds information about sound objects and other hypermedia objects connected to the physical space via the augmentation layer by using metadata. The domain model builds up a virtual acoustic space, in which the location of virtual sound sources and spaces are defined. Stopping in front of an exhibit generates audio information about the art piece. Moving the head and body activates a further audio source, where music deepens the visitor's impressions, or the voice of a commentator talks about the artist or describes the period the painting originates from.

The *User Model* contains knowledge and profile information about the system's users. While the user moves in physical space events are sent to the user model, which is continuously refined. The user model stores knowledge about the visitor like preferences, interests in arts, and the history of the visit. In combination with the visitor's spatial position (delivered by the tracking system), the content of the user model strongly influences the presentation of information according to the current visitor's context (cf., Zimmermann et al., 2003).

The knowledge base described by the number of models needs to be represented in a computer-accessible manner. Furthermore, curators who know best about the artifacts in the exhibition usually are non-experts in computer science, databases and programming. Both problems can be addressed by using information brokering technologies mediating between the curator and the computer system. In this chapter we will therefore focus on the value of information brokering for authoring and managing the huge amount of information pieces in different media types in virtual science centers.

Information Brokering

Information brokering is a value adding process of mediation between information demands and information offers. The added value emerges from the understanding of the domain complexity and the definition of a useful vocabulary. Additional structures and interpretation rules (implicit and explicit) in information exchange simplify the processing and comprehension of exchanged information. Figure 2 illustrates the interrelation between data, information, and knowledge.

Three roles participate in the information brokering process: the *provider* offers information, the *consumer* demands information and the *broker* mediates between the two. The quality of the information brokering process depends to a great extent on the knowledge available to the broker. Knowledge about available sources, the domain, consumers and other brokers is needed. *Source knowledge* is created in the domain representation or maintenance processes and describes the quality of sources and how they can be accessed. *Domain Knowledge* is about the contents of the brokering domain and should reflect the provider's understanding of the domain as well as the consumers' perception of the domain. *Consumer Knowledge* is created in the consumer-oriented process and describes the consumer and his concrete information need. Consumer knowledge has to map onto domain knowledge to fulfill the information need. To ensure

Figure 2. From data to information and knowledge

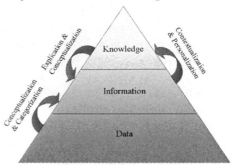

Figure 3. Roles, tasks, and domain models

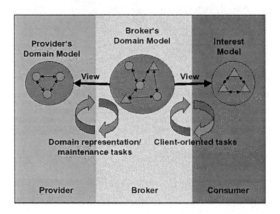

an optimal service to consumers, they should be served by the best broker according to their information need. This assignment task depends on the availability of *Expertise Knowledge* about different brokers.

Information brokering processes make use of and create a number of *information items* that describe single units of information (cf., Klemke & Koenemann, 1999). As Klemke (2002) has defined, each information item is an instantiation of a *concept*, which describes the structure of the brokered items. In order to organize information items, *categories* describe fundamental principles or ideas. In many cases, these categories define hierarchical trees. In Figure 2, *conceptualization and categorization* is illustrated as a procedure for the concretion of raw data.

For authoring valuable information on exhibits and distributing demanded information among visitors in an efficient way, a brokering tool can be understood as a service mediating information items between a Content-Management System (CMS) — maintained by the exhibition's operating company — and the visitor's personal display. Using such a brokerage development and management environment [like the *Broker's Lounge*

Figure 4. The information broker mediates between the application and the content-management system

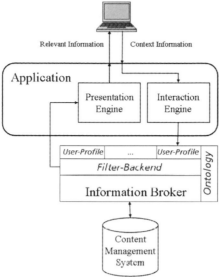

(Jarke, Klemke & Nick, 2001) developed at Fraunhofer FIT], a large variety of scenarios within the general framework of Figure 2 and Figure 3 can be quickly developed, and efficiently and flexibly executed.

For a virtual science center application, the brokering tool can be understood as a server mediating information items between the exhibition's curator and the external application interacting with the visitor. In the role of the information provider, the curator provides the domain knowledge, specifies the context parameters and authors the information items. From the information broker's point of view, the information is consumed by the visitor's application, e.g., a browser, although it finally passes the information to the visitor as the real consumer. In its role of the information consumer, this application defines an interest model regarding to the context parameters like the visitor's location, preferences, interests, and history of already visited objects.

As illustrated in Figure 4, the interaction engine inside the application observes the changed parameters and delivers the information to the brokering software. In contrast to the schema in Figure 2, the information filtering process was outsourced, directly returning the relevant information back to the presentation engine. In information brokering terminology, the application builds an information model and, triggered by events, inquires the broker. Thus, the system logic concerning information filtering becomes independent from the overall application logic. Moreover, the ontology representing the domain knowledge and the information items can be authored independently of the application-internal intelligence. The next section provides an outline of the definition of information items we had applied in our LISTEN project (Unnützer, 2001) for the Kunstmuseum Bonn (Museum of Art at Bonn, Germany). It also illustrates a hierarchical tree to categorize them.

Modeling Information Items

As described, information-brokering processes create and make use of a number of information items that are instantiations of concepts describing the structure of the brokered items. In order to organize information items, categories describe fundamental principles or ideas. In case of a virtual exhibition, the use of a central management component for all information items facilitates the management of items composed of text, sound, graphics, movies and other hypermedia items like interactive flash components.

In recent projects developing guiding systems and electronic art guides, the structuring and internal representation of information appeared to be a central issue for delivering the right information at the right time to a specific user. Even more, for the personalization of presentations and the contextualized delivery of information pieces, the underlying structure of the represented information occurred to be core part of the work. Nevertheless there was a trade off between the efforts of authoring information items into a highly enriched information representation and the daily work of curators and information providers for museum environments.

To find the right balance between user efforts for authoring and creating museum information and metadata in previous projects we often used an object-oriented approach, where the main entities were the art objects as such. This led us to typical presentation and structuring of information like those found in art databases and information systems of virtual science centers or museums.

By detailed user needs analysis of museum visitors and by the analysis of human guides we realized that the event character of museum visits in most cases is much more important than plain information about the art object as such. The presentation of data about artworks is just one style of presentation that can be appropriate, but in practice is rarely used by human guides. Furthermore, through our work with museum curators and artists in the LISTEN workshops we elicited different strategies and methodologies to structure information items and how to combine those information items in a highly flexible way for different users.

The main entities for generating presentations are therefore not the artworks as such, but the parts or chunks of a presentation that can come from a diversity of sources and use a variety of stylistic means to make the museum visit an *interactive* experience. Therefore, for creating interactive audio augmented spaces we decided to choose the sound item as the main entity. Sound items can be classified in a category system with several dimensions describing the sound items technically and from a stylistic point of view. An overview of the dimensions can be seen in Figure 5. This figure illustrates the domain model inside the graphical interface of the information-brokering tool, organized as a tree, and a list of information items of the LISTEN project. The single sound items are independent episodes or chunks of presentation that can be combined flexibly because they contribute to a variety of art objects.

We designed the domain ontology (metadata) for the exhibition to allow for the description of small information items on a variety of dimensions allowing for the connection and the individualized sequencing and presentation of these items. Besides the simple classification of the information items in a tree structure, the information-brokering tool allows for classification of sound items into multiple categories.

Figure 5. Screenshot of the domain-ontology for the Macke Exhibition

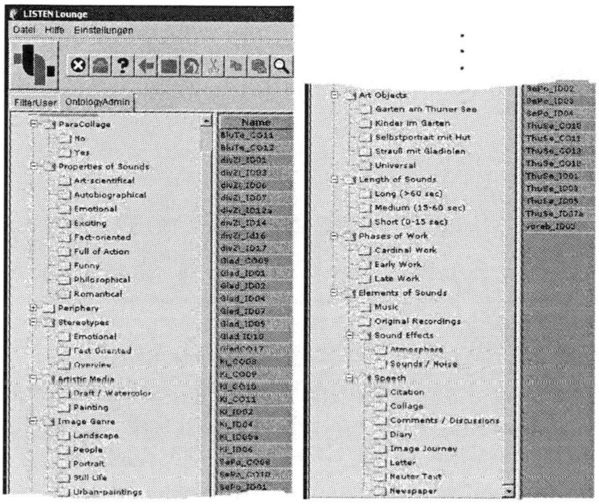

The domain ontology contains:

- technical descriptions of the sound items such as length of item, type (music, speech, sound effects),

- classifications concerning the relation to the physical space objects (art objects to which an item contributes, physical area zones or focuses to which they are connected),

- classifications about phases of work, image genre, or art technical aspects,

- classifications about the preferred target group like the stereotypical listeners for such a sound item, or the emotional impacts or dramaturgy.

In particular, speech sound items could be further classified into subcategories like Citation, Collage, Diary, Letter, Newspaper and others to describe their style of presentation. As described above, the multidimensional classification of the sound items allows for a variety of sequences and presentation styles even combining sound items on several channels, that is, typically music, effects, and speech. An enormous advantage of the description in such a way is that the curator of an interactive experience is not forced to design a complete sequence of information presentation but can combine his/her resources in a collage style or define sequencing rules on the level of possible connection categories and not for single sound items.

In the LISTEN application, the visitor's position and orientation are observed by a tracking system and interpreted within the context of a virtual environment, which connects the real world objects with virtual objects. Since the museum's visitors are mobile in physical space, their spatial positions are translated into virtual positions in the electronic space relative to virtual objects. The virtual environment enables the definition of virtual sound sources and the segmentation of the physical space into virtual zones. Depending on the visitor's location, preferences and so forth, combinations of categories are dynamically selected in order to pre-filter the information items. The next section illustrates how this solution can be adapted to an electronic museum.

Contextualization of Virtual Science Centers

Virtual science centers popularize science and technology among the online public through virtual science museums and exhibitions. In contrast to other museums, which usually host art collections or historical artifacts, virtual science centers display scientific artifacts with emphasis on *interactivity* (Orfinger, 1998). The interactivity (even simple controlling episodes like in http://www.deutsches-museum.de/mum/bildshow/index1.htm) enables online users to immerse into an online visit of an exhibition. The consideration of the personal preferences, interests, and knowledge forms the basis for adaptive information selection and presentation. To create immersive environments for the online visitor, we describe our approach to enhance personalization by considering context parameters and even the visitor's personal style of interaction.

Adaptation of the Existing Approach

In order to transfer our information brokering solution to the domain of a virtual science center, we have to map the physical space to an electronic space (cf., Honeyman, 1998). Without much doubt, the physical environment can be described for virtual reality technology. In such a VR-Environment, the visitor moves through and is tracked as though s/he is moving physically (Tramberend, 1999). In this case, the approach works unchanged because the application receives the same events from the interaction with its users and falls back on the same presentation strategies. Since this approach is not applicable for Web-based systems, and the visitor needs to be equipped with VR-hardware (like glasses, gloves, etc.), we follow an overlay model approach for context-aware information systems as introduced by Gross and Specht (2001). This methodology enables us to apply and reuse the algorithm for information filtering described in the previous section within other domains. Thus, the three concepts tracking, modeling of the information items, and domain modeling must be adapted or extended.

As described in the last section, the domain model connects the objects to categories. If the user enters the zone connected with a certain domain object, the category associated with this object is selected, in order to filter information items according to the user's preferences *and* position. In information brokering terminology, the application builds an information model and inquires the broker triggered by events. In a virtual science center application, the tracking system would deliver the current URI as position and the selected objects (e.g., text links, images, or clicks on maps) as a kind of the visitor's movement. Here, an event is fired when the user visits any Web source that is connected with a category of the broker's domain model.

Currently, the information broker falls back on a set of audio pieces (e.g., spoken text, music, effects) and returns the best matching item according to the inquired interest model of the user. In the domain of a virtual science center, the curator may probably extend this set of information items by any kind of interactive components, multimedia presentations, text fragments or links to Web pages. Therefore, the broker's domain model must be extended also to adequately reflect the changed information architecture. New categories have to be added and the metadata description has to be augmented, which can easily be accomplished with the assistance of the information-brokering tool. As a result, the system provides additional features for the specification of preferences to the users (e.g., users can turn off streamed video presentations) and thereby the user model is extended at the same time.

In contrast to the fixed sound installation in the real world LISTEN environment, a virtual science center has to take into account the abilities and features of the visitor's particular environment, that is, the Web browser. The user's specific settings can be initially inquired and automatically preset the information brokering tool. Based on known environmental properties the information items are filtered in advance and presented in an appropriate format. The ontology model of the information-brokering tool has to be adapted in a suitable way.

The mentioned extensions of the information-brokering tool do not affect its capability of filtering information items that best fits the user's context. The overlay model enables the retention of the presented concepts no matter whether the user moves in physical or in electronic space. The next section shows how the information presentation in a virtual science center can be further personalized and adapted to the visitor's needs and behavior. By means of a user model additional filters can be triggered or refined.

Personalization

A Web museum environment offers the ability to adapt the order of the presented exhibits and the attached information to the visitor's profile and position within the exhibition. In order to provide a personalized adaptation of the environment according to the visitor's context (i.e., interests, preferences and "motion"), the system has to build-up and maintain a user model. To give an idea of the personalization process, we will firstly introduce the approach we have applied to several context-aware applications.

Sensor data and metadata are an important data resource in user modeling and adaptive systems. A growing number of data sources about a user allow for more valid inferences and much more contextualized interactions between users and adaptive systems. The enrichment of information items with significant metadata enables the customization of information offers. By requesting user preferences, different user profiles can be built-up to facilitate the information filtering according to the user's needs. Besides informa-

Figure 6. Context-aware architecture of the LISTEN System

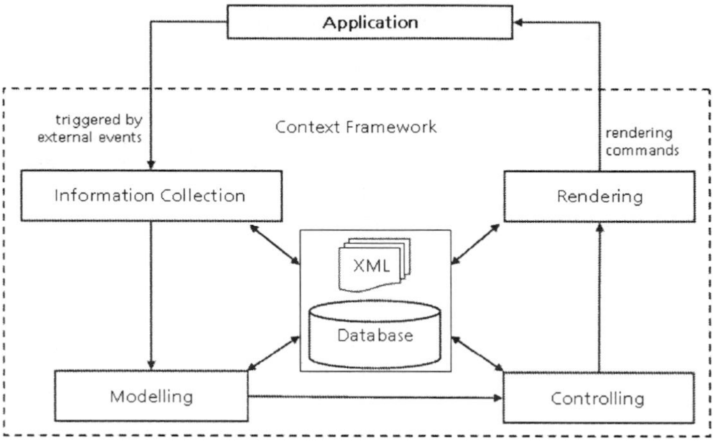

tion presentation, the system can provide recommendations to the visitor regarding to his/her context. On the basis of meaningful user profiles several adaptive strategies can be applied to guide the visitor of the Web museum on a specific (strictly predefined or context-aware adapted) tour.

In virtual science centers users automatically navigate an information space designed as a complement or extension of the real space. The selection, presentation and adaptation of the content of this information space take into account the user's current context. For the flexible and easy combination of the appropriate content and the current context, we chose to provide a centralized Java-based context and user modeling server. In addition, we use XML as the modeling language, which allows for an intuitive design, definition and configuration of the system components. Since this design tool will be used in various follow-up projects, we introduced a new model for the design of context-aware systems (cf., Figure 6).

In the preceding section we have illustrated the general approach for adaptive information presentation: observing the interaction, analyzing the data stream to gain knowledge, determining the information piece to be presented, and finally rendering this information appropriate to the user's device. This approach is true for both virtual science centers and real world exhibitions. Therefore, our personalization approach of virtual science centers is divided into four concrete steps: information collection, modeling, controlling and rendering. Each step fulfils a certain role within the user modeling process. The next subsections describe these modules in more detail.

Information Collection

A virtual science center has to consider the abilities and features of the user's particular environment (e.g., Web browser). Therefore, a network of sensors is placed in the environment and connected to variable parameters of the domain. These sensors are used for recognizing changes within the environment and especially for the perception of the user's interaction with this environment. An observation module receives all incoming events sent by the Web server. These event descriptions are pushed into a database. Thus, an event history for every visitor is saved and an implicit user profile is recorded.

Modeling

In a second step, the incoming sensor values are interpreted by several algorithms (e.g., machine learning or data mining algorithms) based on different models (e.g., overlay, statistic, etc.). Thus, semantically enriched information is extracted and more significant knowledge is gained relating to the user's behavior.

In our user modeling approach for a museum environment we chose to employ our above mentioned adapted information-brokering tool for modeling *user preferences*. Parts of the domain model of the museum environment are mapped to an ontology model (as shown in Figure 3). The users specify their preferences by simply selecting topics s/he is interested in from the displayed categories. From these requests, different visitor profiles can be built-up, which are stored within the tool. Based on these user selections and on significant metadata describing information items the personalization process performs a pre-selection and filtering of information offers and customization according to the users' needs (cf., Pazzani & Billsus, 1997).

Furthermore, we chose to employ *stereotypes* to define the user's *observation type*, which is common in adaptive systems (cf., Rich, 1989). In contrast to Oppermann and Specht (2000), the user's classification into one stereotype is done manually by the user and cannot be changed automatically during runtime. Additionally, the personalization engine is not yet able to perform an automated clustering and derivation of new stereotypes

Controlling

In a third step, a controlling component is necessary to decide what consequences must be taken if certain conditions in the user's context and in the individual user model configuration appear together. Based on these information sources, the control component assembles a sequence of commands in order to adjust certain variable properties of the environment. Different sequences of commands lead to different kinds of information presentation. Meaningful user profiles accurately document the visitors' activity within the virtual science center and can be exploited to adapt the environment in order to support the user or to provide personalized information. In our approach an XML-configurable rule-system controls the presentation of the scenery of a virtual science center. First, a pre-filtering of information items is performed based on the user's current Web location and focus. From this list the best-suited information item is selected referring to the user's history, stereotype and interests.

Rendering

Rendering means handling the connection back to the domain. This engine translates the assembled sequence of domain-independent commands into domain-dependent commands changing variable parameters of the domain (i.e., content of a HTML page) according to the user's behavior. Thus, the decisions taken by the controlling component are mapped to real world actions.

Contextualized Delivery and Information Logistic

Mainly the user's location in the information space describes the user's current context. In a virtual science museum the options for user support are manifold. The user's movements in the virtual world can be tracked and connected to the domain ontology, in order to learn about user's interests. Based on current context parameters available from the augmentation model these movements trigger information requests to the information broker. Such triggers can be the user visiting any Web source that is connected with a category of the broker's domain model or the user entering the virtual zone connected with a certain domain object. By automatically selecting the category associated with this object, information items are filtered according to the user's preferences and position.

In a virtual science center the basis for every kind of adaptation is the presentation of the content with some associated information. Besides the decision of which item is to be shown, the options for modifying their presentation are manifold and with combinations of these possibilities, a wide range of adaptability is already accomplished. Besides the selection of which item is to be displayed, other dimensions influence the presentation: for instance when, with which character (e.g., motion, blinking, …), at which position, and for how long an item is displayed.

Adaptation of the Space Model

In a virtual science center, users enter and leave zones and regions in virtual space. With growing interest and knowledge users tend to navigate through the Web site more quickly. The structure of the Web site (i.e., the space model) may easily and appropriately be adapted to the user's needs

Adaptation to the Social Context

If a group of users is spatially and temporal similar (e.g., two users being on the same Web page at the same time and having similar interest), they may obtain similar information or may be brought together for a chat. Through building such clusters of people, for example a subsequent discussion about seen objects is possible (cf., Zimmermann, Lorenz & Specht, 2002).

Adaptation to the Level of Immergence

Within a virtual environment interest in objects may be expressed by the time a user's focus lingers on these objects. The level of interest corresponds to the complexity, the amount, and the style of already received information about one object and is transferred to succeeding objects. If one of these objects complies with the user's interests, the presentation style directly steps into the right level of interest, and information items are displayed that are classified on the adequate information depth and style.

Adaptation to Movement and Perception Styles

Several kinds of common behavior can be identified with people "moving" through the environment (e.g., clockwise in real museums). Attractor cues (e.g., sounds, blinking or marked links, …) emitted from different sources are used to draw the user's attention on certain objects of the environment. Thus, entire tours through the virtual science center can be recommended. The selection and dynamic adaptation of tour recommendations can be adjusted to the user's stereotypical type of movement and his/her preferred perception style.

Conclusion

Information brokers mediate between information demands and information offers. The information items to be brokered in a virtual science museum are hypermedia pieces. The use of an information-brokering tool facilitates the management and distribution of a huge amount of information within this domain. Through mapping domain properties to an ontology model we benefit from a better understanding of the domain's complexity. In addition, the ontology model in combination with the enrichment of information items with metadata descriptions enables a personalized presentation of multi-media information.

The tools and strategies described in this chapter enable service providers to classify content according to important context parameters and set up the relationship between certain events in the visitor's context and the delivery of content. For that purpose, Fraunhofer FIT is working on a context toolkit that allows an easy management of context data, and the abstraction, generalization and combination of context parameters. In this contribution we present a concept for reusing our experience, which we have gained during the application of the information-brokering tool within audio-augmented environments, in the context of a virtual museum. We followed a modeling approach building on and extending the following three aspects: tracking, information item modeling and domain modeling. The information-brokering tool supports the adaptation of these three concepts, so that applications like an electronic museum or a virtual science center takes advantage of the benefits.

References

Brusilovsky, B. (1996). Methods and techniques of adaptive hypermedia. *User Modeling and User-Adapted Interaction, 6* (2-3), 87-129.

Dey, A.K., & Abowd, G.D. (2000). *Towards a better understanding of context and context-awareness.* In The 2000 Conference on Human Factors in Computing Systems (CHI 2000): Workshop on The What, Who, Where, When, and How of Context-Awareness. Hague, The Netherlands.

Eckel, G. (2001). Immersive audio-augmented environments. In *Proceedings of the 8th Biennial Symposium on Arts and Technology,* Connecticut College, New London, Connecticut.

Goßmann, J., & Specht, M. (2001). Location models for augmented environments. In the *Workshop Proceedings of Location Modeling for Ubiquitous Computing,* Atlanta, Georgia (pp. 94-99).

Gross, T., & Specht, M. (2001). Awareness in context-aware information systems. In H. Oberquelle, R. Oppermann & J. Krause (Eds.), *Mensch & Computer – 1. Fachübergreifende Konferenz* (pp. 173-182). Germany: Bad Honnef.

Honeyman, B. (1998). Real vs virtual visits: Issues for science centers. In *Australasian Science & Technology Exhibitors Network,* Canberra.

Jarke, M., Klemke, R., & Nick, A. (2001). Broker's lounge: An environment for multi-dimensional user-adaptive knowledge management. In *HICSS-34: 34th Hawaii International Conference on System Sciences*, Maui, Hawaii.

Klemke, R. (2002). *Modeling context in information brokering processes.* PhD Thesis. RWTH Aachen, Germany.

Klemke, R., & Koenemann, J. (1999). *Supporting information brokers with an organisational memory.* In 5. Deutsche Tagung Wissensbasierte Systeme - Bilanz und Perspektiven, Workshop Wissensmanagement und Organisational Memory. Würzburg, Germany.

Oppermann, R., & Specht, M. (2000). A context-sensitive nomadic information system as an exhibition guide. In *Proceedings of the Handheld and Ubiquitous Computing Second International Symposium* (pp. 127-142). Bristol.

Orfinger, B. (1998). Virtual science museums as learning environments: Interactions for education. In *The Informal Learning Review* (pp. 1-10). Washington, DC: Informal Learning Experiences.

Pazzani, M.J., & Billsus, D. (1997). Learning and revising user profiles: The identification of interesting Web sites. *Machine Learning, 27,* 313-331.

Rich, E. (1989). Stereotypes and user modeling. In A. Kobsa, & W. Wahlster (Eds.), *User Models in Dialog Systems* (pp. 35-51). Berlin: Springer-Verlag.

Shilit, B.N., Adams, N.I., & Want, R. (1994). Context-aware computing applications. In *Proceedings of the Workshop on Mobile Computing Systems and Applications* (pp. 85-90). Santa Cruz, CA: IEEE Computer Society.

Tramberend, H. (1999). Avango: A distributed virtual reality framework. In *IEEE Virtual Reality Conference,* Houston, Texas.

Unnützer, P. (2001). LISTEN im Kunstmuseum Bonn. *KUNSTFORUM International, 155,* 469-470.

Zimmermann, A., Lorenz, A., & Specht, M. (2002). Reasoning from contexts. In N. Henze (Ed.), *Personalization for the Mobile World: Workshop Proceedings on Adaptivity and User Modeling in Interactive Systems (ABIS)* (pp. 114-120). Hannover, Germany.

Zimmermann, A., Lorenz, A., & Specht, M. (2003). User modeling in adaptive audio-augmented museum environments. In P. Brusilowsky, A. Corbett, & F. de Rosis (Eds.), *Proceedings of the 9th International Conference on User Modeling,* Johnstown, Pennsylvania (pp. 403-407). Berlin: Springer-Verlag.

Chapter IV

Starting With What We Know:

A CILS Framework for Moving from Physical to Virtual Science Learning Environments

Bronwyn Bevan, Exploratorium, USA

Abstract

This chapter examines attributes of learning in informal environments, using a research framework developed by the Center for Informal Learning and Schools. It considers how essential characteristics of learning within science centers can translate and apply to learning in Web-based informal learning environments. It argues that in designing virtual environments, informal science institutions need to build on their particular strengths and pedagogical design principles in order to fill an educational niche in the Web landscape, and not compete with commercial or even K-12 educational agencies similarly engaged in the development of online learning environments.

Introduction

Cultural institutions — historical societies, art museums, zoos, botanic gardens, science museums, and science centers, for example — offer their communities unique sets of subject-matter resources and expertise. They are adept at designing environments that can engage learners at all age levels and prior knowledge. They know something about sparking curiosity and more deeply drawing visitors into the subject matter.

Yet, much of this knowledge is unanalyzed and unarticulated among educator practitioners in these institutions. As cultural institutions move from the development and mediation of exhibit environments to the development of print or Web-based learning tools and environments, it is important that they start from who they are and what they are (Bevan & Wanner, 2003). They need to build from their unique approaches, pedagogies, and collections in order to be the best that they can be, and also to avoid competing with commercial or even K-12 entities on terms that are not their own.

At the Center for Informal Learning and Schools, a U.S. National Science Foundation-funded initiative[1] to strengthen K-12 science education through the generation of new leaders and knowledge in the domain of informal learning, we focus our work with practitioners on deepening their understanding of the environments that they work in and the underlying pedagogical principles that inform those environments. We do this through series of institutes that meet over a two-year period to create shared experiences involving learning science through the exhibit collections, as well as promoting group discourse around a number of ideas and thinkers concerned with science teaching and learning.

CILS is a partnership of the Exploratorium, King's College London, and the University of California Santa Cruz. Since its inception in 2002, CILS has worked with over 100 museum educators, has enrolled two dozen graduate students, and has launched a dozen studies investigating informal learning institutions and opportunities. The purpose of CILS is to strengthen alliances between informal and formal systems of education. These alliances can be leveraged to enhance and expand student interest and understanding of the subject matter taught in schools. They also can ensure that the wealth of cultural resources housed in museums are made accessible to, and shape experiences of, audiences from socio-economic groups who traditionally do not visit museums.

CILS has begun to articulate areas of knowledge informal educators require in order to form effective alliances with schools. We have also developed a research framework for asking questions of these environments and alliances. Drawing on these two areas of focus — on practice and research — this chapter will examine, from a practitioner point of view, some of the particular attributes of learning environments of cultural institutions and their implications for the design of virtual environments, particularly for school audiences.

Looking at Cultural Institutions

While schools, too, are cultural institutions, within this chapter the use of the term "cultural institutions" refers to institutions and organizations that collect, curate, and program public learning environments for visitors of all ages and backgrounds. These include museums, historical societies, botanic gardens, nature centers, science centers, zoos, etc. I use the shorthand "museums" for "cultural institutions" interchangeably throughout this article, because I also seek to avoid the word "informal" and therefore do not want to use the even shorter term "ISI," for informal science institution.

The experiences that many museums aim to promote, as Hein (1998) points out, are not unique to cultural institutions, but fall within a range of educational designs reflective of theories of knowledge and theories of learning. While some museums may follow didactic theories of education, most science centers, as well as growing numbers of other types of cultural institutions, attempt to create discovery-based or constructivist learning experiences consonant with Dewey's progressive theories of education (Hein, 2004) and found in progressive K-12 schools around the globe. By situating museum-based learning on the progressive end of the continuum of learning experiences, we can begin to dispel the formal/informal dichotomy that operates to separate, indeed marginalize, museum-based learning from other types of learning, notably school-based learning. In the U.S., this marginalization adversely affects public funding and utilization of cultural institutions, and results in an imbalance in equity and access issues relating to families and individuals who have the capacity and cultural traditions to draw upon the wealth of cultural resources in a community and those who do not.

Situating museum-based learning in this way also provides museum educators with a firmer grounding, and a more robust research base, upon which to make design decisions, and to find ways to better integrate the museum experience into experiences outside of the museum, including K-12 learning.

Learning in Cultural Institutions

We learn — accrue and assimilate knowledge and experience — at least from the moment we first open our eyes. Our environments, from the beginning, are curated by our socio-economic circumstances as well as by the interventions of those expert adults, our parents and other family figures. Rogoff et al. (2003) point out that from the earliest stages, children learn through active observation and listening. Proficiency in at least one spoken language is generally developed before much formal schooling takes place, and usually without direct instruction by an adult. What might casually be called "informal" learning — in that it occurs outside of school walls or without direct, formal instruction — Rogoff et al. call *intent participation*, meaning that children learn by keenly or intently observing and listening, and with the intention of joining, engaging in, or adopting the actions or skills they observe of others. This mode of learning (not limited to children) can be clearly witnessed at home or on the playground. It is seen less often in classrooms where, at least traditionally, children learn subject matter through direct and usually de-contextualized, or abstract, representations of knowledge. A key attribute of intent participation is that the learner is motivated to observe, listen, and learn because she intends to engage in the activities she is observing.

At school we participate in the process of learning a range of agreed-upon subject matter, ideas, processes, and social conventions. In these settings, expert adults (teachers) supervise the successful acquisition or development of new knowledge and understanding. In most cases, the learners have little choice about what they are supposed to learn. However, they have a great deal of control over what they choose to learn, by either agreeing to or resisting participation. Motivating students to learn, particularly at the higher grade levels, is of primary concern to teachers, and increasing concern to the education reform community. To engage students in school learning — cognitively, emotionally, and behaviorally — school reformers advocate for learning experiences that are seen by students as authentic (relevant), realizable, and that capture the imagination (NRC, 2004).

An evolving vision for school learning is one that provides students with a range of experiences —gained through listening, reading, researching, designing, collaborating, building, and experimenting — both individually and socially constructed. Different contexts for learning — classrooms, homes, museums, Web environments and television — can also support and contextualize student knowledge, interest, and motivation to learn that, it is posited, can create a readiness or capacity to learn in schools.

While there are many highly functioning schools, teachers and students, this particular moment in history — with the advent of the information age as well as advances in the cognitive sciences — poses particularly challenging problems and questions to the school community, having to do with what people need to know and understand, and the most effective ways of achieving those goals. Osborne (2004) explores this question as it regards science education, arguing that in addition to the stuff of science (content knowledge) it is increasingly important for students to understand the nature of science and the processes of science in order to achieve a scientifically literate citizenry (consumers and producers of science and the products of science).

In cultural institutions, communities have access to a range of rich resources, experiences, and expertise usually pertaining to specific disciplines (e.g., physical science, modern art, natural history). These institutions provide highly formalized learning environments, designed through an elaborate process of research, prototype, design, test, and feedback. Over the past half-century, many cultural institutions have sought to redefine themselves from the formal collection and taxonomies of artifacts (still critical in terms of curation and collection, but seen less relevant to engaging learning) into places that celebrate the subject matter, and draw visitors into the associated human experience of either the subject matter or the study of it (as with natural history museums) (Bevan, 2002). They thus attempt to capture the imagination and to make the material relevant to a broad range of visitors.

In addition to supporting the assimilation of a body of knowledge, museums can impart insights into the discipline or medium — kinesthetic, tangible, and evocative experiences that can provide a starting place for deeper inquiries, including traditional academic study and the development of expertise. Entering a room filled floor to ceiling with beetle specimens provides the visitor, at once, with a glimpse into the world not only of beetles, but also of the study of beetles. Indeed, Crowley & Callanan (1998) maintain that the most important learning outcome from children's visits to museums, specifically science museums, may be the opportunities to engage in the core processes of science — asking questions, considering evidence, describing results — rather than learning and retaining sets of facts.

The attributes of learning and of instructional design that we will describe here are not exclusive to cultural institutions, and cross many contexts. Yet, the context for learning can play a critical role in one's interest, motivation, participation, and meaning-making. Some contexts are alienating and might shut down willingness to learn; some are inclusive and encouraging of intellectual community. And different people respond differently, although through the study of culturally and linguistically diverse classrooms, patterns of behavior, comfort, and participation may begin to emerge that suggest that specific contexts and structures for participation may encourage more or less motivation and engagement (Lee, 2001).

The next section will look at ways in which cultural institutions design learning experiences to motivate and engage learners. We do this to extrapolate to virtual environments and the role they can play in enhancing formal K-12 science.

Frames For Consideration

The Center for Informal Learning and Schools (CILS) has developed a research framework that identifies four themes for investigation of CILS questions. They are:

1. Learning environments and their designs
2. Means and structures of participation
3. Explanation, communication, and discourse
4. Systems and structures that support alliances between K-12 and cultural institutions

This framework overlaps with aspects of the framework developed earlier by the Museum Learning Collaborative (Schauble et al., 1997), which is foundational to the work of CILS. The Museum Learning Collaborative framework was based strongly on socio-cultural theory whereas the CILS framework draws on a wider range of theoretical perspectives, as represented by the CILS faculty. These include research and scholarship in developmental psychology, science education, the natural sciences, and museum education. Many of these perspectives are, however, strongly influenced by socio-cultural theory.

The Museum Learning Collaborative (MLC) identified three research themes: learning and learning environments; interpretation, meaning, and explanation; and identity, motivation, and interest. We see these themes as being different (but consonant) with the CILS themes. They are consonant in that the focus on learning and learning environments are shared (Theme 1), as is the interest in interpretation, meaning and explanation (Theme 3). The MLC theme of "learning and learning environments" has been resolved by CILS into two themes of "learning environments and their design" (Theme 1) and "means and structures of participation in informal learning" (Theme 2). Motivation, identity, and interest feature in both of these CILS themes.

Because of the CILS focus on schools, we have, additionally, a research interest in the systems and structures that support informal learning institutions and opportunities and their connections with schools.

This section will consider — from the practitioner perspective — some of the salient features of learning in informal settings using the CILS framework as a thematic guide for analysis.

To do this, the paper will refer to science centers, and more specifically to the case of the Exploratorium. The Exploratorium was founded in 1969 and pioneered, with others, the design of interactive science learning environments.

It is interesting to note that the original Exploratorium exhibit collection was largely a direct three-dimensional translation of a school physical sciences curriculum. Exhibits were organized around physics topics, themes, and activities found in the Elementary Science Study (ESS) kits, produced by Education Development Center in the 1960s. Topics included Light & Color, Sound & Hearing, and Electricity & Magnetism. ESS took an interdisciplinary approach to science learning, particularly through integrating, to varying degrees, aesthetics and the arts. The exhibits thus built upon sound principles of progressive classroom curriculum and instruction, but translated them to a three-dimensional public learning environment, facilitated not by classroom teachers but by museum educators and docents.

Capturing the Imagination: Learning Environments and Their Designs

The design and spatial, as well as thematic, organization of the environment and/or collections is a form of explanation and/or narrative that conveys the meaning of the subject matter to the visitors. It is a physical manifestation of the ways in which the museum thinks about and understands its own collections, and in particular how it relates to the public, and vice versa.

Insights into the subject matter. A critical design element is the identification of compelling subject matter and the development of innovative perspectives on the topic. Indeed experiences of exhibitions that fall flat, despite the bells and whistles, may relate to a lack of insight into the core of the subject matter and what it has to engage our interest — how it resonates with human experience and thus makes a personal appeal to the visitor.

For example, an exhibition that the Exploratorium held in the 1990s, on medical imaging, approached the subject through the lens of how popular culture, dating back centuries, used current technologies — from Renaissance wax anatomical figures to fetal ultrasound images on late 20th century automobile billboards — to shape conceptions about our bodies and human life. This award winning show shifted the focus from simply showcasing a range of medical imaging technologies, to pushing visitors to think about the relationships between the *imaging* and *imagining* of our bodies, and the social, historical and ethical implications of this relationship.

Additionally, museums build on insight into their subject matter to design generative exhibits that can engage at multiple levels of experience. These exhibits — not designed solely for novice or expert — offer opportunities for people, over time, to get to deeper levels of understanding. They become like great books — always offering new insights and experiences, depending on the times in life one reads them.

Personal. Ultimately, the museum experience aims to resonate with the individual, to trigger curiosity, wonder, and interest — to help the learner along a path toward long-term (life-long) engagement with the subject-matter. Our success stories often relate to

Figure 1. 'Sun Painting' exhibit; Photo by Susan Schwartzenberg

evidence from later times that people continued to think about or study the subject matter encountered in cultural institutions.

A key to developing a personal connection is to stimulate a person's curiosity or wonder, and to do this in playful (not intimidating) ways. A classic example of engaging visitors through a visceral **aesthetic** experience is a long-time centerpiece of our light and color collection: *Sun Painting*, by artist Bob Miller. The *Sun Painting* is a shifting, colorful panel that reflects sunlight coming into the building through a shaft and filtered through an array of prisms. Clouds in the sky and visitors manipulating the prisms change the splashes of light and color on the billboard-sized panel. The beauty of the colors, the delight in being able to play with sunlight to create a painting, and the science of refraction draws the visitor into the collections where they can engage in more sustained inquiries into prisms, light, and color.

Many exhibits use the **counterintuitive** or odd juxtapositions to puzzle the passer-by and provoke their curiosity. Why do things behave unexpectedly? A classic example is the *Touch the Spring* exhibit. Visitors see a spring sitting inside a box and are asked to touch it. When they reach for it, they find it is not there. What is going on? By groping around in the box, visitors will find that the spring is lodged upside down in the front of the box, and concave mirrors are used to project an incredibly realistic image standing in the space of the box. This firsthand experience draws visitors into thinking and further explorations of mirrors, images, and light.

Finally, by connecting phenomena with their real world contexts or applications, exhibits seek to situate the science within a **commonplace** frame of reference. At the Exploratorium we have a number of exhibits — dealing with complexity, resonance, vibration, or patterns that use sand as their central medium. Almost everybody has walked on sand, whether in a sandbox or at the beach. We all marvel at the natural contours of sandy landscapes, and when one encounters sand, it is difficult not to mold it, carve it, trickle it into new shapes. The experience, prior and current, of sand is a way into looking at the effects of a vibrating drum. It is a way into thinking about complexity. It is a way into thinking about fluid motion. It is not intimidating. It reinforces the idea that science is all around us, and not confined to textbooks, classrooms, or even science museums.

Encouraging Active Learning: Means and Structures of Participation

Participating in learning in cultural institutions is both personal and social in nature. As Duensing (2004) points out, the nature of the social space varies with the cultural context. In Mexico City, Exploratorium exhibits have more space between them to allow for larger family groups. More manipulatives, as adjuncts to exhibits, were incorporated into exhibits had been in Brazil. Each museum considers its dominant culture or cultures when designing spaces for participation. Here we want to discuss facets of participation in the U.S. that include the role of observation and the relationships of experts and novices in collaborative learning arrangements.

Observation. Rogoff et al. point out that observation is a part of participation (Rogoff, 2002). Observing provides ideas and models for participation and can motivate people to engage more deeply. Museum environments are designed for observation of objects, people interacting with the objects, and the people interacting with each other. (Classrooms, on the other hand, are typically designed for students to observe the teacher, and not each other.)

To encourage observation and participation, museums create a visual horizon that allows people to see where they are coming from and where they might go next. Semper (1998) likens a museum environment like the Exploratorium's to a working town, with different neighborhoods (the biology neighborhood, the electricity neighborhood, etc.), piazza-like gathering and observing spaces, avenues for passage, and a visible infrastructure like power, pipes, and signage. In such a setting, groups assemble and disassemble — this populated environment offers a wide variety of choices for interaction, as well as models for interaction. Because most visitors are actively engaging with exhibits, other visitors can observe them engaged in this meaningful activity — see smiles of delight, watch as people beckon their friends over to "try this," and watch as some visitors become deeply engaged in an activity that clearly has the power to fascinate. It is common to see visitors watching other visitors at exhibits, and then go up behind them and replicate the actions that they have observed. This is as true of adults watching children as of children watching adults.

Experts and novices in collaborative learning. Museum visitors often attend in mixed-aged groups, and take different roles in the process of noticing, manipulating, and discussing objects or exhibits in the cultural institution. At any given exhibit, a different person in the group may be the "expert" (who knows the content or knows the idea, or knows the exhibit) at a given moment.

To encourage interactions, exhibits are often designed to allow for several users to engage at once, and in some cases may require more than one user for the exhibit to fulfill its purpose. For example, with *Resonant Pendulum* the visitor swings a small magnet onto a steel collar attached to a 300-lb. pendulum hung from the ceiling. Because of the weak magnet, the visitor learns that only by pulling in time with the swing of the pendulum (in resonance), can the pendulum be moved. Two magnets are tied to the fence at 90 degrees to each other so the users, if they cooperate, can alter the pattern in which the pendulum swings (circle, ellipse, line, etc.). Visitors need to discuss strategies to get this exhibit to work easily, thus building a sense of learning as a social endeavor.

As Crowley and Callanan (1998) point out, museums provide children settings where they can learn in ways that allow children and adults to equally control what they pay attention to, how they engage and learn, and when the learning is done (and they can move on to a new exhibit). Through the process of shared engagement and discussion, through articulating what they are thinking and finding, children are more likely to remember what they have learned and to have more powerful learning experiences than if they were learning in isolation. The social environment thus affords particular opportunities for children to develop experience and knowledge.

Expertise may sometimes exist outside of the immediate family or social group. People in cultural institutions observe other people in the process of observing, reading, interacting with the collections. They may become interested in objects by virtue of noticing other people's interests, or overhearing other peoples' conversations, including those led by trained museum educators.

Developing Understanding: Explanation, Communication and Discourse

Explanation refers both to the explicit descriptions of how and why things occur — which may be found in exhibit labels, in conversations with or presentations by museum docents, in audio or tour guides, or in conversations among visitor groups — as well as the implicit messages and meanings conveyed by the ways in which the subject-matter is presented.

As both the MLC and the CILS research frameworks note, research has begun to tie the learning of science to the nature of the discourse, both in and out of classrooms. Language is seen as an essential tool for creating scientific explanations, arguments, narratives, metaphors, and analogies. While constructing and communicating (i.e., explaining) their ideas or understandings about what they encounter, visitors externalize, clarify, and restructure their knowledge.

Contextualization. An important part of the implicit explanation of the subject matter relates to how the phenomena are presented from a disciplinary point of view. To generate a broader frame of reference, as well as to provide a gestalt for the subject matter, contextualization helps visitors understand the nature of the discipline and how it is investigated and developed.

In school settings subject matter, and perhaps science in particular, is presented as so many bricks of knowledge (Osborne, 2000). Few school students gain an understanding of the whole, either as a discipline or as an endeavor. Museums have time and space to give visitors a more holistic sense of the subject matter. For instance, the Seeing collection at the Exploratorium raises big questions about the human experience of seeing — In what ways is seeing a subjective activity? How do we actively construct understanding from what we see and notice? How do we interpret the physical phenomenon of light into mental and visual images? — and in so doing, relates the science of optics, cognition, visual perception as parts of an integrated, daily experience.

Science museums can also foster an understanding of how phenomena operate under different conditions and in relationship to one another. One way to do this is through

redundancy, providing multiple entry points for visitors, and multiple modes of representation of the phenomena. Redundancy provides the possibility of multiple perspectives on the same phenomenon. It reveals different aspects and behaviors of the phenomena. It opens broader vistas for understanding. Rather than encouraging narrow conceptualizations, visitors are able to develop more complex understandings, and begin to make connections among phenomena or behaviors of phenomena.

Redundancy also tends toward the use of different media — video, audio, hands-on demonstrations, mediated discussions, open-ended investigations — to provide a range of visitors (different ages and levels of prior knowledge) with multiple ways of knowing and seeing.

For example, one can learn about resonance through a number of different exhibits at the Exploratorium. Some, like *Bells,* provide direct hands-on experiences with aesthetically compelling materials. *Bells* consists of two metal plates which vibrate when they are stroked along their edges with a rosined bow. When each plate vibrates at a specific tone, sand bounces on the vibrating regions of the plates and collects in those areas where there is no vibration, forming intricate, beautiful patterns.

Pipes of Pan, on the other hand, provides visitors with an aural encounter with resonance. This exhibit demonstrates how tubes of varying length act to select and amplify sound waves of different frequencies. Ten glass tubes of different length filter out specific tones from the ambient noise of the museum. Each pipe is resonant at a certain frequency, which is determined by its length. If you listen to each pipe in turn, from longer to shorter, you will hear a progression from low-pitched tones to higher ones.

The process of encountering the same or related phenomena in a range of forms (such as finding waves in sound as well as in light, in water and in sand) conveys the message to the visitor that learning and scientific investigation itself is a process of repeated encounters with the phenomenon. Cultural institutions thus send an important message to their audiences when they present material through the accretion of relevant encoun-

Figure 2. 'Bells' exhibit; Photo by Susan Schwartzenberg

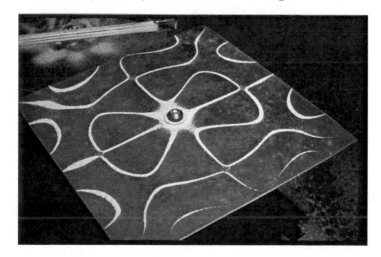

ters and experiences. Returning to Crowley and Callanan's point (1998) — they can lead visitors into the heart of the matter.

Integrating Into Educational Infrastructure: Systems and Structures Supporting Informal Learning

CILS's questions about structures to support stronger integration of the resources and learning opportunities of cultural institutions into the K-12 system of learning focus largely on funding and policy issues — on how these resources are in fact used, what drives their use and what impedes their use.

Programs designed to support K-12 science — particularly teacher development programs and field trips — are designed primarily to support/boost the teaching and learning as envisioned by the school system. That is, they address common curriculum foci, standards, and assessment. They develop teacher and student content knowledge, as well as understanding and relationships with the nature of science and science process skills.

In general, work of museums to support schools or other systems is programmatic and not object-based. In other words, it is the interpretation and facilitation of the use of the collections, and not the collections themselves. The Exploratorium experimented with one large-scale exhibition project where a set of exhibits was designed according to the latest California state science framework themes. While overall there was mixed response to the exhibition, the constraints of the school curriculum did not match the goals of the public environment.

However, museums have the resources upon which to build significant programs that can provide direct and personal experiences with science to nurture interest and engagement with the subject matter (in addition to skill or knowledge levels).

Museums can work well at two levels that schools struggle with — the novice level and the expert level. Both for elementary teachers with little science background, and for students who are unengaged in science, museums can provide programs that motivate inquiry, interest, and the development of self-conceptions as successful science learners. In addition, museums can offer advanced science content experiences for high school teachers as well as for students whose interest in science extends beyond what is offered by schools.

Science museums — dedicated to the discipline of science and to science learning — can provide teachers with content-rich immersions. Here their conversations with colleagues are about science and the teaching of science, and emanate from the doing of science and science investigations. Whereas within the school and classroom environment, much of teacher talk has by necessity to do with administrative and student issues. That is, museums can serve as an oasis for teachers, to rekindle and rejuvenate their passion for their subject matter. They also can use these environments to more deeply engage in thinking about and reflecting on the nature of science — and the importance of understanding the epistemological aspects of science for both themselves and their students.

Summary

Using the CILS Research Framework as a way of identifying attributes of learning in cultural institutions that we need to attend to when building learning experiences for the virtual world, we thus identify the following parameters:

1.   **Learning designs and environments** that provide compelling *human* perspectives on the subject matter and that do so in ways that viscerally and/or aesthetically stimulate the learner's curiosity and interest.

2.   **Participation structures** and social environments that allow learners to browse, to enter and participate at varying levels of active engagement and expertise.

3.   **Explanatory structures** that allow learners to build understanding together, engage in discourse, and that contextualize the subject matter within a broader epistemological framework.

4.   **Systemic structures** and avenues that suggest ways that the resources can be mapped into supporting, and enriching, a broader educational framework.

In the next section, we will explore how two Exploratorium Web sites — Origins and Global Climate Change — reflect these design principles, and look at new design and learning opportunities afforded by the Web environment. Note that both of these Web sites were developed as a learning environment unto itself — that is, not as an extension of a pubic floor experience.

Into the Web Environment

When the Exploratorium launched its Web site in 1993 it was among the first of several hundred Web sites in the world, and one of only a handful that were operated by cultural institutions. Like the cavernous building that beckoned to a mixed group of scientists, artists, and educators who created the Exploratorium, the Internet was viewed as a place of huge possibility — one that had as much potential for active engagement as the physical exhibit floor. It was also imagined that, as with the early Exploratorium, visitors were as likely to stumble into the virtual space as they were to seek it out specifically for science content and learning. The space was therefore developed to be playful, to catch the eye and the whimsy of the accidental visitor.

The Web site was developed for the remote audience, not as a marketing tool but as a learning space. Only recently have we begun to also explore its potential to extend the physical visit to the museum in San Francisco.

Like the museum floor, the Exploratorium Web site is a convergence of ideas and experiences. But there are also many differences.

For one thing, Web visitors usually visit alone; that is, they are not sitting at the computer with friends or family. They are not engaged in discussion and discourse. They also cannot "see" fellow visitors engaged in learning. So the experience is by definition not a physically social one — yet to the extent to which the Web learner engages with the ideas and products of society, it may be extremely social in other dimensions.

Another striking difference is that although Web learners are visiting a carefully curated and designed environment, it is logistically easy and by nature quite likely that visitors can, with a click, depart for entirely different environments and subject matter. That is, one can move quickly from the Exploratorium site to a news site, a shopping site, or any other site on the Web. Entrances and exits can be both quick and unanticipated. Not only can the visitor bypass the gift store near the exit, but they can also remain unaware of entire rooms of knowledge and content just around the corner.

Finally, the Web site is a less sensory experience — there are few sounds, no smells, no touch — and thus the experience becomes primarily visual and intellectual. It is fundamentally text dependent, and requires a kind of verbal and graphical literacy that the physical museum does not — it therefore is less prepared to engage younger audiences, or visitors who do not read English. It thus targets, perhaps, a more literate and sophisticated audience than the physical environment. It goes without saying that it is therefore socially somewhat exclusive. On the other hand, our Web site currently reaches 18 million people annually, while our physical building only receives about 500,000 per year.

Given these differences, how do we — as cultural institutions — design for engaging Web experiences, building on what we know about developing engaging public floor experiences and about engaging science learning? How do we motivate inquiry and learning in the virtual environment? How do the opportunities provided by the Web support the goals of schools for student and teacher learning? Two Web sites will serve as reference points for investigating these questions.

Inroads and Insights to People and Places: *Origins*

The *Live@ Exploratorium: Origins* Web site (www.exploratorium.edu/origins) was developed to explore and communicate cutting edge science related to origins of matter, the universe, Earth, and life on Earth.

Rather than simply telling the story of the science and the current scientific research, the project is conceptualized as a series of virtual expeditions, or field trips, to remote places where fundamental research is being performed in particle physics, cosmology and biology. Sites include a rain forest research station in Las Cuevas, Belize operated by the Natural History Museum in London; McMurdo Station, Antarctica; and an underground particle accelerator at CERN in Geneva, Switzerland.

A total of six research institutions are featured with interactive elements, video clips, articles, images, and live webcasts that enable users to virtually tour the facilities and provide a window into the world of the research as it unfolds. As of 2004, *Origins* has been visited by over 3 million people.

Figure 3. The "Origins" fronts page includes a selection of clickable animations for each of the six primary locations, enabling visitors to the site to get a sense of the research locations before entering the full site.

Learning design. It is difficult for a visitor to experience current science research and get a sense of the enterprise, the endeavor, and the science itself — what the questions are and why they matter; who the people are and what they do; the instrumentation; the implications; the connections with what we know and don't know.

Origins takes two particular, innovative perspectives on the challenge of communicating current scientific research — that of place, and that of the human endeavor of scientific research — who are the people doing this work, and what drives them? How do they live and what tools do they use? What do they worry about and what do they dream about?

To enhance the sense of "placeness" of the six research locations — to fully realize the metaphor of the field trip — the Web site is designed to create a sensory experience through an extensive collection of visual images (taken by photographic artists). These images include the broad landscape of the site — through aerial photos, manipulate-able 360-degree views of the facilities, and maps highlighting the scientific instrumentation and laboratories. They also include peeks into the nooks and crannies, the behind the scenes, close ups on a scientist's hand holding a beetle, or on the scribbled notes in a field notebook.

These images create an aesthetic and textural sense of place that is then organized into structures that underscore the humanness of the endeavor. Images on the Cold Spring Harbor site are organized like Polaroids pinned to a bulletin board, each a snapshot memory of a trip abroad. Clicking on an image takes the visitor to an interview, a quote, some insight into the science or the scientific discussions taking place at Cold Spring Harbor.

Building on another human metaphor, the Las Cuevas images and text are situated in a lined notebook that includes hand scribbled side notes, and links and references to other "pages" in the book. In all of these and other constructs, the human beings engaged in the work, or their scientific forebears, are highlighted, interviewed, photographed, and quoted. We meet scientists and technicians, cooks, and guides.

Like an exhibit floor, this approach to the material and the Web site creates a populated landscape of choices, objects, faces, and ideas. It allows visitors first-hand, if virtual, encounters with a rich tableau of science and scientists and a variety of scientific activities. It creates contexts for understanding how different experiments or research teams interrelate, and what the meaning of their work may be. In some ways it is less playful or whimsical than the museum floor experience — geared as it is to a highly literate, older audience. (Although whimsy is highlighted as it is encountered.) Yet it affords a fundamentally personal journey, driven by what intrigues the visitor, which may be one of the fortunate confluences of Internet-based learning and museum-based learning.

Participation structures. The Web is uniquely suited, through its variety of media-based entry points, to make the journey into a complex environment (such as cutting edge scientific research) a compelling one. The mix of text, photos, video clips, sound, layered images, branching data sets, and Web design tools that can show movement, time, and dimension, creates a huge number of possibilities for how to approach the subject matter, and (in this case) the place, ideas, and tools surrounding the research. Unbounded by the temporal or spatial constraints of the museum visit, these resources, interviews and "field trips" are available for personalized, directed, or exhaustive Web visits that can unfold over time in ways that a museum visit could never provide for.

In a rich Web site like *Origins*, where there are scores of layers of information and images, the subjective feeling of being able to meander along pathways of one's own choosing, creates a sense of infinite choice and the feeling of personalizing one's journey — of following one's own interests and looking more deeply into the thing that catches one's eye. This sense of choice, as well as sudden encounters with new ideas and images, parallels the personal experiences one has on the museum floor in choosing pathways and engaging directly with phenomena.

But unlike the museum experience, this journey is undertaken alone, albeit in collaboration (or complicity) with the Web site curators, educators and developers. This is a difficult transition for museum educators — steeped in a social environment and culture designed to support social interactions — to grapple with. To experiment with social interactions and different kinds of participation structures for Internet visitors, the *Origins* team soon began to experiment with how emergent webcast technologies could be used to create more discussion and collaboration among visitors.

Over the course of the project, the team developed over a hundred live webcast components that connected remote scientists at the research labs with live Exploratorium visitor audiences as well as with live Internet audiences. While visitors were less able to interact with each other, they were able to interact directly with scientists at the research sites and at the Exploratorium, through e-mailing questions and observations — and indirectly by following up with questions and comments from previous visitors.

These webcasts — a form of broadcasting over the Internet that can be watched live or later retrieved on demand — enabled remote audiences to connect in real time to scientists and activities as they were happening. Mediated by Exploratorium scientists, working before a live audience in the museum's webcast studio facility, these 30- to 60-minute programs allowed for time-based material to be introduced to both remote viewers and studio audiences.

Examples of the most compelling live time-based events were the unveiling of a new Hubble image, looking inside a bat researcher's bag after a day of collecting, and speaking with scientists at CERN after they announced that they had produced anti-matter. Many such events lose some of their luster when re-packaged into exhibition or magazine/Web site-like entities. They lose their 3-D-ness and they also lose the essential connections to the people involved in the science: the person holding his breath before opening the bag of bats.

Webcasts also were structured so that Exploratorium science educators could build and demonstrate, on the museum floor, 3-D models of various events, thus providing visitors with opportunities to conceptualize the location of the Hubble in relationship to the earth and the moon by having educators, holding tennis balls, stand at a distance to one another that represented the distances between the telescope, Earth, and Moon. Audiences were able to predict and direct where the scientists would stand before they moved into position, showing how close the Hubble is in relationship to Earth.

Thus Origins experimented with ways to mix novices and experts, learners and teachers, through a variety of live interactions. The archival value of these interactions is not clear – in some cases, such as the unveiling of the Hubble, later visitors might be interested in viewing the dated material. In some cases, the webcasts served to create an historical record — interviews with scientists and others engaged in the process of discovery. One day in the future, these video archives may prove a valuable piece of history. But, from the point of view of designing engaging learning experiences, their real value was in the live, unexpected, and interactive format. They allowed people to interact with other people interested in the science, and they also allowed people to interact with significant scientific events in real-time.

Such encounters, like chat rooms, and to a lesser degree bulletin boards, begin to suggest active social engagement that allows people to develop new questions or insights not only from the material presented by the museum, but from the ways in which others are engaging with and interpreting that material.

Explanatory structures. To encourage visitors to engage with the material, the site seeks to develop explanatory and narrative structures that establish the broader framework into which the individual stories and insights fit. It organizes the material, first, by research site/topic, and, second, by cross-cutting themes of people, places, tools, and ideas. The sites are indexed in ways that will bring them up quickly in a Google search on the specific topic. But it is also anticipated that people who find their way there more serendipitously will be drawn in by the stories of the places and people.

The site index is organized to provide visitors with a kind of visual horizon within which they can develop their own pathway. It allows them to "see" what the site has to offer in terms of people to meet, tools to learn about and representations of the science to be explored. It presents the stories as layered landscapes of people and ideas — vertically, by research enterprise, and horizontally, by human endeavor. It thus creates a context that both makes connections and illuminates pathways.

Figure 4. Site index for Origins allows visitors to navigate either by research location or by topic area. For example visitors to the site may "tour" all of the tools across research locations sampling those used for collecting particle data and those used for sampling biodiversity. Or a visitor may examine each of the locations as a unit. Clicking on an image takes you directly to the that section of the sub-site.

Once you choose a starting point, your path takes you to pages with a plethora of links to new stories and people and questions. One story leads to another story. For example when we click on the question "Why are there so many living things?" in the Las Cuevas site, we meet botanist Nancy Garwood. By clicking on her name, we encounter images of her work in the rain forest, which leads us to links back to London's Natural History Museum where further research is conducted on the collected specimens. Whole new worlds of visual images and places — from the collection, study, and preservation perspective — are opened up.

The narrative in *Origins* is about the interlinked activities that together make up scientific process, endeavor, and discovery. That the activities are driven by human passion to know and understand, by human obsession with the minute and the cosmological, is at the heart of the narrative. This provides the context for further exploring the science that the site details.

The looped nature of the narratives serves in its own way as a pedagogical redundancy. While specific ideas may not be encountered in many forms, in the way that science phenomena are on the museum floor, aspects of the scientific enterprise as a human one are repeatedly revisited. This is done through the complex layering of information and insight — through the branching and looping of the Web site, and through the narrative construct, as described above, of organizing the material in terms of location as well as people, tools, and ideas.

Systemic structures. *Origins* experimented with ways of connecting the science research Web resources to the school system, although it did not develop Web resources specifically for this audience. However, for each research institution, the team developed targeted webcasts for specific classroom groups from middle schools or high schools. Classrooms attended these live webcasts as part of a field trip to the museum and interacted beforehand in a special program warm-up. They would then join the live studio audience at the museum during broadcast and have more opportunities to converse with and question scientists at the remote location during the webcast. Teachers would prepare students beforehand by using the Web site or project representatives would visit the classroom.

In research we conducted on the outcomes of these experiments, we found that the nature of the live webcast — where many events, some technological, some human — intervened to delay, repurpose, or redirect the transmissions, did not match well with the nature of the K-12 classroom, where teachers have plans and goals for their students. For example, in one case a class was prepared to interact with and ask questions of a particular scientist in Belize. However, due to technical difficulties, the webcast had to be conducted in a one-way stream from Belize to the Exploratorium and the Internet.

This was an important learning experience for us, and has redirected our examination of the material for K-12 audiences away from webcasts (as a live event) and toward webcasts as archived resource material along with other Web site resources. The *live event* of the webcast is inherently opportunistic and open-ended, and cannot easily be slotted into a prescribed curriculum.

On the other hand, when the technology worked — as with our webcasts of a scientist at the Licancabur Volcano in Chile — the webcast allowed students opportunities, and a sense of excitement — seeing, hearing, and talking with somebody who they had previously only read about — that traditional means of communication (letters, books) would not. In the Chilean instance, a collaboration with a professional development program called Project ARISE focusing on "research at the extremes," students were able to interact directly with a scientist whose work they had been studying for two years. In addition to these interactions, Exploratorium science educators developed a series of teacher professional development workshops for Project ARISE, building on the science from *Origins* to support the program and its teachers.

This connection between schools and science research, and the mediating role that places like the Exploratorium can play, is an area we continue to grapple with. For the Mars Rover Web site, we developed a series of hands-on classroom activities to support teacher use of the resource. We intend to do the same, post-hoc, for *Origins*.

The other system that the Web site may allow cultural institutions to tap into is the system of scientific conferences. The Exploratorium has had some experience webcasting conferences and meetings from Cold Spring Harbor, and has expanded this way of connecting a system of scientific enterprise — through conferences such as the annual American Association for the Advancement of Science, or the American Geophysicists Union annual meeting.

Interpreting Current Research:
Global Climate Change Research Explorer

The Exploratorium's *Global Climate Change* Web site (www.exploratorium.edu/climate) offers another approach to science learning in a Web environment by exploring the potential of using the Web to provide access to and insight into the world of research through a very current and somewhat (in the U.S.) politically controversial topic: global climate change.

The study of the phenomenon of global climate change involves a variety of scientific disciplines. Data featured on the site are organized into four research areas: Atmosphere, Hydrosphere, Cryosphere, and Biosphere. Data are mostly short-term or near real-time — meaning constantly current — with some long-term data where available. The data come from a variety of institutions, including the Massachusetts Institute of Technology, University of Wisconsin, Florida State University, Boston University, the National Science Foundation, NASA, and the National Weather Service.

Deep linking to data ensures regular updates of the site. Each data set is accompanied by analysis and commentary about the data collection methodologies and the implications for climate change. A comprehensive selection of links allows visitors to access the source of data and search in areas of interest.

Figure 5. Home page of the Global Climate Change Research Explorer. Animation displays the variety of data accessible through the site. In this example the cryosphere is shown.

The *Global Climate Change* Web site allowed the Exploratorium to experiment with the mediation of complex, text-heavy, value-laden science by bringing to bear pedagogical approaches inherent in the museum floor. These principles include both isolating (in the instantiation of one exhibit or one visual graph) and contextualizing (within an exhibit collection or a Web page) the phenomena. They also include building visual representations of complex ideas, and providing accessible explanatory texts to support learners.

Learning design. The *Global Climate Change* Web site approaches the subject of global climate change in two ways. First, it provides visitors with first-hand experiences with the data themselves. Second, it stresses the inter-relatedness and complexity of the data, measures, and interpretations — and how scientists accrue these different data to suggest trends and probabilities. It thus provides Web visitors with experiences (the data) that provide a context for understanding the dilemmas facing scientists. In so doing, it provides insights into the scientific process of trying to understand a complex phenomenon.

Understanding global climate change involves examining and interpreting a wide range of data sets within many different disciplines. In addition to an introductory text, the site is divided into five sections — one for each of the four "spheres" (hydro, atmo, bio, and cryo), and one that focuses on overall effects of changes in these spheres.

In each section, visitors can explore data collected in the study of one of these spheres. The site represents this complex array of data through a series of images, graphs, charts, and maps, accompanied by explanatory texts. This highly visual environment is designed, like an exhibit floor, to offer the visitor a variety of visual prompts and choices that effectively break down the complexities of this topic into bite-sized pieces. Links lead to more detailed descriptions and data, as well as to other research domains found on the site.

Like scientists engaged in understanding global climate change, visitors soon understand that the science is highly inter-related. For example, data in the biosphere section show how coral systems are being impacted by increased seawater temperatures, which are represented in data found in the hydrosphere section. In the cryosphere section, data exploring decreasing ice mass is linked to changes in water levels explored in the hydrosphere section. The visitor comes to understand that in a complex system like climate, evidence must be assembled slowly and carefully, and pieced together to form a picture of the whole.

Participatory structures. The site is developed in ways that allow for wide-ranging navigation and an egalitarian approach to what is important to notice or know. Like a browser in a library or in a museum, visitors to the *Global Climate Change* site are confronted with a large number of possibilities and choices, indexed in ways that allow personal interests or proclivities to control the learning. No one "sphere" is deemed more important to examine than another. No one piece of data takes precedence over another.

Although each page has an explanatory text that provides an overview to the data, the text does not dominate the page, but is placed in a column on the left-hand side. The dominant images are the colorful series of visual representations of the data. The learner can decide where to start — with images and data, or with context and narrative. Drawn to an image, caption, or headline, visitors can click on screen items and go deeper into the subject matter. Or they can go back. Information is layered in ways that inquiry-rich

Figure 6. The e-mail form for submitting questions about one of the atmospheric data sets

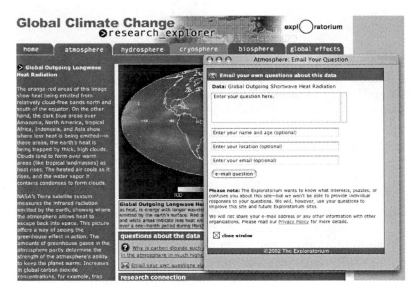

exhibits might be layered, with ever deepening possibilities. As new questions are raised, new links lead to new answers or consideration of the questions.

Although they can't observe the learning of others, the materials are organized in ways that allow the visitor to observe the many different directions that science and scientists themselves are taking in order to better understand the science of climate change. They can observe on the Hydrosphere page, for example, that there is being data collected on sea temperatures, sea currents, and precipitation. And that this relates to the data being collected in the Cryosphere related to ice on the earth's surface, and measurements of glacier growth or decline.

Thus, there may be something about the Web environment that allows visitors not to participate with their peer and family groups, as a museum environment does, but rather with an enterprise, such as scientific research. Museums also seek to do this in their physical spaces, but the Web may be more effective in engaging people socially with the social and cultural products and processes of a field or discipline.

There is also a mediated environment for posing questions, where visitors can ask questions about a specific data set, and get responses from Exploratorium staff, who consult with science advisors, as required.

Explanatory structures. The Web site is structured to not only share or explain the process of scientists grappling with complex questions and systems, but in fact to allow visitors to engage in that process themselves in order to better understand both the science and the scientific investigation of global climate change. In essence, the Web site is structured to support a visitor's trajectory through the processes of starting from curiosity or a question, moving to the selection and examination (although not the collection) of data, to considering (through mediation provided by texts as well as the data) the evidence, to thinking about predictions — how one would make them, what they might be.

Within this overarching explanatory framework, the *Global Climate Change* Web site seeks to make the complexities manageable, as well as compelling. The site has an extensive six pages of text introducing the visitor to the big picture, and the questions that are being asked. The Exploratorium's decades of experience making science, including scientific research, accessible and digestible mean that even in the more formal and didactic narratives, the tone and the terms are carefully made accessible to audiences of many ages and levels of prior knowledge. Each page has a glossary section (which can be printed in toto) that gives definitions for words ranging from *anomaly* to *phenology*.

Each of the four "sphere" sections is organized into "what we know" and "what we don't know." The "what we know" section is the data, presented in visual as well as textual form. The "what we don't know" is generally text that considers the evidence (from the what we know part) and the uncertainties that confront the scientific community, as well as the society in general, in interpreting the data.

The narrative thus allows the learner to experience and consider the processes of science — of trying to make sense of a mass of data, from a variety of sources, to consider the uncertainties, and to think about what types of predictions are possible. A section called *Global Effects* looks at the results of the data in the four "spheres" and explores how scientists go about predicting global climate change.

The site also reinforces an understanding of science as data-driven. By taking different looks at different elements — not only at the cryosphere and the biosphere, but also at the golden toad and the polar bear — scientists piece together their best understanding, explanations, and predictions of the world they are investigating.

Systemic structures. There were no specific efforts to link this site to K-12 science education. However, in online user surveys, 44% of respondents stated that they were teachers, professional developers, or students.

The site is so data rich that it can serve as a dynamic (albeit upper grades) textbook for studies of the environment, as well as of specific topics or subjects.

Like with *Origins*, we have assembled a team of staff to think about ways in which we can make these resources accessible and used by school audiences.

Conclusion

This examination of two Exploratorium Web sites, *Origins* and *Global Climate Change*, point out perhaps that many of our approaches to the learning design — which fundamentally involve *humanizing* science and the process of science (whether it means using common materials like sand in exhibits or putting faces and names to the researchers exploring the ends of the earth) — can cross the boundaries between physical and virtual. While the hands-on nature of the presentation is slightly different, in that it is not tactile, clearly it can be hands-on, including use of online exhibits where people can manipulate variables with keystrokes instead of with buttons or levers.

While Web-based participation structures are by definition less social in nature — in that it is currently much more common to travel alone when visiting Web sites than it is when visiting museums — they need not be any less learner-driven. Indeed the choices are manifestly greater with the Web's ability to link to a myriad other ideas or places. Rather than building on local social groupings, the Web may offer a greater connection between the learner and the designer or the learner and the cultural products of science and scientists — creating new inroads and insights into the world of science that is different than what can be provided in a museum, but perhaps no less important. In time, talk of networked learners and distributed learning environments may become more real than projected, and may even begin to reach across the socio-economic ranks to include a broad array of learners.

Explanations offered through Web-based media can go far deeper and far wider than what can commonly be offered in physical environments. The explosion of space and time, in that the learner can spend hours and hours — in one sitting or stretched over multiple visits — interacting with enormous quantities of ideas and people is substantially different, and is an important way in which people learn, augmenting the vital socially dynamic relationships that people have learning with other people in shared physical environments and discourse. Because they are predominantly text-dependent, these sites currently operate best for more advanced learners, more literate learners, and, in these two cases, English language speakers. However, there is much work underway to explore how to expand Web-based explanatory structures to be compelling and non-text dependent, and there are plenty of examples of powerful Web sites that do this — although perhaps not so many that do this in science and get to core science content or process domains.

As explored above, the connection of the Web resources into systemic structures continues to challenge us. Is it enough to build it and let them (whoever "them" may be) come? This is of great interest to the Center for Informal Learning and Schools. Many cultural institutions have been built by the ideas and work of groups of creative, rather iconoclastic and independent thinkers. This may (or may not) be truer for the younger field of science centers. It may be especially true for new media designers working in science centers! For many of these thinkers and designers, schools represent a stifling approach to learning. Thus, engaging them in designing for school audiences and uses is not easy. On the other hand, as I hope this chapter begins to suggest, the learning designs that these creative iconoclasts can build in the Web environment may be tremendously powerful learning tools — *especially* for kids who don't come to museums, and who might not naturally seek out science resources on the Web. With the iconoclasm often comes a sense of egalitarianism, but schools — despite the fact that they are our communities' most ubiquitous and democratic institutions — fail to be perceived as such by most of us.

Schools are an important audience for cultural institutions (Delacote, 2003). Working with schools is one way in which a museum can pay attention to issues of access and equity for its community's citizenry. It may be the beginning of a long-term relationship with future adults. It may support the nature of how schools think about and present knowledge and experience related to the field (Bevan, 2002). For these reasons, thinking about how schools use our resources is important for cultural institutions to grapple with, including the use of Web resources.

Like museums, which are designed for public audiences, these Web resources may be designed for the general public of all ages, but we may need to think about specific ways in which to adapt or mediate the virtual environments so that they can be accessed and used by school audiences (teachers and students).

It is useful to think about the special tools, guides, and docents that many museums provide for field trip visitors. These facilitators seek to augment and/or focus the museum experience in ways that can reinforce school learning goals.

In addition to better understanding how the field trip is incorporated into the school experience — what attributes of engaging with scientific phenomena, the process of inquiry, and learning science support the content goals of schools, and how? — we need to examine how virtual resources, designed by museums, can similarly move beyond the straight content goals of many K-12 and commercial Web resources to examine how we can support student understanding of the nature and processes of science, and build motivation to learn more about science. By building on what we do best — which includes but goes beyond science content knowledge and encompasses developing understandings of the epistemology and nature of the subject matter — cultural institutions can play an important role and fill a niche in supporting science learning in K-12 and beyond.

References

Bevan, B. (2002). Windows onto worlds. In J. Amdur-Spitz & M. Thom (Eds.), *Urban network: Museums embracing communities*. Chicago: The Field Museum.

Bevan, B., & Wanner, N. (2003). Science centre on a screen. *International Journal of Technology Management, 25*(5), 371-380.

Crowley, K., & Callanan, M.A. (1998). Identifying and supporting shared scientific reasoning in parent-child interactions. *Journal of Museum Education, 23*, 12-17.

Delacote, G. (2003). Apoptosis: The way for science centres to thrive. *International Journal of Technology Management, 25*(5), 371-380.

Duensing, S. (2004). Culture matters: Informal science centers and cultural contexts. In *Learning in places: The informal education reader*. New York: Peter Lang Publishers.

Hein, G. (1998). *Learning in the museum*. London: Routledge.

Hein, G. (in press). John Dewey and museums. *Curator*.

Lee, C. (2001). Is October Brown Chinese? A cultural modeling activity system for underachieving students. *American Educational Research Journal, 38*(1), 97-141.

National Research Council and the Institute of Medicine. (2004). *Engaging schools: Fostering high school students' motivation to learn*. Committee on Increasing High School Students' Engagement and Motivation to Learn. Board on Children Youth, and Families, Division of Behavioral and Social Sciences and Education. Washington, DC: The National Academies Press.

Osborne, J. (2000). Science for citizenship. In M. Monk & J. Osborne (Eds.), *Good practices in science teaching: What research has to say*. Buckingham, UK: Open University Press.

Osborne, J. (2004). What "Ideas-about-Science" should be taught in school science? A Delphi study of the expert community. *Journal of Research in Science Teaching*, *40*, 692-720.

Rogoff, B., Paradise, R., Mejia Arauz, R., Correa-Chavez, M., & Angelillo, C. (2003). *Firsthand learning through intent participation*. Annual Review of Psychology.

Schauble, L., Leinhardt, G., & Martin, L. (1997). A framework for organizing a cumulative research agenda in informal learning contexts. *Journal of Museum Education*, *22*(2&3).

Semper, R.J. (1996). The importance of place. In *About learning: A field guide for museums* (pp. 3-8). Association of Science-Technology Centers Newsletter. Washington, DC: ASTC.

Endnote

[1] CILS is funded through the U.S. National Science Foundation as one of 13 Centers for Learning and Teaching, each of which is funded to address some critical aspect of the national education infrastructure in K-12 science and mathematics. CILS is the only center to focus on informal learning. Other centers focus on issues such as student assessment, rural education, equity in education, mathematics curriculum, technology integration, etc.

Chapter V

Weaving Science Webs: E-Learning and Virtual Science Centers

Susan Hazan, Israel Museum, Jerusalem

Abstract

Drawing on Bruno Latour, this chapter argues that science and technology need to be replaced into social context dissolving the artificial boundaries between art, culture and science inscribed in institutional activities.

Introduction

Through a series of projects from museums and over online architectures, this chapter explores innovative systems that harvest data across electronic highways, online collaborations between museums and their public, the production of bottom-up narratives that invigorate community knowledge and novel ways to simulate and visualise science discourse.

In these examples science and art work together to produce meaningful narratives, replacing them back into their social context, and dissolving the familiar dichotomy inscribed in the rationalizing project of modernity. Combining the weft and warp of culture and science into interwoven discourses no longer results in our intellectual life being out of kilter as these discourses become one and the same.

Science is feasible when the variables are few and can be enumerated; when their combinations are distinct and clear. We are tending toward the condition of science and aspiring to do it. The artist works out his own formulas; the interest of science lies in the art of making science. *Paul Valéry (1871-1945)*

History, ethnography, art and anthropology museums weave stories around social and cultural narratives and lead us to compare our own narratives with those on display, and in doing so distance ourselves from, or connect with specific threads that bind geographically and temporally distanced cultures. While these kinds of exhibitions serve to reaffirm our own histories and social affiliations, these cultural mappings are not the only kinds of knowledge that inflect the meta-narratives of society. Science museums on the other hand, like their sister institutions, the university-based, science faculties are concerned with the systematized knowledge of the physical or natural world and in doing so purport to display finite knowledge, knowledge that can be demonstrated through controlled experimentation. This separation may be seen as artificial, as all of scientific practice is articulated in social space and their ramifications culturally inflected.

Bruno Latour (1993) re-places science and technology into its social context, blurring the boundaries between nature and science, between human and thing, while dissolving the familiar dichotomy inscribed in the rationalizing project of modernity. The webs woven across science channels, science museums and natural history museums, as well across the Internet all work together to construct the rational and empirical knowledge base of scientific discourse. While scientifically determined projects tend to be seen as distilled from anything as serendipitous as culture, this chapter will argue that they do in fact construct a wealth of culturally invigorated narratives. In an investigation of the social and cultural messages inscribed in these practices and conventions, it becomes clear that these kinds of narratives impact the ways we think about our lives and our environment and serve to transform society no less than the cultural variety.

Weaving Science Webs is divided into four sections. The first section, *Locating Science, Locating Culture,* draws on Bruno Latour's assessment of the "Modern Constitution," and the dichotomy between nature and culture to problematize the practices that artificially separate science from culture. These practices are reflected not only in the traditional articulation of these discourses in schools, (science lessons, art classes, etc.) and over the media, (history channels, science channels, and art magazines, etc.), but are also firmly entrenched in the ways that museums have traditionally located science and culture. Institutional articulations of these discourses are not only represented through separate display strategies in the gallery, they also have become arbitrarily divided into separate physical buildings; the science museum, natural history museum as distinct from the social history museum and the art museum.

The second section, *Interacting with Science*, describes the body "as source of experimental knowledge" (Barry, 1998, p. 99) observing Frank Oppenheimer's significant contribution to the display of science and the pedagogic activities he instituted in the Exploratorium in San Francisco. This section also introduces two specific exhibitions from the Natural History Museum and Science Museum in London that are as much about art as they are of science. The third section moves *Beyond the Museum Walls* to explore how the Internet has extended and enabled museums to move beyond their traditional, physical constraints, and discussing the *Moving Here* Web site and the Exploratorium online. As illustrated in these projects there are now numerous new opportunities for museums to extend themselves online through:

- innovative systems that harvest data across electronic highways;

- online collaborations between museums and their public;

- production of bottom-up narratives that invigorate community knowledge; and

- novel ways to simulate and visualise science discourse.

Through four case studies, *Virtual FishTank, Walking with Woodlice, Country Cures and Journey North*, the fourth and last section, *Real (Time) Collaborations* investigates synchronous and asynchronous interactions over science Web sites. The projects illustrate how Internet architectures articulate and greatly enable large-scale collaborations as they disseminate them across electronic networks. In the same way that the culturally invigorated, gallery-based, interactive displays demonstrate culturally determined narratives, the online activities illustrate that scientific discourses when hosted and disseminated by science and natural history institutions are also confidently engaging with artistic and cultural discourse. As Oppenheimer had expressed in his philosophy, and demonstrated in his practice, "science has an aesthetic dimension" and art and science are united in the human quest for understanding (sic. in Hein, 1990, p. xvi; quoted in Barry, 1998, p. 103). These brief descriptions of museum activities serve to demonstrate how cultural messages are inscribed in scientific discourse and directed in meta-narratives across society. As these discourse are read together as the weft and the warp of the science web, the artificial boundaries between science and art are blurred and visitors may discover not only instructive scientific narratives but also golden opportunities to reset their cultural compass.

Locating Science, Locating Culture

The Modern Constitution, according to Latour (1993) has divided nature and culture, and with this separation, the knowledge and discourses that flow from language, society, and the arts have been detached from empirical scientific practice and theory; the same universally recognised systems that are embedded in the disciplines of academic practice that separate universities into different faculties. Museums have similarly maintained the status quo, keeping these two kinds of knowledge mostly distinct. This distinction is evident not only in their display of science and arts objects separated in different kinds of exhibitions, but also in the meta-narrative of these disciplines reflected in the institutional taxonomies of the institutions themselves; where science museums, and natural history museums are thought of as producing very different kinds of expectations and experiences from those of art and social history museums.

According to Latour, the distinction between scientific objects and the cultural objects of social discourse is historically rooted in the seventeenth century practice of demonstrating scientific proposals in the laboratory. "The facts are produced and represented in the laboratory, in scientific writings; they are recognized and vouched for by the nascent community of witnesses. Scientists are scrupulous representative of the facts" (Latour, 1992, p. 28). It was there, and then that the mute, inanimate or inhuman objects were given voice at a time when impartial scientific experiments gave rise to new forms of knowledge. "Yet the scientists declare," according to Latour "that they themselves are not speaking; rather, facts speak for themselves. These mute entities are then capable of speaking, writing, signifying within the artificial chamber of the laboratory or inside the even more rarefied chamber of the vacuum pump" (Latour, 1992, p. 29).

Latour is referring here to Robert Boyle's air pump. Boyle, as a member of the Royal Society of the English Restoration of the Experimental Way of Life, and known as the father of chemistry, conducted his experiment in the public laboratory where the practice of witnessing was the prerogative of a special community; clerics and lawyers. Donna Haraway points out that "these privileged zones of 'objective' reality were never the prevail of women, who were understood as 'covered' persons, subsumed under their husbands or fathers, women could not have the necessary honour at stake" (Haraway, 1997, p. 27). Subsequently, according to Haraway, as empirical science began to emerge, the modest, self-visible, transparent, non-polluting gentleman, she argues, could have never have been a woman.

Donna Haraway challenges situated knowledge as constructed in the Boyle laboratory through a feminist critique, and actor-network theory, (Haraway, 1991, 1997; Latour, 1996) and in her Manifesto (Haraway, 1991, pp. 149-181) describes the cyborg, the hybrid of organism and machine that resists the historical separation that has been set up between nature and culture. Like Haraway, Latour recognizes how "our intellectual life is out of kilter. Epistemologies, the social sciences, the science of texts, all have their privileged vantage points provided that they remain separate" (Latour, 1993, p. 5). Once divided artificially, scientific practice that has been separated from cultural discourse becomes detruncated and problematic. "Half of our politics is constructed in science and technology," Latour reminds us, while "the other half of Nature is constructed in societies. Let us patch the two back together and the political task can begin again" (Latour, 1992, p. 144).

The narratives that museums present to their audiences are first and formally separated by their location in totally different physical spaces. Visitors either pop in to their local science museum to explore the latest developments in genetics or biodiversity, and the natural history museum to measure themselves against dinosaurs. It is unusual to find a preference for artistic intervention in these kinds of locations although there are several, welcome exceptions that bring both kinds of discourse together under one roof.

Interacting with Science

In the summer of 2003 the Natural History Museum in London opened CleanRooms, billed as a temporary exhibition where *Art Meets Biotechnology* featured a collection of artworks and performances that illustrated the ethical issues of biotechnology. In a video installation called Silvers Alter, Gina Czarnecki presented a series of life-size human figures and visitors were asked to select and create new beings from the people they encountered on the screen in front of them. After getting to know the screen people through the biographies presented, using the "human-menu," visitors made their selections and were encouraged to activate their choice to create a new generation of virtual people. The process took place through an innovative interface triggered by sensors under the carpet and visitors determinately walked across the carpet, they were able to see the fruits of their action materialise before them on the giant screen in the darkened room (http://www.nhm.ac.uk/cleanrooms/crhome.htm). Interacting with the installation, this artistic gesture offered novel ways to think about human relationships and through the creative and empowering visualisation enabled visitors to enter into dialogue, not only with the scientific processes of procreation, but also with the cultural and social implications resulting from their own symbolic intervention.

Interactivity, granted by this kind of symbolic intervention, according to Andrew Barry, can be of particular significance in the museum of science "drawing together concerns with, for example, both public 'participation,' 'empowerment,' and 'accountability' and with more specific questions and anxieties about the proper way to bridge the gulf between popular culture and the esoteric world of science and technology" (Barry, 1998, p. 98). Barry argues from the *political anatomy* of the museum visitor (Foucault, 1979, p.138; in Barry, 1998, p. 99) "not only in relation to the development of communications theory and technology [...] but also in relation to a rather more longstanding concern with the body as a source of experimental knowledge" (p. 99). Barry reminds us how the incorporation of interactives in the gallery was not intrinsically new, particularly in the context of the science museum. He describes how, in 1969, the nuclear physicist Frank Oppenheimer, who had been blacklisted from practicing as a scientist by the House Un-American Activities Committee had established the Exploratorium in San Francisco as an alternative to the traditional science museum (1998, p. 102). Empowering visitors in this way through "interactive pedagogic technique" (Hein, 1990) could turn visitors into active participants rather than passive observers. This was the political and intellectual message inscribed in the Exploratorium, and expressed by Oppenheimer as "the whole point of the Exploratorium [...] to make it possible for people to believe that they can understand the world around them. I think a lot of people have given up trying to comprehend things, and when they give up with the physical world they give up with the social and political world as well" (Oppenheimer, quoted in Hein, 1990, p. xv).

It is not only the public who interacts with science in the science museum — sometimes it is the artist who instigates the dialogue between art and science. In the *Digitopolis Gallery* at the Wellcome Wing at the Science Museum in London, several artistic interventions are included in the narrative including Gary Hill's video projection *HanD HearD* — Variation, (2001) presented on seven projectors from DVD players. According to the Science Museum Web site, visitors will discover "intriguing art-pieces, poetry and opportunities for you to have your say on the technology affecting your life" (http://www.sciencemuseum.org.uk/on-line/wellcome-wing/digitopolis/about.asp). Thinking about science through Valéry's "artistic formulas" opens up not only new ways to interpret the rigor of scientific practice but also articulates scientific discourse across cultural and social spaces.

In May 2004, The Science Museum in London invited visitors to the Dana Centre to come to the museum and engage in an evening focusing on genetic testing in embryos to discuss whether we should …"be able to select for certain genetic characteristics in our children? Or is this a step too far?" (http://www.sciencemuseum.org.uk/).

Where does science end and culture begin? Well, in fact it doesn't and with this provocative invitation the public is already beginning to understand how our personal and social lives are impacted by genetic testing and what could be described as a purely scientific process. When experts in the museum enter into dialogue with the public over critical issues such as whether to treat a defective gene or even to abort a genetically impaired foetus, scientific processes may be mobilized to illustrate how we see ourselves, our [genetically produced] flaws and the imperfections of others. Clearly it is artificial and arbitrary to try and detach science from culture and to objectify scientific practice as the exclusive domain of scientists. Reinserting these narratives back into culturally directed social discourse can only produce richer interpretations and a fuller understanding of scientific practice.

Latour notes "[…] but when we find ourselves invaded by frozen embryos, expert systems, digital machines, sensor-equipped robots, hybrid corn, data-banks, psychotropic drugs, whales outfitted with radar sounding devices, gene synthesizers, audience analyzers, and so on, when our daily life newspapers display all those monsters on page after page, and when none of these chimera can be properly on the object side or the subject side, or even in between, something has to be done" (Latour, 1992, p. 49). A lot is already being done in science museums and natural history museums around the world, as well as through their online surrogates in the virtual museum.

Beyond the Museum Walls

The Internet has opened new highways to disseminate cultural and social discourses and in doing so offer new forms of links, nodes, networks, and paths that Barthes described as *ideal textuality* in *S/Z, An Essay* (1975), a textuality that George Landow likens to hypertext denoting "an information medium that links verbal and nonverbal information" (Landow, 1997, p. 4). Across this *ideal textuality*, where once the written text privileged discourse, now science and natural history museums promote and disseminate their own narratives across electronic highways, and in doing so enthusiastically take up the tools available to them; online databases, VRML, (virtual reality modelling language), environments, Flash animations, QuickTime VR and Web casting.

As museums develop their activities online, they not only extend their mandate they also offer novel opportunities for remote visitors to take on an active role. This echoes Oppenheimer's philosophy about knowing the world through experimentation and potentially sets up new relationships about the ownership of knowledge that is the legacy of museum tradition. (For debates of museum and the making of knowledge see, for example, Bennett 1995, 1998; Fahy, 2001; Hooper-Greenhill, 2001; Knorr Cetina, 1999; Macdonald, 1998.) Through the active collaborations and contributions of remote visitors, the authorship and mediation of knowledge is subtly shifted. This demands new adaptations by the museum as gatekeepers not only of tangible collections but also of the institutional knowledge systems that has been inscribed in the objects. While traditionally it has been the museum that actively collects, collates and categorises not only collections but knowledge on behalf of the public, new online interactions may reverse the traditional creator and receiver of institutional knowledge in novel scenarios. The intertextuality of the museum now freely disseminates educational scenarios, inspired and authored by museums. (See, for example, *Best Museum Web Site Supporting Educational Use,* http://www.archimuse.com/mw2004/best/educational.html.) These creative architectures are opening up new opportunities both for institutions and visitors to think about the role of the museum and the articulation of scientific and cultural knowledge and social experience.

The projects described below in fact illustrate new kinds of relationships where museums inspire remote visitors to take on an active role. In new symbiotic relationships the public is encouraged to contribute their own experiences to common narratives, or their own scientific findings to community-based, scientific collaborations. Looking at online activities from natural history and science museums that extend the museum mandate across communities, this chapter will explore how these kinds of narratives inflect the way we think about ourselves and the world around us. As science museums weave their webs across culture, art and aesthetic creativity, we begin to make connections between bio-diversity and an understanding of cultural diversity and, as we consider genetic engineering processes, we then take into consideration the socially inflected implications.

When artistic practice is almost exclusively exhibited in the art museum and pure science the prerogative of the science realm, there is little opportunity to contest the status quo and to bring both science and culture together into social discourse especially where knowledge is under the institutional management of an academically orientated museum (see Bennett, 1995, 1998). Now that new Web-based applications are appearing and evolving online, museums are taking up these opportunities and turning them to formulate new kinds of narratives — those that are not disseminated to the public but intrinsically emanate from the public. Connecting people across cyberspace from different cultural backgrounds, these networks enable remote visitors an opportunity to exchange ideas and stories and to explore whatever they wish from the comfort of their armchair — may it be flint arrowheads, Van Gogh sunflowers, or wood lice. These networks already span the globe, slip silently across geographical and cultural boundaries, and are available 24 hours a day at the click of the mouse.

The traditional mandate of the museum is to preserve, display and interpret extraordinary objects. The online variety becomes a space where people can contribute their own knowledge, data or experiences — even their own interpretations. A virtual network does not intend however to displace the museum when it comes to the singular and extraordinary objects, but it can be instrumental in collecting digital artefacts or scientific data, give meaning to them and thus building on and enriching shared knowledge and community narratives. Reversing the traditional relationship of museum and visitor, museums can build horizontal fraternities allowing leadership to shift from one participant to the next where the relationship is predicated on the offering (and accepting) of each other's interpretations, experiences and histories. This becomes a space that can be accessed by all, and while museums have traditionally taken the lead in producing knowledge, and with it the institutional narratives, the Internet is a space where new relationships may unfold with the remote visitor, with his or her own role to play, and his or her own narratives to contribute.

When it comes to the sharing of what we understand as cultural narratives, museums are already drawing from their online constituencies as resources for different kinds of explorations and participatory activities. *Moving Here*, a partnership of 30 heritage institutions across the UK, explores records and collates why Caribbean, Irish, Jewish, and South Asians came to England over the last two centuries (Moving Here, http://www.movinghere.org.uk/). Via an intuitively designed Web-based interface this evolving fraternity reverses the role of curated and curate. *Moving Here* grants everyone (at least those who have themselves journeyed to the UK with a story to share) an opportunity to contribute his or her own narrative. The idea that this is a space for anyone to participate in is reiterated by the invitation — "Your life is history. Your experiences are history. Your story is history."

Benedict Anderson (1991) describes three institutions that contribute to the formulation of the national imagination: the map, the census and the museum. Each one enables the citizen to imagine the parameters of his or her nation, and each institution, in their own way, sanctions a national history. In projects such as *Moving Here,* new narratives are formed as they filter up from the field from the community — it then falls to the institutions to organize and make sense out of the new knowledge and to represent these narratives coherently to the public. *Moving Here* opens up many questions about who may participate in the narration of history, who might be authorized to document the national narrative[s], and who would be responsible for compiling them for the national memory. Museums, libraries and archives, the acknowledged compilers and preservers of national narratives for society will need to embrace these new kinds of authorship. Distributed authorship is becoming more ubiquitous online with blogs and personal homepages are all predicated on individual agency without being subject to local or international law, ethical codes, or other critical mechanisms as is evident in professional journalism. All these kinds of narratives reflect a more flexible presentation of history and with them new opportunities for individuals to author micro-histories and narratives bypassing institutions. Like histories, knowledge is potentially a valuable and powerful commodity, especially once it is free from the restrictions of the traditional tethering of institutions like archives, libraries and museums.

Building on the Oppenheimer tradition, in 1993, the Exploratorium was one of the first science museums to go online with a comprehensive Internet presence. Clicking on the Exploratorium Web site reveals their focus, described as "investigating the science behind the ordinary subjects and experiences of people's lives. The topics themselves provide 'hooks' that get people excited about science. Then, when we investigate these topics, we can also look at the historical and social issues surrounding them, thus providing a context for scientific exploration" (http://www.exploratorium.edu).

One of the goals of the online museum is to provide an extension of the gallery floor and, according to the Exploratorium, does not in any way claim to displace the physical experience. The staff describes these activities as "'real' experiences for our online audience, not 'virtual exhibits.'" Internet access therefore offers surfers entrance to the museum at a click of a mouse and remote visitors can now access the exhibitions at home and in school from the comfort of their chair all over the world.

One of the earlier online demonstrations was *The Cow's Eye Dissection* (Hazan, 1995), which according to the museum "for many years it has helped people satisfy their curiosity about what is inside an eye. The material presented to the public is meant not to replace the act of dissecting a cow's eye, but rather to enhance the experience." (http://www.exploratorium.edu/learning_studio/cow_eye/index.html). Clearly the virtual, bloodless activity is sufficiently persuasive to provide for an exchange of scientific knowledge without the necessity for corporeal interaction. In fact avoiding the physical vivisection of *the real thing* may possibly be seen as a highly preferable in this case for most students who might have a loathing of dissections.

If we are to believe the SimSuite promotional material, the practice of simulating surgical operations is gradually displacing not just experimentation on laboratory animals, interns honing their theoretical knowledge on unknowing patients, or even on the experience gained from operating on cadavers but represents a major revolution — a computerised and totally bloodless simulation. According to their Web site:

> SimSuite® Centers feature Simantha™, a state-of-the-art tactile-force-feel simulator, which produces distinct and unique experiences for each user. Their experience is affected by response time, actions and decision-making abilities in real-time interventional cases. The realistic clinical environment also includes other visual and tactile elements that contribute to suspending the user's disbelief and enhance the opportunity for learning (http://www.simsuiteed.com).

Not only do these simulations appear to be instructive, efficient, and persuasive, but unlike the carbon variety, the patients of the surgical interventions would be instantly retrievable after failed surgery, and presumably highly advantageous with a view to market potential when, in addition to all this saving of in viva or post mortem experimentation, Simantha™ is limitlessly cloneable

For those students, however, who do choose to forgo the virtual, bovine experience for the real thing they will find instructions on the Exploratorium site to be able to order a cow's eyes at a butcher shop or be directed how to purchase one directly from a slaughterhouse. (Further information is available online with the recommendation to "try to get eyes with the muscles and fat still attached. If possible pick up the cow's eyes the day of the dissection; eyes are easier to cut when they are fresh."). One would think that the availability of the virtual experience in itself would more than justify the entire project of an online museum of this nature, but it seems that some students of biology would still privilege the real over the virtual.

The Exploratorium in San Francisco promotes itself as is a museum of science, art, and human perception and is proud of their over 650 interactive "hands on" exhibits. According to Web site information from June 1997, each year more than 660,000 visitors come to the Exploratorium, over 67,000 children come on field trips, and more than 500 teachers attend professional development programs which focus on inquiry-based teaching and learning in the K-12 classroom. Today (summer 2004), the Web site boasts over 15,000 Web pages exploring hundreds of different topics, including instructions for over 500 simple experiments, many sound and video files, exploring hundreds of different topics and currently serving fifteen million visitors a year (nearly thirty times the number of visitors who come to the museum in San Francisco). One wonders why they don't simply close up their physical site and promote themselves exclusively online, but, as the cow's eye dissection proves, it seems that people, wherever they are, still seem to prefer the real over the physical.

Real (Time) Collaborations

Computer mediated communication (CMC) takes place both across synchronic and asynchronic architectures. Synchronic activities (which may be between person and machine or between person to person over electronic networks) take place in real time, (i.e., all participants are communicating in a shared space simultaneously) and in this kind of conversation, participants need to be identified in some way. Often, performing identity may be less than straightforward, with surfers invoking a "handle" or purported name commonly used across IRC (Internet relay chat) discussion. In an online VRML environment players may choose to present themselves with a realistic name or an accurate textual or visual representation of the person or, in certain spaces, players may be represented by an avatar. The promotion of self through an assumed name is often confusing — both for the chatter and the *chatee* and neither can truly know whether the person they are relating to on the "other side" is really the person he, she or it claims to be. Sherry Turkle explored these alternative models of identity in her book *Life on the Screen*, in her discussion of MUDs (Multiple User Dimension, Multiple User Dungeon, or Multiple User Dialogue). In MUDs, players soon discover that the presentation of self is a flexible negotiation and that the idea that they are a unified self is simply fiction. This is not a space that necessarily inspires trust, but museums, whose institutional mandate do inspire a sense of trust may appear more attractive for users who wish to seek out a social or culturally directed narrative when hosted by a publicly funded institution with a tradition of integrity.

Virtual *FishTank*™

None of this is particularly relevant however when your avatar happens to be a fish. *Virtual FishTank*™ introduces a 3-D, real-time, interactive environment delivered directly into the museum. *Virtual FishTank* was developed by Nearlife, Inc. (http://www.nearlife.com) and by the MIT Media Lab (http://www.media.mit.edu/) primarily for the Museum of Science, Boston (htp://www.mos.org), and more recently with tank access at the St. Louis Science Center (http://www.slsc.org/). Both in the Boston Museum of Science or St. Louis Center, visitors can create their own designer fish, in a specially constructed interactive kiosk, save them and then release them into the Museum tanks.

Figure 1. Virtual FishTank™

The Boston Museum of Science describes the activity…

> As visitors venture inside The Virtual *FishTank*, they are immersed in a 2,200-square-foot virtual undersea world. Lights shimmer on the waves of perforated metal above as if sunlight were hitting the water's surface. Bright blue gravel covers the virtual ocean floor spilling out onto a carpet of deep sea blue. Twelve large projection screens form windows (each measuring 12 square feet) into a spectacular 400-square-foot central tank, populated by nearly 100 bold-colored, cartoon-like, mechanical fish. The windows are edged in metal plate riveted together with nuts and bolts, simulating a wall built to hold back the aquatic environment. The sounds of waves, water dripping, bubbles, and fish swimming by and being gobbled up by predators infuse the exhibition space.
>
> (http://www.mos.org/exhibits/current_exhibits/virtual*FishTank*/vft_walkthrough.html) © *Nearlife Inc.*

Illustrating complex biological behaviour becomes a lot more immediate and a lot more enjoyable through the *Virtual FishTank*, but the social aspect of taking part in a shared activity also greatly contributes to the pleasure of the activity. In the same way as your own fish is designed, formed and released into the aquarium through the gurgling pipes among all the other fish, it is tempting to make a connection between the aquatic world

Figure 2. Fish building kiosk, © St. Louis Science Center

Figure 3. © *Nearlife Inc.*

and the world around those visitors who participate in the activity together. Apart from the fun side of things, there is meaningful engagement in this activity where visitors learn how complex patterns can emerge from individual interactions among simple objects. In Boston, visitors can "dive deeper" into six stations that demonstrate to visitors other implementations of the phenomena they have observed in the central tank. Through reiterating the rules they learned from the fish tank, visitors may also understand how other complex systems work, such as insect colonies, highway traffic, and market economies and to go on to make connections from the world of science to the world of culture — breaching the artificial divide that can no longer be maintained as separate.

While not every visitor might be able to go to Boston to dive into this undersea world and while others may not be able to even set foot in the United States for the whole of their lifetime, remote visitors may still access the tanks via the Web version. The Web

Figure 4. Screen grab of the online aquarium © *Nearlife Inc.*

version of **The *Virtual FishTank*** purports to turn your computer into an aquarium. This requires that visitors first log on to the *Virtual FishTank* and once connected to the digitally, enhanced aquatic system may do a number of things including:

- Launch your fish into a Personal Tank on your own computer and see how they behave.

- Send your fish to the Museum of Science ***FishTank*** and receive reports on their life stories.

Forging the connection to the physical museum across the invisible science web the activity extends into another a further dimension. Fish-makers who may wish to follow the fate of their newly built fish may also:

- Come to the Museum, retrieve and release your fish, and watch them interact with other cyberspecies in *The Virtual FishTank.*

- Build a fish at the Museum of Science and send it to your home PC.

Sharing both fish and a pleasurable moment, this project breaches the walls of both institutions as newly created fish are propelled from one museum to the other - users can send fish to either destination in Boston or St. Louis — albeit some 1,179 miles apart as the crow flies but a mere second's swim across cyberspace. The sense of the porous museum wall is further inferred by the fact that you can take part in this activity from the comfort of your own home or school, making connections beyond the previously impermeable walls of the museum. The activity is further extended when the fish data (identified by the user's unique login details and fish name) is gathered and collated on the backend database. Subsequent interrogation of this data, the fish's life story (recorded depth data, food eating events, etc.) may be retrieved by visitors in the form of a graph as he or she follows the life cycle of his personal fish. Not only is this a charming and playful way to engage in what would otherwise be a complicated process for young children and adults to grasp, but this also reiterates the sense of being part of a biosphere, albeit an aquatic one, but one that is not too dissimilar to the one we all live in.

Walking with Woodlice

Obliquely referring to our prioritising of some creatures over others, the Natural History Museum, London, asks their Web visitors — "why woodlice? You may think that once you've seen one woodlouse, you've seen them all." The museum urges young scientists to search at home in order to help the museum-based scientist find out how many different kinds of woodlice live in their neighbourhood. Out of the 37 species of woodlice in the UK, remote visitors are requested to identify their very own woodlice using a printable key and to send in their results to the museum. The Museum explains how "scientists

Figure 5. Screenshot of the Walking with Woodlice *key,* © *Natural History Museum*

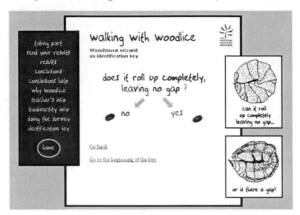

still don't know everything about woodlice — and this is where *you* can help" (http://www.nhm.ac.uk/interactive/woodlice). *Walking with Woodlice,* inspired by the BBC series (at least nominally), *Walking with Dinosaurs*, aims to get children to take part in a nation-wide scientific study to use the Internet to share scientific information, to develop their scientific skills through challenging the online results, and in doing so nurture an enthusiasm for bio-diversity.

Young scientists of biodiversity are requested to find out:

- Where do woodlice live?
- How many different kinds of woodlice live near you?
- Which are the most widespread UK woodlice?
- Do different kinds of woodlice live in different place?

Once the study area has been chosen students are required to collect woodlice, and using the key identify the species they have found.

Following the [virtual] woodlice around the online questionnaire, participants are instructed how to identify the woodlice they find then in their own homes, schools and gardens and to upload the results on an especially designed Web interface. The results in turn need to be interpreted and the questions asked are provocative. Terms such as biodiversity, taxonomy and systematics are explained online and students are encouraged to interrogate the data that had been gathered from 2001-2004.

The lessons learned on biodiversity not only serve to understand the meaning of the term but the museum also reminds students how the miniature life they discover in their own homes demonstrate the principles of biodiversity without having to visit far away lands...." The mention of biodiversity commonly conjures up images of steaming tropical forests, and these are without doubt some of the most diverse areas on the planet," *instructs the museum,* "but did you know that the UK is also an area rich in biodiversity?" (http://www.nhm.ac.uk/interactive/woodlice/biodiversity.html).

Bringing scientific narratives close to home not only allows for active participation on the part of every single one of us — after all we all have woodlice wandering around our homes — but also serves to focus on the potential of even the smallest of creatures. Although unable to compete with the mega-stars of the animal kingdom (the dinosaurs) in science channel television programs and museum exhibits, the un-trendy woodlouse, for all of its modesty is still able to teach children not only a good lesson in biodiversity but also to help us recognize the potential of all creatures, large and small.

Country Cures

Reaching out in a similar way to the community to gather diverse data, the Natural History Museum also gathers information on herbal remedies, "not those from ancient books or old scientific journals, but those kept alive by word of mouth between generations" according to the *Country Cures* Web site (http://internt.nhm.ac.uk/cgi-bin/country_cures). Visitors to *Country Cures* learn about home remedies that keep gnats away — as long as the person is prepared to carry a sprig of basil (*Ocimum basilicum*), which evidently when eaten produces a foul smelling sweat that insects don't like in the least. This particular contributor, from Glamorgan, also claims that basil has been traditionally used to keep away mosquitoes as well.

The plants and shrubs that produce these home remedies are available to visitors who may wish to visit the wildlife garden that is open to the public each day from April until the end of October, from 12.00 to 17.00 (http://www.nhm.ac.uk/museum/garden/index.html). Visitors who may not be able to come into the urban garden in the heart of London may access the garden through the Web site — and view a series of photographs taken at weekly intervals over a period of two years. The site also demonstrates the seasonal factors that affect the garden — the length of light and dark hours over the different seasons and the maximum and minimum temperature of any given day. For visitors who do not have their own garden, this is a delightful opportunity to follow the waxing and waning of nature and, even if remote visitors are unable to take pleasure in the smells that waft up from the herbs and flowers, they still may gather a wealth of knowledge about garden life.

Journey North

The examples of public institutions extending activities to draw in their online constituents into new opportunities for community participation still replicates the traditional educator/educated scenario to a certain extent, but they also illustrate the potential to extend the new paradigm as remote visitors take on all kinds of roles, including the full authorship of the project. *Journey North* is a project that engages students in a study of wildlife migration and seasonal change across North America. According to the Web site, "students track the coming of spring through the migration patterns of monarch butterflies, bald eagles, robins, hummingbirds, manatees, whooping cranes — and other birds and mammals, the budding of plants, changing sunlight and other natural events. (http://www.learner.org/jnorth).

Now in its 10th season, a project such as this was not possible before the Internet as it is built on and is disseminated across the electronic networks that connect geographically distant classrooms. The thousands of participants in this science education program tracks migration and seasonal changes and share their own observations with others up and down the country. *Journey North's* spring program begins each year around Groundhog Day (February 2nd) when about a dozen different migrations and signs of spring are first apparent. Today, over 11,000 schools, representing more than 490,000 students, are participating in the 2004 Journey North Program from all 50 U.S. States and seven Canadian Provinces.

Journey North, an independent 501(c)3 organization based in Washington, D.C. was established in 1991 with a grant from the Annenberg Foundation to the Corporation for Public Broadcasting. While this is obviously not a museum-based project, it is an impressive one that contributes vital threads to the webs of media and communications that serve to improve math and science education for the nation's 44 million school children. According to the *Journey North website,* collaborating in this way students are brought together across 17 time zones, four states, two continents and two countries. In the same ways that museums reach out to their public, *Journey North* not only produces a rich data set of vital material on the migration habits of many kinds of species but also represents an inspirational project that encourages community participation and instigates network collaboration.

Conclusion

Weaving science webs has looked at several innovative systems that harvest data across electronic highways, subtly shifting the traditional role between museums and their public through online collaborations between the institutions and their geographically dispersed audiences. The production of bottom-up narratives in this way invigorates collective knowledge and establishes a sense of community through collaboration and the public sharing of micro-experiences. The novel ways that serve to simulate and visualise science discourse also open up new possibilities for formal and informal learning scenarios, where the public become agents of the production of knowledge and add a further dimension to institutional activities that now extend beyond the museum walls.

The in-house exhibitions described here in the Science Museum, the Natural History Museum and the Exploratorium not only activate the visitors through these kinds of gallery-based interactions but also illustrate how art and culture may be reinserted into the realm of science. In this way, science and art work together to produce meaningful narratives, replacing them back into their social context, and dissolving the familiar dichotomy inscribed in the rationalizing project of modernity. Combining the weft and warp of culture and science into interwoven discourses no longer results in our intellectual life being out of kilter as these discourses become one and the same.

Embedded in archaic mythologies, these combined narratives and discourses reinforce narratives of kinship, national configurations and cultural affiliations, and through the telling of stories, whether these stories are the rational science based wefts, or culturally directed warps; they all impact and enrich our world both directly and indirectly.

References

Anderson, B. (1991). *Imagined communities*. London: Verso

Barry, A. (1998). On interactivity: consumers, citizens and culture. In S. Macdonald (Ed.), *The politics of display, museums, science, culture*. London; New York: Routledge.

Barthes, R. (1975). *S/Z: An essay*. New York: Hill and Wang.

Barthes, R. (1977). The death of the author. In *Image, Music, Text*. Essays selected and translated by Stephen Heath. London: Fontana.

Bennett, T. (1995). *The Birth of the museum*. London; New York: Routledge.

Bennett, T. (1998). *Culture: A reformer's science*. London: Sage Publications.

Fahy, A. (2001). New technologies for museum communication. In E. Hooper-Greenhill (Ed.), *Museum: Media: Message*. London: Routledge.

Foucault, M. (1979). *Discipline and punish*. Harmondsworth: Penguin.

Haraway, D.J. (1991). *Simians, cyborgs, and women: The reinvention of nature*. London: Free Association.

Haraway, D.J. (1997). *Modest_Witness@Second_Millenium. Female_Man_ Meets_Oncomouse: Feminism and technoscience*. New York: Routledge.

Hazan, S. (1995). Museums and art on the Internet. In D. Bearman & J. Trant (Eds.), *Conference Proceedings, Hands on Hypermedia & Interactivity in Museum, ICHIM '95 - MCN '95*. Pittsburgh, PA: Archives & Museum Informatics.

Hein, H. (1990). *The Exploratorium: The museum as laboratory*. Washington, DC: Smithsonian Institution Press.

Hooper-Greenhill, E. (2001). Museums and communication: An introductory essay. In E. Hooper-Greenhill (Ed.), *Museum: Media: Message*. London: Routledge.

Knorr Cetina, K. (1999). *Epistemic cultures: How the sciences make knowledge*. Cambridge, MA: Harvard University Press.

Landow, G.P. (1997). *Hypertext 2.0: The convergence of contemporary critical theory and technology* (2nd ed.). Baltimore; London: Johns Hopkins University Press.

Latour, B. (1993). *We have never been modern* (C. Porter, trans.). New York; London: Harvester Wheatsheaf.

Latour, B. (1996). *Aramis, or, The love of technology* (C. Porter, trans.). Cambridge, MA: Harvard University Press.

Macdonald, S. (1998). Exhibitions of power and powers of exhibitions: An introduction to the politics of display. In S. Macdonald (Ed.), *The politics of display: Museums, science, culture*. London: Routledge.

Turkle, S. (1995). *Life on the screen: Identity in the age of the Internet*. New York: Simon & Schuster.

Valéry, P. (1970). Analects. Moralités. In J. Matthews (Eds.), *Collected works* (vol. 14). (Original work published 1932).

Chapter VI

Resource-Based Learning and Informal Learning Environments:
Prospects and Challenges

Janette R. Hill, University of Georgia, USA

Michael J. Hannafin, University of Georgia, USA

Denise P. Domizi, University of Georgia, USA

Abstract

Recent changes in the role of resources have not only transformed how we think about resources, they have distributed production of and access to digital resources while altering fundamentally how, when, and for what purposes resources are created and used. The metamorphosis has been propelled by the exponential growth of information systems such as the Internet and the Web, and the ubiquitous presence of enabling technologies in classrooms, libraries, museums, homes, businesses, and communities. These changes portend exciting educational opportunities, particularly in resource-rich environments such as science centers and museums. In this chapter we explore RBL and ILEs, providing examples of how an RBL approach might be implemented in an ILE and describing the opportunities and challenges associated with such an endeavor.

Introduction

During recent years, the definition, role and uses of resources have undergone a metamorphosis. The changes have transformed how we think about resources, the distributed production of and access to digital resources, and how, when, and for what purposes we create and use them. The metamorphosis has been propelled by the exponential growth of information systems such as the Internet and the Web, and the ubiquitous presence of enabling technologies in classrooms, libraries, museums, homes, businesses, and communities.

While increasing the numbers of and access to resources is energizing, realizing the educational potential of these breakthroughs may prove daunting. This is particularly true in formal learning settings (i.e., school), where current practices do not emphasize optimizing available resources or preparing individuals to learn in resource-rich environments. Informal learning environments, in contrast, offer considerable promise for resource-based learning (RBL). Science museums and centers, for example, provide a variety of resources to investigate, as well as learning from and with, as visitors explore exhibits. Informal environments offer freedom not available in formal environments, where instruction usually focuses on established curriculum goals, sequences, resources, and activities. Informal learning environments provide an opportunity to exploit resource-based learning alternatives, expanding both the materials and methods used in teaching and learning.

The purpose of this chapter is to provide an overview of resource-based learning (RBL) and its applications within informal learning environments (ILEs). We begin by providing an overview of resource-based learning, describing RBL components, what they are and how they work. Next, we discuss RBL examples in science centers/museums. Finally, we describe both opportunities and challenges associated with RBL in informal learning environments.

Overview of Resource-Based Learning

Resource-based learning "…involves the reuse of available assets to support varied learning needs" (Beswick 1990). Several factors make resource-based learning (RBL) viable: 1) increased access to resources (print, electronic, people) in a variety of contexts not previously available; 2) resources are increasingly flexible in their manipulation and use; and 3) economic realities dictate that resources become more readily available, manipulable, and shareable across a variety of contexts and purposes.

Increased Access

The raw amount of information we are exposed to on a daily basis increases exponentially. Recently, researchers at the University of California-Berkeley estimated that information grew 30% annually between 1999 and 2003. According to Lyman & Varian (2003), the volume of information on the Web alone is 170 terabytes — 17 times the volume of collections at the Library of Congress — a tripling of information since 1999 when the last survey was conducted.

We also have significantly greater access to information. We interact face-to-face, on the phone and via computer-based technologies such as e-mail and Instant Messenger®. We obtain a multitude of print-based resources, ranging from books to magazines to newspapers. We acquire seemingly unlimited amounts of electronic information — some traditional (i.e., books, newspapers), many non-traditional (e.g., video clips, sound bytes). According to Lyman and Varian (2003), the telephone and Web have now become leading technologies for accessing and sharing information.

Another profound change related to increased access are the multitude of perspectives those resources represent. We are no longer limited to local town newspapers or newscasts for our information. We obtain news broadcasts from around the globe (e.g., Channel Africa, the Australian Broadcasting Corporation), newspapers from any number of groups (e.g., *The Times* (London), the *New York Times*), books in almost any language imaginable — all with the point and click of a mouse and an Internet connection. Resources are seemingly limitless in scope and depth.

Flexible Manipulation and Use

In addition to exponential increases in the number and type of resources, we can now manipulate and use resources in ways heretofore unimaginable. The technology of learning objects (see Wiley, 2000) has provided new ways of thinking about the very fabric of resources. What was once an entire video is now a series of clips that can be split apart and used in multiple ways. A book is now a collection of words, charts, figures and images that can be used as a whole or as individual objects.

Use is also influenced by the capacity to identify resources as something other than intact, "whole" units. The capacity of a resource to be broken down into its component parts allows users to appropriate individual pieces for their own purposes. For example, an individual Web page containing images, video and text provides a multitude of individual resources for the individual user. The very fibers of the fabric are now moveable and malleable, enabling meaningful uses in similar or entirely different contexts.

Economic Realities

Learning initiatives of all types and contexts have experienced radical budgetary constraints in the last 20 years (NEA, 2004). If ever a surplus of supplies and resources existed (many would argue there never was, but might at least agree that things were better at the end of the 20th century), it certainly does not exist today. Increasingly, teachers need to buy their own supplies and fund their own copying. Science centers and museums face similar constraints with reductions in funding for displays and supplies.

Budgetary restrictions in all educational sectors demand that resources become more readily available and shareable. Further, availability needs to be ensured across a variety of contexts to enable resource use for a variety of purposes. Resource-based learning, particularly in digital environments, offers considerable promise, enabling educators in a variety of contexts to provide access to an ever-expanding library of digital resources.

Components of Resource-Based Learning

Resource-based learning (RBL) features four basic components: enabling contexts, resources, tools, and scaffolds. Taken together, these components enable educators to create and implement learning environments of considerable diversity and flexibility. Table 1 provides an overview of key characteristics. Each of the components will be briefly described in this section [for a more in-depth description, see Hill & Hannafin (2001)].

Enabling Contexts

Enabling contexts supply the situation or problem that orients learners to a need or problem, such as recognizing or generating problems and framing their learning needs. By creating an enabling context, meaningful learning can occur with and through the resources provided or obtained (Brown, Collins, & Duguid, 1989). Enabling contexts can be imposed, induced or generated. *Imposed* contexts — frequently specific problem statements or questions — clarify expectations explicitly and guide teacher and student strategies implicitly. Teachers may use state-mandated objectives; science centers may adopt standards established by nationally recognized organizations (e.g., National Science Teachers Association).

Table 1. Components and characteristics of RBL

RBL Components	Key Characteristics
Enabling contexts	*Imposed*: teacher/curator or external agency determines goal *Induced*: learner or learner and teacher/curator determine goal
Resources	*People, things or ideas* that support the learning process
Tools	*Objects* used to help facilitate the learning process. Range from processing to organization to communication tools.
Scaffolds	*Support* that is faded over time. Includes conceptual, metacognitive, procedural and strategic scaffolds.

Induced contexts introduce a domain where problems or issues are situated, but not specific problems to be addressed. A typical scenario enables multiple problems or issues to be generated or studied based on different assumptions, topical relevance, and the context of use. For example, a science center might situate learning about electricity by providing a story for the user of the Web site to help guide building understanding.

In *generated* contexts, specific problem contexts are not provided; rather, the learner establishes an interpretive context based on his or her unique needs and circumstances. In a generated context, the user of the science center Web site becomes responsible for situating the electricity information in their environment. Similar to induced contexts, the learner activates relevant knowledge, skill, and experience to guide their activities as they define and evolve new circumstances.

Resources

Resources are "raw materials" that support learning, such as electronic databases, print textbooks, video, images, original source documents, and humans. Resources may be provided by a more knowledgeable other (e.g., a teacher, a curator) to assist others in extending or broadening knowledge or understanding. Resources may also be gathered by the learner as questions and/or needs arise.

Exhibits might include multiple resources, ranging from historic artifacts, to manipulables, to video, to detailed explanations of the objects being viewed. Electronic resources, physical or virtual, are often included as well. They provide a theoretically unlimited library of RBL source materials supporting different educational goals. Given varying contexts of use, the utility of a resource may change dramatically from situation to situation. The Web, for example, enables access to millions of source documents, but their integrity and usefulness is judged by the individual and in accordance with the context of use. As resources become both increasingly relevant to the learner's need *and* accessible, they assume greater utility.

Tools

Tools enable learners to engage and manipulate both resources and ideas. Tool uses vary with the enabling contexts and user intentions; the same tool can support different activities and functions. Eight types of tools are used in RBL: processing, seeking, collection, organization, integration, generation, manipulation, and communication.

Processing tools help students to manage the cognitive demands associated with RBL. Processing tools, such as self-directed learning systems, for example, enable learners to work with ideas, extending their cognitive abilities and reducing the need to "remember" or engage in unnecessary mental manipulation [see Jonassen & Reeves (1996) for a discussion of cognitive tools].

Seeking tools (e.g., keyword searches, topical indexes, search engines, etc.) help to locate and access resources. Popular search tools like Google® (www.google.com) identify the location of and direct access to a variety of Web-based resources. Seeking tools can also be specific to a particular context. For example, the State of Georgia has created *GALILEO*, an extensive database of search engines and electronic resources available through libraries via the Web. *GALILEO* users have access to traditional resources (e.g., academic journals) as well as other resources (e.g., multimedia) (www.galileo.usg.edu).

Collection tools, ranging from paper-based worksheets to high-end PDAs, aid in amassing resources and data for closer study. Learners might use collection tools as they explore a learning space or after completing a tour. Organization tools are used to represent and define relationships among ideas, concepts, or "nodes." Like collection tools, organization tools range from electronic to non-electronic devices. Concept mapping tools (e.g., Inspiration®, www.inspiration.com) are powerful devices that enable users to demonstrate relationships and links between and amongst ideas.

Integration tools help learners to relate new with existing knowledge, which helps to both organize and integrate ideas. Integration tools might range from a word processing program to a Web page. The depth and breadth of what is represented by a single tool or set of tools vary according to the needs and abilities of the user. Generating tools, as simple as a Web page or as sophisticated as a modeling tool (e.g., SimCity® or SimEarth®), help learners to create "objects" of understanding. Manipulation tools, which also range in their complexity, are used to test or explore beliefs and theories-in-action. The Virtual Solar System®, for example, enables learners to hypothesize, construct, test and reconstruct a solar system as they test their beliefs related to the relationships between and among objects in the solar system (lpsl.coe.uga.edu/live/vss).

Finally, communication tools support efforts to initiate or sustain exchanges among learners, teachers, and experts. Asynchronous tools such as e-mail and listservs enable fast and easy access. Higher-end synchronous communication tools such as videoconferencing enable access to people "down the hall" through "down under."

Scaffolding

Scaffolding — support provided to assist learners initially and subsequently faded (see Vygotsky, 1980) — varies with problem(s) encountered and the demands of the enabling context. Four types of scaffolding are useful in ILEs: conceptual, metacognitive, procedural and strategic.

Conceptual scaffolds guide learners in *what to consider,* identifying knowledge related to a problem or making organization readily apparent. Worksheets have traditionally been used in formal and informal learning settings to help guide students as they explore a new concept or topic. Conceptual scaffolding might be extended through communication tools in the form of leading questions or scenarios that set a context for the learners on a Web site. Problem-based learning makes considerable use of conceptual scaffolding to help guide learners as they explore new areas and build understanding [see Knowlton & Sharp (2003) for a collection of articles related to problem-based learning].

Metacognitive scaffolds support the underlying cognitive demands in RBL, helping learners to initiate, compare, and revise their approaches. Scenarios or cases are often used to focus and guide the learner as they explore and attempt to understand. Scenarios or cases can present ideas for learners to consider as well as checkpoints where learners examine their understanding, seeking to uncover what they do and do not know or understand (see Kolodner, 1993; Kolodner, 1993/1994).

Procedural scaffolding aids the learner while navigating and emphasizes *how to utilize* a learning environment's features and functions. WebQuests, for example, use procedural scaffold extensively and have been used in a variety of contexts and content areas. According to Bernie Dodge, the primary creator, "WebQuests are designed to use learners' time well, to focus on using information rather than looking for it, and to support learners' thinking at the levels of analysis, synthesis and evaluation" (Dodge, n.d.). By focusing on *how to*, procedural scaffolds free up cognitive resources for other important learning activities (e.g., problem solving, higher-order thinking).

Finally, strategic scaffold provide ways to analyze, plan, and respond, such as identifying and selecting information, evaluating resources, and integrating knowledge and experience. Several models have been particularly useful in selecting and evaluating resources. Alexander & Tate (1999), for example, proposed several techniques for evaluating Web pages ranging from advocacy to informational Web pages (see www2.widener.edu/Wolfgram-Memorial-Library/webevaluation/webeval.htm for several online resources). The I-Search process (Joyce & Tallman, 1997) strategic scaffolding focuses on integrating knowledge and experience. I-Search enables learners to select a topic of personal interest, then guides through the process of finding and using information and developing a final product.

Resource-Based Learning in Science Centers and Museums

Informal learning environments are provided in a variety of settings and ways. Examples can be found in everyday environments, ranging from learning about economics in the grocery store to exploring physics on the playground. Informal learning also occurs in more formal settings such as science centers and museums (Borun, Chambers, & Cleghorn, 1996; Paris, 2000). While many science centers strive to engage their visitors through interactive multimedia, they sometimes emphasize "bells and whistles" over the inherent educational value of their holdings and exhibits (Aldrich, Rogers & Scaife, 1998; Cairncross & Mannion, 2001). Resource-based learning, properly implemented, encourages visitors to explore beyond the scope of traditional material or exhibits, challenging them to question, pose "what-if" scenarios, and seek answers with available resources.

Given the vast array of resources available to science centers and museums, they are natural environments for RBL approaches. In this section, we describe several informal science and museum settings in which RBL has been implemented (see Table 2 for a summary of how the museums and exhibits exemplify RBL components). The Lemelson Center provides an example of how learning science can happen through virtual play. *The Smithsonian Institute* and the *Seeing Exhibit* exemplify how blended approaches to

Table 2. Examples of use of RBL components in museums and exhibits

Museum/ Exhibit	Enabling Context	Resources	Tools	Scaffolding
Lemelson Center for the Study of Invention and Innovation/ *Invention at Play*	Several, including puzzle blocks and building mazes	Examples of inventions, print-based information	Sketchpads, "hardware" for building mazes	Help buttons, stories, guiding and integrating questions
Smithsonian Institute	Several, dependent upon specific exhibit	Multiple, print and electronic	Several, dependent upon specific exhibit	Several, dependent upon specific exhibit
Exploratorium/ *Seeing*	How do we see the world?	Field trip pathways, companion Web site	Worksheets with pathways for processing, collecting, organizing and integrating	Guiding and integrating questions
National Geographic Education Guide	How do we understand a culture?	Multiple electronic resources	Search, manipulation and generation tools	Lesson plans, scenarios

learning (face-to-face and online) can extend and enhance the user's experience. Finally, the *National Geographic Education Guide* demonstrates how a completely online experience can support and enhance the learning of science.

Invention at Play: Lemelson Center for the Study of Invention and Innovation (www.inventionatplay.org)

Science centers and museums are often associated with "playtime." This does not, however, suggest that learning does not occur. Indeed, researchers suggest that children can, and often do, learn through play (see Rieber, 1996). *Invention at Play* incorporates play into all its exhibits. There is an extensive traveling physical exhibit, but for those who cannot go to the places where the exhibit will be displayed, the virtual center provides several opportunities for exploring science concepts.

One of the hallmarks of this center is the unique nature of the contexts in which the learner is placed for exploring science concepts. Problem solving skills are centered within the context of a puzzle. Learners are challenged with creating a bird, boat or man by putting together various puzzle blocks. As stated on the site: *Many inventors can see patterns that aren't obvious to others. Changing or manipulating patterns is one way to generate new ideas.*

Resources and tools on the *Invention at Play* site vary depending on the science concepts being explored. Examples of creations from inventors are provided to assist learners in extending their understanding of how invention occurs. Several print-based resources are linked to the site to assist with extending the use of the online center.

Tools range from general idea capturing sketchpads to specific pieces of "hardware" (e.g., pipes, hammers, springs and sockets) for "tinkering" to get a ball into a cup. According to the site: *Exploratory play is about asking questions: "What happens when I do this?" "What if I did it this way?" Experimenting with materials and pushing their limits encourages us to consider a wide range of possibilities when problem-solving. Playing around with objects and ideas helps us see that there may be more than one solution.*

Invention at Play provides several forms of scaffolding as the learner works with various experiments. When interacting with the puzzle blocks, learners can get assistance by clicking on a "help" button. The shapes are then outlined in the empty form (e.g., boat) so the learner can see where the pieces fit. Additional scaffolding is found in some of the resources, such as the stories by inventors and guides for the exhibit.

Smithsonian Institute: Science and Technology (www.si.edu/science_and_technology/)

The Smithsonian Institute has hundreds of online exhibits related to science and technology that allow for learning and exploration to co-exist. Each exhibit is indexed by main fields in science (Animals, Astronomy, Aviation and Transportation, Computers and Communications, Ecology and Environment, Evolution and Paleontology, Geology, Health and Human Sciences, Industry, Machines and Electricity, Marine, and Plants), enabling access to almost anything one might be seeking for exploration related to science and technology. Within each of the main science fields, further contextual information is provided for specific areas. For example, in the Animal exhibits, visitors are taken around the planet to explore a variety of animal species, ranging from turtles in the Galapagos to camels in the Australian Outback to frogs in North America.

Resources within the overall Smithsonian Institute are extensive, enabling educators, families and students to find information related to specific topics from the museum (see http://smithsonianeducation.org). Resources within each exhibit are just as far ranging as the contexts in which the individual exhibit is centered. Print-based worksheets and vocabulary lists are available to guide initial exploration of many areas. In the Astronomy exhibits, learners can view pictures of stars in far-away galaxies or listen to Neil Armstrong, Michael Collins and Buzz Aldrin reflect on their Apollo 11 mission to the moon. In the Marine exhibits, extensive databases of fish to floor plans to enable virtual exploration of physical exhibits are amongst the resources made available to learners.

Tools range from traditional to more interactive materials. Worksheets can be printed; a "Polar Pairs" matching game in the Artic Wildlife virtual exhibit can be played multiple times; and a message board enables visitors of Ocean Planet to share information with other learners. In an exhibit titled, *Mirror Molecules*, learners are provided with directions for experiments so they can learn about molecules — how they are formed and shaped.

Scaffolding in this museum is enabled by various means. Online tools are used to some extent (e.g., the discussion board described earlier) to provide access to others. Step-by-step procedures are provided in some exhibits (e.g., how to conduct experiments) to assist learners with exploring concepts. Extensive explanations of concepts are also obtainable so that more advanced learners can extend their understanding. For example, in the *Resource Guide to Paleoanthropology*, advanced users can get detailed information on the overall process of evolution as well as the process of human evolution.

The Seeing Exhibit at San Francisco's Exploratorium (www.exploratorium.edu/seeing/index.html)

The Seeing Exhibit at San Francisco's Exploratorium incorporates a strong tie between the resources in its physical exhibit hall and those in its accompanying online virtual exhibits. The collection's exhibits are grouped according to the phenomenon they illustrate: light and the eye, seeing in context, seeing and attention, interpreting images, seeing color, and seeing motion. The purpose of the collection is to illustrate the many aspects of visual perception. In the "seeing in context" exhibit, for example, an eye-tracker monitors the movements of a visitor's eyes as they pore over an image. This illustrates not only where on an image an individual concentrates, but how individuals search the same image in a completely different manner.

The exhibit induces a context by asking, "How do we see the world?" from both a biological perspective (how our eyes and brain interpret light) and a subjective perspective (the way our own identity shapes our interpretation of what we see). Additional imposed contexts are provided to teachers through Field Trip pathways. Open pathways suggest creative ways to engage students, and guided pathways provide a specific course for the students to follow by giving them "big questions" to consider, such as "Is it possible to look at something and not see it?"

The companion Web site (www.exploratorium.edu/seeing) serves as an additional resource that includes exhibits and interactions demonstrating principles of seeing, and provides links to other Web sites containing additional relevant information. The fieldtrip pathways include worksheets that serve as both resource and as a variety of tools. By asking guiding questions and allowing for different interpretations per both a biological and subjective perspective, the worksheets can function as a processing, collecting, organizing, and integrating tool.

The exhibits, especially those containing field trip pathways, offer conceptual scaffolding by suggesting what to look for and consider. The information and smaller "guiding questions" contained in the pathways help students to interpret, interact with, and process the information they encounter at each exhibit, while allowing for free exploration. These metacognitive scaffolds can be extremely helpful to students who are asked to think using an unfamiliar process or new perspective. Once they finish their visit, students are asked an "integrating question." This question helps them to synthesize and internalize what they have learned and draw conclusions based on their own life experiences beyond the exhibit.

National Geographic Education Guide (www.nationalgeographic.com/education/)

National Geographic's Web site offers a rich array of resources with a large area of their Web site dedicated to both children and educators. The content ranges from online adventures and homework help for kids, to lesson plans and a community resource for teachers.

The Web site offers both exploratory and guided contexts. Each lesson plan, targeted to teachers, introduces an imposed enabling context by giving an overview of the lesson, materials required, student objectives, and a suggested procedure. For example, in *Two Ancient Cities — Machu Picchu and Chichén Itzá*, students in grades 9-12 are asked to compare the two civilizations by researching "the surrounding geography, the main structures and their purposes, the inhabitants, the building materials, the general layout, and any other revealing information or artifacts from the city." Students are then asked how the culture and geography of each city influenced their inhabitants' lives. The site also offers a range of induced enabling contexts. Visitors are invited to explore dynamic maps, atlases, and over 100 online adventures.

The resources of the National Geographic Web site are many. In addition to both dynamic and printable maps, the site has articles, photos, and games (for both in the classroom and for children at home), lesson plans, science experiments, and crafts. A "homework help" section offers pictures, maps, and information on subjects including animals, history/culture, maps/geography, photos/art, places, and science/nature. A click on *Alaska* in the "places" section leads to an invitation to explore "Alaska's land, wildlife, history, and people." This section also includes a travel guide, a discussion forum, resources, and links.

The site offers an assortment of tools to help engage and guide, including extensive seeking tools. Educators can search for educational materials by subject (geography, science, social studies, reading/writing), resource type (lesson plans, books/workbooks, kits/overheads, maps/globes/atlases, online adventures, videos, software), grade (K-12), or keyword. There is also a one-stop search for more popular subjects.

In one interactive feature, as visitors learn about natural disasters, they are asked to predict when and where natural disasters will occur, and to generate their own natural-hazard map based on their knowledge of plate tectonics, historic earthquakes, and volcanoes. This generating tool helps visitors to both integrate and organize their new knowledge of natural disasters into a product. In a less structured activity, visitors learn about great moments in architectural and engineering history (for example, Stonehenge, the Panama Canal, the Pyramids of Giza) and are then asked to design a theme park devoted to these great achievements using anything from modeling clay to pictures pasted on a sketch board to computer modeling.

The site offers scaffolding at several levels. The lesson plans provide conceptual and metacognitive scaffolding for both the teachers and the students by supplying a contextual overview of the problem and leading questions or scenarios that help guide the learners through the experience. In *Crack the Code,* for example, first-graders learning about latitude and longitude are presented with a scenario where they must decode a clue based on latitude and longitude to find thieves who stole "armfuls of priceless maps" from the Royal Geographical Society in London. Once they have an understanding of latitude and longitude through the scenario and find the thieves, the site offers lesson extensions based on age and understanding. The lesson plans also offer procedural scaffolding for the teachers through suggested procedures for how to facilitate the lesson.

Opportunities and Challenges with RBL

RBL creates opportunities for ILEs heretofore unavailable, optimizing the affordances of available and future technologies across a range of diverse settings. While increased availability and access to resources are necessary for RBL, they are not, by themselves, sufficient to promote effective learning. Unprecedented opportunities abound, but formidable challenges loom.

Opportunities

RBL enables access to a multitude of perspectives on a given phenomena. One of the most compelling characteristics of RBL is the ability to view a variety of resources from a potentially unlimited number and range of perspectives. This is currently apparent in how textbooks are used in formal learning settings, but the explosion of digital resources and the capacity to access and use them has increased dramatically. Textbooks are often written from a particular perspective to promote a specific view of events or processes. Digital resources may also be written from a particular perspective, but ready access and easy cross-referencing enable extended access to more resources in online science centers and museums — and, therefore, multiple perspectives.

Increased access also has implications for use in online science centers and museums. RBL places more control (and responsibility) in the hands of the end-users — instructors and/or individual learner — increasing the flexibility of the resource and expanding its potential utility to events and activities beyond the immediate context of its creation. This portends unprecedented opportunities to re-use and re-purpose available resources, and to redefine their meaning as the context of use changes.

RBL can be implemented in a variety of contexts. RBL approaches change both the nature and also the role of traditional resources (i.e., instructor, book, articles), as well as the contexts in which they are used. RBL frameworks can be applied in multiple contexts, ranging from formal to informal, electronic to physical, specific to distributed locations, and at particular through unlimited dates and times. While access to multiple resources may be difficult in a specific set of circumstances, in most online environments access can be enabled in several others simultaneously. It is important to note that the *number* of resources used in an online science center or museum may not be as significant as the variety of sources. This creates unique opportunities for diverse resources to be re-purposed and used across environments.

RBL facilitates learner-centered approaches. While RBL tends to focus on individual approaches to learning versus teacher or large group approaches to learning, it is not inherently limited to one-to-one interactions. Learners (individually, in small groups, or classes) can access a multitude of electronic, print and physical sources to assist with their learning in an RBL context within an online science center or museum. While the individual needs may be addressed, it does not necessarily follow that student work is isolated or without guidance. Learners may receive guidance or direction from an instructor, use the same resources identified by their colleagues, or consult with a more expert peer [e.g., scientist who is connected into the online science center or museum via Internet communication technologies (e-mail, discussion board)] as they focus their goals and seek resources needed to attain them. The key RBL focus is what the *individual* learner needs to facilitate growth in knowledge and understanding, not simply the group size or ratio; thus, learner-centered approaches are not only supported but encouraged through RBL.

RBL cultivates transferable 21ˢᵗ century knowledge and skills. The knowledge and skill needs of 21ˢᵗ century learners are different from those of generations past. With the explosion of knowledge, resources, and challenges, learners need more strategic approaches to identifying what is important and the depth of knowledge or skill needed in different contexts. Increasingly, learners need to discriminate when "knowing that" versus "understanding why" is appropriate or necessary. Given the prevalence of inaccurate, questionable, and contradictory evidence, assertions and propaganda expands geometrically. It is no longer sufficient for learners to simply master what they encounter; they also need to demonstrate greater critical thinking, problem solving, reflection, and self-direction than past generations. The use of open questions provided in *National Geographic Education Guide*, for example, stimulate an investigation rather than simple answer-seeking and engages the user in critical examination, reflection, and manipulation of multiple resources, thereby cultivating needed information seeking and evaluation skills. Similarly, children investigating the Kid Club's Ancient Egypt Exhibit are scaffolded in the use of tools as they seek and evaluate clues, helping to hone promote greater facility with increasingly pervasive digital tools. In effect, 21ˢᵗ century strategic knowledge and skill are both prerequisite to and cultivated by RBL.

Challenges

Standards are not available, used, or applied consistently. Several groups continue to create and refine standards and technologies for Internet-based resources [e.g., World Wide Web Consortium (W3C), Internet Society (ISOC), Sharable Content Object Reference Model (SCORM)]. This has great promise for the growth of RBL in a cross-platform environment. Yet, the standards and codes used to "tag" resources for subsequent access and use are still emerging. Those developed have not been consistently adhered to across developers or users. Supporting innovative creation is important, but establishing and maintaining consistent standards and structures is essential for the sustainability and growth of resource use. It is essential that we achieve and implement a needed balance so that RBL approaches can be extended within online science centers and museums.

The importance of context is not well-understood. It is important to recognize the role of context in any learning interaction or event (Shambaugh & Magliaro, 1996). Within the RBL framework, context specifies where the learning occurs (physical or virtual), the perceptions of meaning and understanding shared by users, and the situational referents of learning and performing. The learners, instructors, and values of a given culture all contribute to how contexts are interpreted and acted upon. The opportunity for even more diverse contexts is enabled when working in an online science center or museum. To assist learners and users of these sites, we need to better understand how RBL works in different settings with diverse domains and participants — how resource meaning is (or can be) redefined, how the values and mores of a given culture help or hinder different uses, and how to scaffold differently based on specific situational needs.

Resources remain largely unregulated. Recent legislation has called for restricted access to some Web-based resources in certain contexts (e.g., schools and public libraries). The genesis of the resources themselves remains largely unknown and unregulated on the Web and in other formats (e.g., video, print, people); likewise, content sharing remains unregulated. As perhaps the most democratic of forums, the Internet has enabled unprecedented access to all forms of resources independent of location, time, gate-keeping, and the like. This is exciting, yet not without challenges, particularly in terms of resource integrity, reliability, and access. When RBL applications define boundaries (i.e., pools of images, "approved" URLs, etc.), greater control is afforded but the organic capacity to grow through use and spontaneously inquire based on emerging interest is constrained. This can be particularly troublesome for RBL approaches in online science centers or museums, where the use of resources can be very dynamic and ever changing. It is vital that we find both structural (e.g., directories of "approved" images, better inquiry tools and user scaffolding) as well as human solutions (e.g., developing critical thinking, information literacy, and evaluation skills).

Directed approaches continue to dominate teaching and learning practices. Despite long-term discussions and research related to the value of learner-centered approaches, instructor-centered models dominate learning contexts at all levels. Direct approaches to learning are efficient for reaching large populations of learners who share the same learning expectations or requirements, but are largely ineffective and inappropriate when learning needs, rate, and/or goals vary. Online science centers and museums provide unprecedented opportunities for learner-centered exploration, which can lead to accomplishing individual learning needs and/or goals, and at a pace that works best for the individual learner. While the continued need for directed approaches to harness technology for defined needs remains undisputable, we also need to emancipate and unleash individuals to support unique intentions and needs of individual learners to meet *their* purposes (Hannafin, Hill, & Glazer, in press; Hannafin & Land, 2000).

Conclusion

The potential of resource-based learning for instruction and learning is considerable. Whereas conventional instructional approaches address known learning goals using well-organized sequences, resources, and activities, methods for supporting context-specific, user-centered learning have been slower to develop. Increasingly, individuals evaluate vast numbers of digital resources located in expanding information repositories. Tools and search engines typically (but in most cases only generally) help to locate potential resources; they do not help an individual to determine meaning or relevance. Individuals must recognize and clarify learning needs, develop strategies to address these needs, locate and access resources, evaluate their veracity and utility, modify approaches based on learning progress, and otherwise manage their teaching or learning. RBL enables teachers and learners to take advantage of the information systems we now have available, expanding the resources they use to enhance the teaching and learning process.

References

Aldrich, F., Rogers, Y., & Scaife, M. (1998). Getting to grips with "interactivity:" Helping teachers assess the educational value of CD-ROMS. *British Journal of Educational Technology*, *29*(4), 321-332.

Alexander, J., & Tate, M.A. (1999). *Web wisdom: How to evaluate and create information quality on the Web*. Mahwah, NJ: Lawrence Erlbaum.

Beswick, N. (1990). *Resource-base learning*. London: Heinemann.

Borun, M., Chambers, M., & Cleghorn, A. (1996). Families are learning in science museums. *Curator*, *39*(2), 124-138.

Brown, S., Collins, A., & Duguid, P. (1989) Situated cognition and the culture of learning. *Educational Researcher*, *17*, 32-41.

Cairncross, S., & Mannion, M. (2001). Interactive multimedia and learning: Realizing the benefits. *Innovations in Education and Teaching International*, *38*(2), 156-164.

Dodge, B. (n.d.). *Webquest*. Retrieved April 6, 2004, from *http://webquest.sdsu.edu/*

Exploratorium. (2001). *Electronic guidebook forum*. Report from Electronic Guidebook Forum, San Francisco. Retrieved March 27, 2004, from *http://www.exploratorium.edu/guidebook*

Hannafin, M.J., & Land, S. (2000). Technology and student-centered learning in higher education: Issues and practices. *Journal of Computing in Higher Education*, *12*(2), 3-30.

Hannafin, M.J., Hill, J.R., & Glazer, E. (in press). Designing grounded learning environments: The value of multiple perspectives in design practice. In G. Anglin (Ed.), *Critical issues in instructional technology*. Englewood Cliffs, CO: Libraries Unlimited.

Hill, J.R., & Hannafin, M.J. (2001). Teaching and learning in digital environments: The resurgence of resource-based learning environments. *Educational Technology Research and Development, 49*(3), 37 - 52.

Jonassen, D. H., & Reeves, T.C. (1996). Learning with technology: Using computers as cognitive tools. In D. H. Jonassen (Ed.), *Handbook of research on educational communications and technology* (pp. 693-719). New York: Simon & Schuster Macmillan.

Joyce, M.Z., & Tallman, J.I. (1997). *Making the writing and research connections with the I-Search process*. Neal-Schuman.

Knowlton, D.S., & Sharp, D. (Eds.). (2003). *Problem-based learning in the information age: New directions for teaching and learning*. New York: Jossey-Bass.

Kolodner, J.L. (1993). *Case-based reasoning*. San Francisco: Morgan Kaufmann.

Kolodner, J.L. (Ed.). (1993/1994). Goal-based scenarios. *The Journal of the Learning Sciences (Special Issue)*, *3*(4).

Lyman, P., & Varian, H.R. (2003). *How much information*. Retrieved March 27, 2004, from *http://www.sims.berkeley.edu/how-much-info-2003*

NEA. (2004). *No child left behind?: The funding gap in ESEA and other federal education programs*. Retrieved March 27, 2004, from *http://www.nea.org/esea/images/funding-gap.pdf*

Paris, S.G. (2000, January). *Multiple perspectives on children's object-centered learning*. Retrieved March 27, 2004 from *http://www.nsf.gov/sbe/tcw/events_000121w/1.htm*

Rieber, L.P. (1996). Seriously considering play: Designing interactive learning environments based on the blending of microworlds, simulations, and games. *Educational Technology Research & Development*, *44*(2), 43-58

Shambaugh, R.N., & Magliaro, S.G. (1997). *Mastering the possibilities: A process approach to instructional design*. Boston: Allyn & Bacon.

Vygotsky, L.S. (1980). *Mind in society. The development of higher psychological processes*. Boston: Harvard University Press.

Wiley, D. (Ed.). (2000). *The instructional use of learning objects: Online version*. Retrieved March 27, 2004, from *http://reusability.org/read/*

Section II

Design Considerations

Chapter VII

Interactivity Techniques:
Practical Suggestions for Interactive Science Web Sites

Michael Douma, Institute for Dynamic Educational Advancement, USA

Horace Dediu, Handheld Media, USA

Abstract

This chapter shares our observations, research, and experience with creating interactivity. We explore useful techniques for creating interactive science-oriented online displays, and describe a series of occasions and methods for making exhibits interactive. For each technique, the design issue is described, the methods for addressing the issue are summarized, and there is a discussion of the approach. We explore what kinds of interactivity have proven to work well online, and, perhaps more importantly, what does not work. Generally, technical solutions are prescriptive rather than descriptive, leaving the actual implementation up to the programmers involved in the project.

Introduction

The modern Web browser allows visitors to interact and explore. Over the last decade, a number of techniques have evolved to convey scientific information, and engage visitors.

At time of writing, the Internet has become commonplace, and most museums are committing resources to online exhibits. The Internet is so pervasive, that major museums, such as the U.S.'s *Smithsonian*, have twice the number of virtual visitors than in-person visitors (McMillan, 2004). It is likely that this trend in museum attendance will continue, with more and more content accessed online. It therefore behoves the curator to consider the quality of their online presentations, and make the best use of interactivity to teach and engage visitors.

This chapter is based on experience garnered over the last five years with the *WebExhibits* online museum of science and art exhibits. This site receives over 24 million page views per year from 8 million visitors (webexhibits.org). For comparison, in 2004, *WebExhibits* had slightly less online traffic than the *Exploratorium*, and slightly more than the *London Science Museum*, the *Ontario Science Center*, or *the Museum of Science* in Boston. Over the last two years, while online science museums overall have drawn a smaller fraction of Internet users, *WebExhibits*' traffic ranking is increasing (Alexa Data Services, 2004). We periodically add new exhibits, and make extensive use of interactive technologies.

This chapter shares our observations, research, and experience (Douma & Henchman, 2000a, 2000b). We explore useful techniques for creating interactive science-oriented online displays, and describe a series of occasions and methods for making exhibits interactive. For each technique, the design issue is described, the methods for addressing the issue are summarized, and there is a discussion of the approach. We explore what kinds of interactivity have proven to work well online, and, perhaps more importantly, what we have learned does not work well. Generally, technical solutions are prescriptive rather than descriptive, leaving the actual implementation up to the programmers involved in the project. This is due to the multitude of tools available to designers, none of which is necessarily ideal.

Comparing Data or Images

Data analysis is fundamental to the scientific process. The e-learning environment can allow students to make their own discoveries, and thus participate in the process of discovery. Good e-learning designs are often based on constructivist and learner-centered principles which put the student in control, encourage the student to look at issues from multiple perspectives, and create his own meaning (Wilson & Lowry, 2000). Science inquiry is central to scientific literacy, and has been shown to help students gain skills like questioning, explaining, and making predictions (Songer, Lee, & Kam, 2002; Songer, Lee, & McDonald, 2003). Research shows that this type of inquiry process and support has a strong impact on students' understanding of scientific concepts and content (Bransford, Brown, & Cocking, 2000).

A virtual science center should provide students and visitors with opportunities to engage with science by doing — not just reading — and thereby increase their understanding of how science works, and how it is connected to other social science disciplines. In that spirit, the following are techniques for promoting inquiry.

Virtual Devices and Experiences

Consider the need to explain how a process works by devising a virtual experience for visitors. For example, create a working model of an automobile engine that visitors can take apart, or let visitors perform dissections. Virtual experiences are particularly appropriate for processes, such as flying a jet aircraft or performing a bypass surgery, that visitors would find it extremely difficult to do in real life. Simple drag-and-drop actions are

Figure 1. Virtual devices and experiences

A. Panda habitat (National Zoo)

B. Color saturation and brightness (WebExhibits)

C. Knee surgery (Ed Heads)

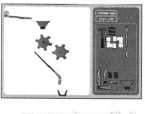

D. Gravity effects (Smithsonian Lemelson)

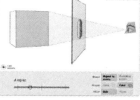

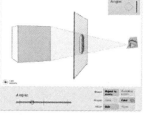

E. Flight of swallow (style.org)

F. Perspective (rotating cube) (WebExhibits)

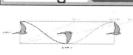

G. Ocean wave periods and amplitudes (National Geographic) (Volvo Ocean Race)

H. Ocean waves (right) (Pearson / Prentice Hall)

surprisingly engaging when the visitor is picking up tools, sawing and cementing, and hearing sound effects immersing them in the procedure (Bort & Wheatley, 2003).

Do not require visitors to do steps just for the sake of interactivity (like turning on the lights or opening doors). Interactivity should always be purposeful, not decorative. Do not add extra animations, or require visitors to drag unnecessarily (e.g., "drag a specimen to the microscope to see details"). It is generally better to indicate "point" than to "click," and "drag" only when necessary. For example, a good instruction is, "point at a specimen to see details." Superfluous interactivity leads to boredom quickly.

Interactive games and quizzes that test the visitor's understanding are sometimes desirable, but such games and the quizzes should be optional, lest visitors be annoyed or put off if they perform badly.

The only exception to the principle of simplicity is when designing for young children, who often find clicking and dragging to be entertaining, and may enjoy superfluous animations. For example, "drag fruits into a blender to make a smoothie."

Comparing Views

Consider the need to compare views of the same item, or from the same viewing angle. For example, compare X-ray and normal views of a human body or a flower; or simulate views of color blindness. To compare views, superimpose the set of images, and allow the visitor to make comparisons. This can be accomplished with three methods:

Method 1: Swapping. Provide buttons next to an image that allow visitors to swap the images. For example, the buttons could be marked "X-ray" and "Regular." Use the indication: "Point at a view."

Method 2: Spyglass. Have a movable window that visitors can move around a scene. The visitor can drag the spyglass, showing and hiding the alternate view (Douma & Henchman, 2000a). Use the indication: "Drag the spyglass." Implementation: Mask the top image, and move the mask in the opposite direction of the cursor. Versions of this technique have variously been called the *spotlight, toolglass, magic lens, looking glass,* or *x-scope* (where x is the name of what is revealed, like "bone-scope" or "pollution-scope") (Bier et al., 1993). This approach can also be used to reveal translations, or permutations of a scene (Goldstein et al., 1997). For example, in a view of cities on a map, a movable filter could reveal which cities have high population densities (Fishkin & Stone, 1995).

Method 3: Slider. Provide a knob that visitors can slide back and forth. The knob should be graphically dominant so that it is obvious to the visitor that they should move it. To further indicate that the figure is interactive, some designers animate a sliding knob for two to five seconds when the page loads. Comparing two images involves a linear slider; comparing three images involves either two

Figure 2.

A. *Color blind simulation (Web Exhibits)*

B. *Light sources (Web Exhibits)*

C. *X-ray, infrared, and other views of painting (Web Exhibits)*

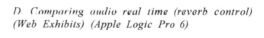
D. *Comparing audio real time (reverb control) (Web Exhibits) (Apple Logic Pro 6)*

sliders or a three-dimensional slider control. Use the indication: "Adjust the (attribute)." Implementation: Track the slider position, and vary the alpha (opacity) of a top layer.

Swapping is generally the easiest to implement, and is best to use if the images are similar. In common with all predators having binocular vision, humans are especially good at detecting differences, which are perceived as motion. By swapping different views, it is immediately obvious to the eye how the two images relate and what is different. The swap should be immediate, with no flicker in between.

The spyglass is helpful if the images are quite different, or viewers need a reference. For example, an X-ray can be confusing without the visual image surrounding the spyglass for reference.

A slider is helpful when it is useful to see in-between views. This is like adjusting the brightness of a television. For example, compare how a scene would have looked to an aging artist suffering from cataracts and color blindness. There could be a photograph of an everyday scene, with a knob on a slider beneath. The visitor moves the knob left and right, adjusting the scene.

Figure 3.

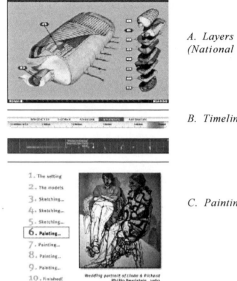

A. Layers in an Incan mummy bundle
(National Geographic)

B. Timeliness (various sources)

C. Painting a painting (WebExhibits)

Viewing a Sequence

Consider the need to display a time-lapse sequence, or a sequence of images. For example, before and after views of deforestation, stages in a rocket launch, development of an embryo, a tide flowing into the beach, the erosion of a canyon, a sunrise, steps in a chemical reaction, a flower opening, an explosive chemical reaction, workers building a skyscraper, micro-time photography (bullet through glass), or MRI or PET slices of the brain. Use a timeline to choose a frame in the sequence.

> *Method 1:* Slider. Use a slider with a knob to choose a point in time. The track for the slider should have markings to indicate that there are discrete steps. Use the indication: "Move knob to view (event)."

> *Method 2:* Filmstrip. Display a strip of thumbnails, like a sequence of film. The visitor points at the thumbnail and sees the larger frame. Use the indication: "Point at a view."

This is similar to the previous technique, but the relationship between the frames is emphasized. Good interfaces have measurements on them, with a labeled axis. For example, if exploring a tide, label the axis with times (low tide, medium tide, high tide); if looking at MRI slices, label the location within the brain. If the sequence is a loop, such as a repeated dance of a honeybee, a horse running around a track, or a rotating planet, instead of a linear slider, use a circular slider, and track the knob in a continuous fashion.

Contextual Captions (Qualitative)

Consider the need to label an image with captions while avoiding clutter. There is too much information to fully label the image without becoming illegible. For example, a group of people, portions of a painting, anatomy of an insect, taste sensitivity of the tongue, components of a circuit board. Display contextual information about the region of the image at which the visitor is pointing.

> *Method 1:* Dynamic caption. Change the caption to correspond to what the visitor is pointing at. For example, in an illustration of the planets, the visitor points at Saturn, and the vital statistics of Saturn are displayed. Use the indication: "Point at a planet."

Figure 4.

A. *Microscopic views of a painting (WebExhibits)*

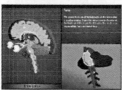

B. *Brain anatomy (BBC Science)*

C. *Excavation of ancient Egypt (Theban Mapping Project)*

D. *Eye anatomy (National Library of Medicine)*

Method 2: Contextual pop-ups. Put detail into floating windows. Use the indication: "Point at a building."

If there are only a few important regions of the image, delimit the "hot" regions by outlining — marking them with labels or numbers, or some other means. It must be obvious where the visitor should point their mouse.

The floating pop-up can be opaque, semi-transparent, or have time-dependent transparency whereby it fades after five to 10 seconds so you can see what is behind it.

If there is information for dozens (or hundreds) of sub-regions spanning the full image, it is not necessary to delimit regions because the visitor can point anywhere. This is especially powerful for annotating images where the interpretation is not obvious (e.g., an X-ray of a suitcase, or micrograph of a cancer cell). Implementation: Create an image map. A simple image map can be created as an html <map>. A more complex mapping requires creating a bitmap. An algorithm is listed in the Appendix.

Another extension of this technique is to conceive of the contextual information through a viewing filter. This is similar to the spyglass or toolglass mentioned above. Different kinds of information are displayed within the toolglass. For example, passing over a map of China, within the toolglass, color-coding could indicate population density. Note also that more than one toolglass can be overlapped (Bier et al., 1993). For example, with an anatomy exhibit, one filter could reveal muscles and another could reveal nerves. In their overlap, both muscles and nerves are shown.

Animated Keys (Quantitative)

Consider the need to display quantitative information about an image. For example, carbon dioxide output per country, number of voters per region, elements in minerals, calories of foods, diets or habitats of species.

Display a fixed key alongside the image, and display or highlight relevant data as the visitor points at the image.

Method 1: Highlight key. If the values are Boolean (present or not-present), display a key with names or icons, and highlight the icons on and off as the visitor points at the image. For example, identify the paints used in a painting. Display a large painting, and next to it are a series of pigment names (Ultramarine, Ochre, Charcoal, etc.). As the visitor glides the cursor over the painting, the pigments in the legend highlight on and off, indicating what paints are used. Use the indication: "Point at the painting to see what pigments are used."

Method 2: Dynamic graph. If there are varying amounts, for example, dietary information about foods. Use the indication: "Point at the foods." The visitor points at a series of foods. Animated bars next to the words "fat," "sugar," and "protein" indicate the amounts of different components in the food.

This technique funnels information by showing the visitor only partial selections of the overall data. Clarity is perceived in simplicity. Traditionally, this would be implemented in a book by displaying an image, with labels (A, B, C, etc.), and a series of bar or pie graphs, each labeled A, B, C, and etcetera. In the Web site, the visitor points at the image and only one graph is displayed. This makes visitors feel less overwhelmed, and also creates a sense of discovery since visitors need to point at an item to reveal its values. It also helps visitors compare regions in an image.

The idea is to allow the visitor to call up additional information (details-on-demand) as their interests guide them. For example, the visitor might expand from a small collection of objects (e.g., tropical rainforest insects) to reveal more about their variables (e.g., habitat or predators). Displaying data on an as-needed basis prevents clutter and allows more of the variables to be mapped to the visualization (Card, Mackinlay, & Shneiderman, 1999). Having a visitor move among a set of linkages, either moving among Web pages, or pre-set views of data, is sometimes called a "direct walk" (Card, Pirolli, & Mackinlay, 1994; Furnas, 1997).

Views of Data

Consider the need to have multiple dimensions of data to display without being confusing. For example, a city map, onto which you want to show the roads, the public transportation system, population densities, and demographics. Reduce the amount of data shown to the visitor by offering different views.

Method 1: Predefined views. Provide a series of buttons on the edge of the city map. The visitor can choose a predefined view, for example, all the roads, or just the parks. Use the indication: "Point at a view."

Method 2: Choose your view. Break down the data into simpler categories, such as streets, highways, demographics, housing density, and etcetera. The visitor can toggle these categories on and off, to compare what interests him. Use the indication: "Click on the categories."

This technique reduces multidimensional data by allowing the visitor to choose only the dimensions of interest. By displaying less information, the visitor is better able to see what interests him. Dynamic queries let the visitor change his view in real-time, for example, movie fans could conduct a real-time search among different movies by choosing among actors, directors, and other qualities.

The origins of these ideas can be traced to a technique called "worlds within worlds," which displays a visual structure of multidimensional data by overloading different axes and by placing each coordinate system within another (Feiner & Beshers, 1990). Another technique called "parallel coordinates" explored the notion of choosing subsets of data (rather than subsets of dimensions) and using the parallel placement of axes in 2-D (Inselberg, 1997). For example, an interactive graph could simultaneously show the

Figure 5.

A. *Alligator habitats (National Geographic)*

B. *Custom map (Museum of Sydney TimeMap)*

Iowa Caucus Results for John Kerry

C. *Political results of 2004 Democratic primary (style.org)*

D. *Word count (Jonathan J. Harris and Number 27)*

amounts of vitamins in different foods, and the visitor could dynamically choose different subsets of foods. This would immediately convey, for example, how both leafy greens and milk are high in calcium, but that both milk and eggs are high in protein. It is often helpful to have several views, and allow a visitor's actions in one view to affect the others (brushing).

For additional inspiration for how good visual representations can reveal hidden structures, while keeping the details understandable (see Tufte, 1990, 1997).

Color Coding

Consider the need to explain the differences between major regions of an image. For example, a painting is the hand of two artists: who painted what? Which parts of an aircraft are responsible for lift propulsion, control? The tastes detected by different regions of the tongue?

Provide a button that color-codes the image into clearly defined regions, so that visitors can see the major sections. Use strongly contrasting colors, overlaid semi-transparently (Douma & Henchman, 2000). This technique is related to the previous two techniques, and is very helpful for identifying regions of an image. It is a way to provide context for an image.

Interacting with Objects

The previous techniques concern analysis of data or an image. Scientific inquiry also involves dissecting phenomena into vital components. Many young engineers had their first tutorial taking apart a motor or a television; so too, a Web site can lend the sense of taking something apart or understanding how it works. Wherever possible, such a presentation should not be linear, like a slide show; rather, visitors should be able to freely interact with objects, and potentially make mistakes.

Independent and Dependent Variables

Consider the need to explain relationship between two variables, where one variable is dependent on the other variable, and the values are continuous; for example, the speed of a horse running around a track, the hours of summer daylight at different latitudes, times in different time zones. Here are three solutions, all detailing the boiling point of water.

Method 1: Show a normal graph, where the x-axis is independent, and the y-axis is the dependent variable. In the case of boiling water, the x-axis is the height above sea level, and the y-axis is the boiling point of water.

Method 2: Animate values with an abstracted scale. In this case, the visitor moves a knob to select different elevations, and another knob animates in real-time to show the corresponding boiling points.

Method 3: Animate values with a realistic scale. In this case, the visitor moves a small icon of a person up and down a mountain. In real time, a ruler scale adjusts to show the current height, an animated thermometer shows the boiling point of water, and an animated timer shows how the baking time varies for cookies.

Figure 6.

A. Latitude vs. length of daylight over the year
(Web Exhibits)

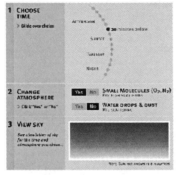

B. Time and atmosphere vs. colors of sunset
(Web Exhibits)

C. Parents genetics vs. childs genetics
(Tech Museum of Innovation)

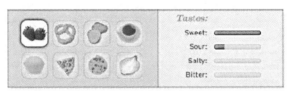

D. Food vs. taste components
(Web Exhibits)

Although a normal line graph actually shows the most information, with the boiling points for all elevations shown simultaneously, line graphs are not intuitive for many visitors. In contrast, an animated figure is much more effective and fun. With Method 3, the visitor has the sensation of discovery, as if he is personally climbing the mountain to conduct the experiment. Additionally, you can add contextual information in your illustration, such as nature of the terrain at different elevations. By coloring the figure, the visitor better intuits the difference between a desert-like climate below sea level, and the thin cold air at alpine heights.

Visual examination of graphs helps visitors acquire insight into the complex relations embodied in a model (Tweedie et al., 1996). This technique helps reduce the number of dimensions of data you are showing. Rather than show a two-dimensional graph of elevation versus boiling point, you show only one dimension (elevation), and one boiling point at a time.

If you have three-dimensional data, such as elevation vs. time vs. temperature of cake batter, it can be difficult to illustrate. However, by separating elevation, and animating it, you can display a line graph of time versus batter temperature, and visitors will understand that it takes longer to bake a cake at high elevations because the batter heats more slowly.

Modeling and Altering Reality

Consider the need to show dependencies between components, or the key themes in a system: for example, an ecosystem consists of many species; an orchestra has multiple instruments; fractals have common arrangement; ocean waves have periodicity; certain color combinations have more contrast than others; the path of a comet depends on its speed, trajectory, and the mass of nearby planets. Simulations are opportunities for learning by doing, and help illustrate abstract, or complex concepts. Simulations are dynamic and task-driven, allowing visitors to experience a concept (Hanna, Glowacki-Dudka, & Conceicao-Runlee, 2000). Present a visitor with a model of the system, and allow them to manipulate it:

Method 1: Multiple instances. The model is a short animation, 5 to 30 seconds long, and the animation depends on the starting conditions. Either randomly set the starting conditions each time the model is run (e.g., change the position of the seed in a fractal attractor), or let the visitor set the starting conditions.

Method 2: Real-time changes. The model is a continuous animation, such as air flowing through a whistle, and the visitor can adjust parameters in real-time, such as the position of the air entry and the dimensions of the air chamber.

This is related to the previous technique, but involves more complex systems such as taking an engine apart. By breaking or modulating a working system, the visitor gains a strong appreciation for how it works, and how the components interact.

This technique is particularly powerful for basic physical phenomena, such as trajectories, orbits, and vibrations. Building the mathematical underpinnings into a complex model, and illustrating the model on a Web site can be very time consuming. However, it is one of the most effective teaching tools.

Implementation: Note that virtual reality can be simulated. For example, you do not need to calculate the fluid dynamics of airflow in real-time. Rather, you can have a series of rendered frames, and recorded sounds, to create the illusion of real-time reality.

Interactive Appliances

Consider the need to decide between making generalized tools that all work similarly, and inventing several unique tools. Add just the controls necessary to yield a functional information appliance, but do not add unneeded controls (excessive features) or portions (like gears) that a real appliance might have. Many of the examples discussed earlier in this chapter, such as animated graphs with moving knobs, are all dedicated information appliances (Norman, 1999).

Keep the user interface widgets common. For example, always use bright green knobs. However, the design of your interactive figures should always match the purpose. Again, only present the minimum complexity to achieve your purpose.

For millennia, tools have been invented to achieve just one purpose. In the digital age, Jef Raskin first promoted the idea of a simple device designed around the information needs of a single function in 1978 at Apple Computer. The concept of focusing on a single function has evolved into a design strategy for interactive tools hoping to break through the complexity barrier imposed by such general purpose computing devices as the personal computer (Norman, 1999).

The stopwatch is not useful for telling time, but it is both useful and highly usable for timing the duration of events. The stopwatch is a highly specialized information appliance that does one thing — track elapsed time — very well.

Real-Life Experiments

A virtual science center should touch the real world. Consider the need to have visitors experience a topic first-hand: for example, making butter at home; folding origami animals; creating a pin-hole camera; or building a small ship out of plastic soda bottles (Wrigley, 2001).

Devise real-life experiences that complement the online exhibit. Provide detailed instructions, materials, and photographs to explain what to do. The text should be easy to follow, and convenient to print.

If the experiment is intended for use in a classroom, provide simple instructions that can be printed for students, and a more detailed version for teachers including additional notes on implementation and the science concepts being taught.

There is extensive research into experiential learning, but ideas for experiments are sometimes hard for curators to find. By working with science educators, classroom activities (with extra background information for teachers) can be incorporated into the design of the online exhibit. Examples of classroom activities include building and testing models of sailing vessels, simulating the effects of wind and waves; replicating Galileo's experiments; and studying waves by using long strings (Wrigley, 2001).

Figure 7.

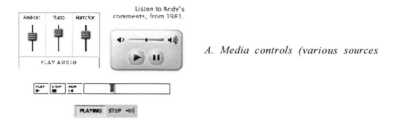

A. *Media controls (various sources*

Gaining Additional Insight and Context

An engaging online exhibit should maintain a sense of realism and narrative. Adding multimedia, such as commentary from a noted astronomer, can engender realism. Narrative is sometimes maintained by removing extra information from the core presentation, for example, keeping images as small thumbnails that can be enlarged.

Vignettes and Clips

Consider the need to lend a sense of context and mood. Add a short audio clip, including a personal recount, or reflection. Include a volume control, play/pause, and restart. The audio augments the textual content with selected examples. We discourage narrating the whole exhibit because the visitors become too passive, as if watching television.

Personal comments are invariably more interesting and engaging that factual monologues. A simple phrase like, "we were so surprised at the results of this experiment. It revolutionized our understanding of the atom," comes to life if delivered by an engaging scientist or doctor. Use voice-over to provide context, not to instruct. It can be difficult to write concise copy for a script. Keep in mind that real dialogue is very colloquial, and if two speakers are talking, they rarely say more than two sentences without interruption.

The following are some examples of how different media types can effectively lend depth and interest (Sutcliffe, 2003). Audio clips can illustrate the setting for a story, sound of rain and storms, sound of skiing, sound of a starting gun, noise of a tornado, echoes in a cave, sound of snoring. A video can show documentary of a person talking or an interaction, can reveal body language or subtle movements. If using video to illustrate a phenomenon, such as an explosion, rocket flight, a dancer twirling, or a storm cloud, allow the visitor to control the timeline, so they can dissect the action frame by frame, as described in *"Viewing a Sequence"* .

Zooming In

Consider the need to let visitors zoom into an image. A screen has much lower resolution than a book, and visitors cannot see enough detail.

Method 1: An enlarge button. An easy method is to put a small "Enlarge" button next to the image. In current screen displays, zoomed images that fit within 750 pixels wide and 650 pixels high are sufficient.

Method 2: Zoom into and around details. Provide a small thumbnail that the visitor can click on and see what detail they are examining. Also, let the visitor pan from within the zoomed view. Implementation: As of 2004, an excellent Flash software package for creating zooms within a Web page is zoomify from Zoomify, which allows visitors to quickly zoom into an image by streaming only the relevant image data. The technology uses a mixture of encoding, avoiding redundancy in data, data compression, pre-fetching, and streaming to create a very responsive experience (Zoomify, 2003). We discourage the use of Java applets.

Zooming is critical, and virtually all images in your exhibit should be displayed relatively small in line with the text, but have an option of a full-screen view. It is rare that visitors really need to zoom far into an image, so an enlarge button with one state of zoom will suffice for most instances.

Organizations concerned with the copyrights of their images can either watermark them with a faint tint (a watermark with curved lines, and both white and black is virtually impossible to manually remove), or by slightly warping the zoomed image so that the high-resolution original cannot be re-assembled.

Three-Dimensional Views

Consider the need to show a 3-D object or scene. Allow the visitor to rotate, spin, or navigate a 3-D world on screen. If he drags to spin the object, have the object continue to spin for a few moments, slowing down as if by friction, after the visitor releases the mouse button.

Method 1: Pre-render. You can make a pre-recorded animation, in the form of movie clip, or use an interactive 3-D object, such as Quicktime 3-D, or make a 3-D view in Flash. If using flash, keep in mind that the client does not need to calculate the 3-D view natively, typically all the views are rendered as bitmaps or vectors. We discourage special plug-ins other than Quicktime or Flash. Implementation: Swift3-D offers tools for rendering 3-D objects from LightWave and other engines as Flash vectors.

Method 2: Real-time graphics. Simple polygons, and images can be manipulated to appear 3-D. Represent the objects using 3-D points, manipulate them in 3-D space, and have a rendering engine for display on-screen. An excellent set of Flash tools is available as Open Source from levitated.net (Tan et al., 2002).

Most three-dimensional objects come to life when animated on-screen. The illusion of three dimensions is captivating, and visitors will enjoy the virtual reality experience. It is more interesting to navigate a scene or rotate an object, than to simply see a few selected views.

Pre-rendering allows you to have highly ornate images, but since you need a frame for every view, file sizes balloon quickly. Real-time graphics can potentially offer smoother motion, and also allow more interactivity, and degrees of motion. The Virtual Human project from the mid-1990s explored building dynamic query and previewing techniques for rendering voxel images of a human body. The visitor was able to view portions of the human anatomy to view, and the angle from which to view it. By pre-rendering many views, the system was able to send to the viewer only the relevant information, only downloading a small sample of the full dataset; larger selections could then be downloaded of the areas that the visitor was interested (North, Shneiderman, & Plaisant, 1996).

Navigating Among Many Items

Consider the need to let a visitor choose among a set of items, such as pages or thumbnails.

Method 1: Ordered grid. Thumbnails quickly appear disordered, but if you lay them out with a strictly ordered grid, the eye perceives structure, and they will appear more manageable.

Method 2: Animated row or column. The thumbnails are arranged in a linear shape, and the row flows from side to side, showing a subset of the thumbnails, analogous to a train moving on a track.

Method 3: Cylinder display engine. As the visitor moves or drags the mouse, a ring of thumbnails rotates about the vertical axis (into the screen). This brings some thumbnails to the front, while others recede and fade into opacity. Clicking the images both enlarges the image panel and adjusts its outward radius. Spin the rotational device using the horizontal position of the mouse. Properties of the rotational device allow for the precise placement of each image, distance from center, rotation speed, size, and more.

Method 4: Spherical magnification. As the visitor moves the mouse, individual elements are displaced and magnified around the edge of a transparent sphere.

The system works by defining a reactive object in continual watch of the mouse. Pressing the mouse button down creates a momentary increase in the amount of magnification (7x vs. 4x). When the mouse comes within a specified distance of the object, the object is first pushed away, and then attracted to the mouse as it draws nearer. In addition, the object is magnified by a sinusoidal relationship to the distance from the mouse. The effect is a reactive field of objects that appear to be magnified by a lens controlled by the mouse (Tan et al., 2002).

Method 5: Hyperbolic magnification. Similar to spherical magnification, a hyperbolic browser is inspired by the "Circle Limit IV" woodcut of M.C. Escher. The hyperbolic tree view was first introduced in modern computing by the visualization studies at Xerox (Lamping & Rao, 1995), and later commercialized by Inxight. Using hyperbolic trees, large tree structures are compactly displayed by projecting a tree onto a hyperbolic plane. The effect of the projection is that components appear to be diminishing in size and radius exponentially the further they move from the center of the diagram. The hyperbolic tree is patented by Inxight, so if your project is distributed in the U.S., you will need to use a type of the projection that is not patented.

Using an animated series of thumbnails can be eye catching and engaging, but you have to be careful that the motion is not too computationally intensive and slow on lower-end machines. Whenever animating motion, try to use physics-based algorithms for motion, with acceleration, gravity, and friction, so the motion seems more natural. A good source for basic motion algorithms is Bourge's *Physics for Game Developers* (2001).

The notion of zooming can be abstracted from zooming into a photo, to narrowing a data set (Card, Mackinlay, & Shneiderman, 1999). For example, from a set of all the fundamental physical elements, visitors can zoom in on the metals, and then zoom in on the heavy elements.

The User Interface

Much has been written about human-computer interaction, and the importance of the user interface. Many of these concepts appear throughout this chapter. The following points touch upon just a few aspects of interfaces. The overall key is to be "user-centered." Imagine the experience of several visitors, from young students, to experienced Web visitors, to the elderly. The interface must be immediately obvious for all cases (Watzman, 2003; Norman, 1990; Nielson, 2000; Constantine & Lockwood, 1999).

Figure 8. Choosing items and designs for icons and buttons

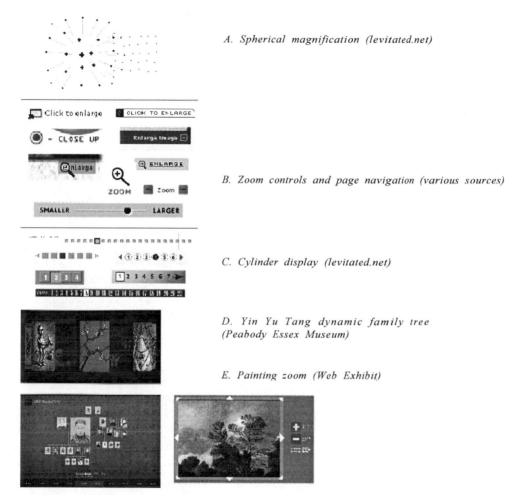

A. *Spherical magnification (levitated.net)*

B. *Zoom controls and page navigation (various sources)*

C. *Cylinder display (levitated.net)*

D. *Yin Yu Tang dynamic family tree (Peabody Essex Museum)*

E. *Painting zoom (Web Exhibit)*

Icons

Consider the need to make *icons* immediately obvious. Except for a few concepts where icons are part of our common visual vocabulary, use text labels rather than icons. The icon or thumbnail can augment the text, but not replace it.

Icons are not useful when they are unfamiliar, and you should not rely on unusual icons for navigation. There are very few icons that are wholly obvious.

However, for those few concepts where the icons are part of our visual vocabulary, judicious use of icons can be a helpful part of your navigational scheme. For example, an arrow icon can help a "continue" or "next" link considerably, augmenting the meaning in a nonverbal way. Similarly, a magnifying glass helps a "zoom" link, and the now ubiquitous "shopping cart" icon is helpful. If you are spawning new windows for help screen or magnifying images, an icon made of overlapping squares indicates that a new window will appear.

Icons should be used when the visual clue *augments* the textual message. For example, a department store might pair a tuxedo icon with the word "men's wear" to suggest formal attire. Similarly, pairing a small thumbnail with a text label would help visitors choose among different size atoms, or climates.

Buttons

Consider the need to make *buttons* immediately obvious. Use a consistent style for all buttons. Thousands of designs appear around the Internet. Two common styles are to create a slightly embossed shape, or to have a small thumbnail with a label.

Buttons are objects on the screen that the visitor can click. If the buttons are highly obvious, it is not necessary to have a mouse-over effect to highlight it. If the buttons appear subtler, a mouse-over effect can be very helpful to draw the visitor's attention to the widget's interactivity. If there is a default button on the screen, consider drawing attention with a slow (2-3 second), subtle pulsing, changing the brightness or size.

Knobs

Consider the need to make *knobs* immediately obvious. Knobs should have consistent styling. There are many styles of knob, but the best designs are reminiscent of real-life knobs and switches, are convenient to use, and are reminiscent of other widgets used in the operating system. Successful styles include brightly colored gumdrops and embossed blocks.

Sometimes the knob should take the form of an object. For example, the visitor might move a rocket up and down, to see how the pull of gravity changes with distance from the earth. In such as case, the knob is not self-evident, and it must be clearly indicated, "drag the rocket to different heights" or "change the position of the rocket." Consider using false color to highlight the icon.

E-Mail or Print This Page

Consider the need to encourage visitors to share your site with others, or archive it. Add an "e-mail this page" link to every page in the exhibit, allowing visitors to add a brief note. Append an identifier, such as "?e-mail" to the end of the URL if you want to track how many hits come from referrals using server logs. Visitors may feel more comfortable using this feature if you allow them to "preview this e-mail" before sending. Also assure them the e-mail address will be used for no other use. Similarly, it is helpful to have a "printable version" of pages with white backgrounds and no fixed margins.

Helping Visitors Find and Access Information

As the amount of information online continues to mushroom, finding information has become increasingly complex, and the metaphors for finding information have shifted from hierarchical structures to searching. Searching has become a critical task for Web users and for personal computer users

Visitors expect to be offered several ways to find information. In addition to making the content of your virtual science center available through remote and local search engines, several other concepts will help visitors find information, and make the exhibit more relevant to their lives.

Editorial Decisions in Choosing Content

Consider the need to choose what to put online. Designing an effective exhibit is as much a matter of what you don't say as what you do say. This is not a matter of dumbing down the material, but seeking epiphanies. Focus on topics that are hard to explain using only words alone, but come to life with interactivity and multimedia.

Good interactive exhibit design helps visitors get their minds around a subject by encouraging active participation. "Visitors do not come to an exhibit to read. They come to learn and do," said an exhibit developer at International Spy Museum in Washington, D.C. A content expert might know the subject inside and out, but that encyclopedic knowledge can be a problem if not reined in (McMillan, 2004). Extra information should be placed on secondary pages within the exhibit.

It is often effective to design an exhibit by working backward. Assemble a collection of ideas for interactive figures and experiences, and then arrange them conceptually.

One method of obtaining content for an exhibit is to adapt the proceedings of an academic conference. Create an overall conceptual structure to encompass the papers, adapt the language and figures from presentations, and interlink topics. But be prepared that using conference proceedings as source material is only a starting point; the material must be digested to become a coherent and engaging exhibit.

There are different definitions of the term "content," and different approaches to building an e-learning module. These include: (1) Information: facts, concepts and procedures to be learned. (2) Objectives: a collection of learning objectives or strategic goals. (3) Media: all the text, graphics, videos, and other multimedia assets. (4) Experience: the overall experience and interactions which the visitor experiences (Allen, 2002).

Site Organization

Consider the need to organize the exhibit. Design the exhibit so that it can be navigated several different ways, but choose an overall organizational model that fits the subject

(Douma & Henchman, 2000a; Baxley, 2000). If the visitors are highly motivated, the content can be organized from the subject-matter expert's point of view, often with a pattern of declaration and elucidation. However, just as mystery novels are more interesting than textbooks, so too should most exhibits avoid the temptation to be content-focused. Instead, focus on the experience of a learner, and assume that most visitors will not take the same path through the site (Allen, 2002).

Method 1: Organize by problem or event. Capture the visitor's attention with a series of case studies or situations. Well-devised imbroglios can be fascinating and anomalies can be intriguing.

Method 2: Organize by conceptual topic. For example, examine various aspects of a painting, such as perspective, color, and motion. Examine van Gogh's life in terms of his family, painting, and illness.

Method 3: Organize by chronology. Use a timeline as the main navigation. Exhibits of this sort generally have a wide timeline displayed at all times. For example, explain the stages of embryonic development, the maturation of a caterpillar, or the evolution of the earth.

Method 4: Organize by structure. For example, explain the workings of the New York subway by examining all the pipes and tunnels beneath the streets of the city.

Method 5: Organize by first-person account. For example, tell the story of the discovery of DNA from the point of view of Watson, Crick, and contemporary scientists. If possible, augment the narrative with audio clips.

Method 6: Organize hierarchically. Start simple and get more complicated. For example, if you are explaining how a complex device works, start by explaining the basic principles and theory that govern the workings of the device. Give examples of similar, but simpler devices.

Method 7: Organize by pigeonhole. Present numerous little facts, and let the visitor synthesize them in their own mind, and apply their knowledge to their own life. Effectively, the whole exhibit is a collection of small sidebars.

Method 8: Organize by question/answer. A combination of Methods 1 and 7 is used in the common "Frequently Asked Questions (FAQ)" format. Focus on narrow questions, such as "Why are days longer in the summer?" Note, that you should not ask mandatory questions that require a correct answer, like a quiz. Visitors prefer to browse your exhibit and pose their own questions.

Maintain a narrative thread to help the visitor grasp the big picture. You are making an *exhibit* not an *encyclopedia*. The difference is that an exhibit has a flow. An encyclopedia is merely cross-referenced information.

Consider using more than one navigational scheme. For example, an archive of Charles Darwin's observations could be organized by subject in a hierarchical table on contents, or in forward or reverse chronological order.

Adapting Real Exhibits to the Virtual World

Consider the need to adapt a real exhibit into an online exhibit. Rather than adapt the exhibit as a whole, instead plan on repurposing much of the material into a different structure. You may augment certain topics online, and skip others. Approaches for design exceed the scope of this chapter, but exhibits are generally arranged in a hierarchical manner (chapters, sections, pages), with a layout similar to a magazine. Most images should be enlargeable, and there should be an emphasis on interactivity rather than textual content. There can be an appendix to the exhibit with more scholarly discussions and references.

During the formative years of the Internet, print, radio, and television were quick to adapt to online publishing. However, museum exhibits, and other in-person experiences, were harder to adapt to the Web browser. Indeed, online and physical exhibits are very different media, and physical exhibits rarely translate directly into online exhibits. Many styles of layouts and presentations were tried in 1998-2002, and by 2002; a number of successful design motifs had evolved. Some of these ideas are discussed elsewhere in this book.

Classroom Guides

Consider the need to help teachers use the exhibit in the classroom. Provide an online teacher's guide for the material. Depending on the nature of the topic, provide guides focused on specific age groups and school level. These guides can be Web pages or printable PDF format.

The guide should include an outline of the curriculum and themes, specific teacher-focused overview of the sections of the exhibit, and additional sources. It should be obvious to teachers how they might use the exhibit in their own classroom.

A lesson plan can vary widely in structure, but may include: a summary of major understandings, essential questions, key knowledge objectives, key skill objectives, a listing of primary and secondary sources (including portions of the exhibit) and several suggestions for assessments (assignments and quizzes). For example, in studying the adventures of Lewis and Clark mapping the Western United States: "Students will take the role of a tribal chief and write a three-to-five-sentence response to the speech. The response must include at least one reason to accept or reject what Lewis has written" (Grove et al., 2003).

The major audience for exhibits is often students in a classroom, and the likelihood of teachers using materials increases dramatically if you provide teacher's guides. Failing to produce a teacher's guide is like failing to advertise a new business, or failing to make a press release for a scientific breakthrough.

Many teachers lack robust knowledge about science, engineering and technology (Lewis et al., 1999). Therefore, include professional development resources with the teacher's guide to provide additional academic background for those teachers who want to be brought up to speed on some of the underlying concepts. Professional learning is critical because teachers must feel confident with the material if they are to teach it well. Like students, teachers will learn by using the exhibit and related real-life activities you include. This is sometimes called a *dual-purpose curriculum*, teaching students and teachers at the same time (Wrigley, 2001).

Web Browsers are Constraining

Consider the need to overcome the constraints of a Web browser. Browsers have limited screen space, limited navigational conventions, low resolution, and limited bandwidth. Expect to compromise and innovate. Producers must massage the content, appearance, pedagogy, and organization. Rather than be concerned with the shortcomings of the Web browser, we encourage you to focus on the opportunities it presents. Virtual exhibits offer an opportunity for interactivity — which is not available in-person — and which the savvy curator should indulge to the fullest extent possible.

Navigation

Consider the need to let visitors move to neighboring pages. The visitor should have a sense of the scope of the site, understand how the contents are organized, and where they are in the site.

In addition to the other navigational tools on the page for chapters, sections, and etcetera, add previous and next links. Add a small arrow, or angly bracket, such as ">" or ">>." These links should be obvious, and placed on the bottom of every page.

If there are several pages in a long series, also add "Page: 1-2-3-4-5-6" links, in which the current page in the series is visually distinctive. Alternatively, you can label the pages as a sequence, "First," "Second," "Third," and etcetera, or by name.

Local Search Engine

Consider the need to find content among a sea of hundreds of thousands of pages. Add a search engine to your site, placing a "Find:" search box on the edge of every page. Also add a search link to the navigation.

Search engines are only helpful if your site has over a hundred pages, so that common search terms will yield meaningful hits. If you install a local search engine, you will need to provide informative page titles, so that the engine returns useful results.

Customize the search results template so that its design is integrated with the rest of the exhibit. The search engine should do full-text searches, and also let you add keywords to your own pages to manipulate the rankings within your site. Implementation: At the time of writing, leading commercial search engines for local servers are Verity Ultraseek and Google appliance. A common outsourced engine is AtomZ.

Search-Engine Robot and Human Accessibility

Consider the need to make pages accessible and indexable by search engines, and have URLs that are short enough that visitors can write them down. Keep your URLs short, and do not use frames. If pages are database driven, do not use a GET query with a question mark in the URL. Rather, cloak the arguments by using the PATH environment variable. Add *alt* tags to important images.

These techniques do not assure high rankings in search engines, but they will ensure that the pages will be easily accessible to search engines, and easily shared and typed by visitors.

Dynamic Navigation and Personalization

Consider the need to meet the needs of different visitors. For example, you need a version of the exhibit for young children, teens, adults, and teachers.

> *Method 1:* Initial interview, or preferences. Ask a few questions of the visitor at the beginning of the exhibit; for example, "What is your school level?...I am in elementary school; I am in middle school; I am a teacher." Depending on the responses, present appropriate navigation leading the visitor to pages appropriate to them.

> *Method 2:* Embed branch points or choices. Present a series of pages, with questions interspersed within the pages. Answers to the questions affect the content and navigation on other sections of the site. For example, in a site that offers movie recommendations, ratings on one page affect the recommended movie on another page.

If your exhibit contains a lot of content that is only applicable to certain types of visitors, it is helpful to cater the content to the unique needs of the visitor.

Broadly speaking, there are two ways to implement dynamic navigation, both representing the content in the site as a tree of nodes or objects. For simple branching, it is generally easier to base the logic on the parent node. For example, if visitors note that they are female, they could be directed towards women's issues. However, for complex logic, it is better to base the logic on the child node. In this case, the site has a user database, which compiles characteristics about the visitor, and each node has a set of rules (e.g., activate this node if the visitor is a woman with cancer and diabetes; or deactivate node if visitor is a student over the age of 14).

Dynamic Content and Localization

Consider the need to meet the needs of multiple audiences which each need different versions of the same information, or different translations (localization). For example, Spanish and English versions of each page, or versions geared towards different education levels.

Method 1: Static tree. Prepare several versions of the same set of pages, geared towards different audiences. Create a mirror version of the site for every version. This is similar to the technique above.

Method 2: Dynamic content. Prepare one set of pages, but embed different versions of the text in each page. Use a scripting engine and a template system to dynamically choose the appropriate text to display.

The advantage of the first method (static) is that it does not require any complex programming. The advantage of the second method (dynamic) is that it is easier to update the Web site because the layout and the content are separated. Foreign language versions may not be completed by the translators until the last days of production, so separating content and layout is critical.

If you are localizing the content, include an option on every page to switch between languages. Use a small icon of the parent country, if appropriate. If you are creating different versions for different educational levels, have a brief introductory page where visitors can identify themselves.

Interactive Encyclopedias and Browseable Databases

Consider the need to make a large amount of data interesting.

Method 1: Provide views. Produce a large series of pages with pre-selected fields, with a clear layout, and illustrations (if possible). Example: A database of hundreds of insects, and their traits. For every insect, there are four views with the most popular data fields — a page each for habitat, anatomy, ecosystem, and history.

Method 2: Include multimedia. Whenever possible, include in the database design fields for appropriate audio (clips from famous persons, animal noises), or visuals (photos of the item or its context).

Method 3: Graph or illustrate the data. For example, in a weather database, rather than display the temperature, barometric pressure, and wind speed as mere numbers, create simple graphs or diagrams to illustrate the data. These can be generated in real-time, or you can pre-generate all possible conditions and display the appropriate one as an image. Similarly, the habitat of insects can be illustrated with a map. This is the basis for most geographic information systems, which display data on a map rather than in tables.

Method 4. Related items. Show connections and provide links to multiple types of related items. For example, in an insect database, provide links to related insects in the similar geographical area, insects in the same taxonomic classification, competitors, common predators, anthropological history (used in ceremonies, as jewelry, etc.), and the alphabetically adjacent.

Method 5. Hierarchical browsing. Using a series of columns, help a visitor navigate a hierarchical tree. This technique was championed in NeXT's file browser, and recently in Apple's iTunes music browser, in which thousands of tracks are quickly narrowed down. The visitor clicks on a genre, then an artist, then an album. In seconds, the visitor reduces data down to a few relevant items.

There are typically two ways to use a database, searching and browsing. Searching is intended for visitors who want to quickly find something by searching for keywords. Browsing allows the visitor to wander through the database and discover items they did not anticipate.

The browsing experience is about visitors' process of discovery, revealing connections, and creating their own sets. For example, in Peabody Essex Museum's ARTscape, after making an initial query, the search engine presents a set of thumbnails in a results strip along the bottom of the interface. Visitors scroll through hundreds of results, scanning the cropped thumbnail representations and rolling over them for titles. When something is of interest, it can be clicked and loaded in the center of the interface where the title, date, creator, origin, medium and donor are revealed. This approach can work for any large classification system. You can also add the capability to "save" or "bookmark" a custom view, or to email it to a friend (Johnson, 2004).

Web sites vary in their degree of flexibility — ranging from static and fixed presentations to dynamic database-driven offerings — and their degree of intermediation — ranging from a high degree of curatorial control over the presentation to giving visitors uninhibited control. Static presentations generally make it easier to create rich, compelling and immersive interactive content; but in static exhibits, the visitor is strictly an observer, there is no technological extensibility, no editorial extensibility and very little design flexibility in fixed features. Dynamic features, by contrast, are fluid, flexible and keep the components (content, presentation and technology) transmutable. But too much flexibility can be daunting; visitors need some guidance and hand-holding.

One approach to guided browsing is for visitors to select a case, and then searches for cases with similar attributes (attribute walk). The technique was developed in a system for searching databases called Rabbit in the 1980s (Card, Mackinlay, & Shneiderman, 1999) and is now often used by stock photography services that let visitors find similar photos by choosing among attributes of a current photo. Music and book retailers use attribute walks to find related items by a musician or author. In a science Web site, the visitors might be exploring minerals and gems, and move from gem to gem first based on color (more "greens"), by hardness (more "soft" minerals), by crystal type (more "monoclinic" crystals), and so forth.

Drawing on the Knowledge and Experience of Visitors

The dynamic and database capabilities of Web servers allow a virtual exhibit to also encourage a sense of community. However, this must be done carefully. Many approaches lead to abuse or misuse, including irrelevant and otherwise inappropriate content.

Visitor-Generated Content

Consider the need to expand your content by drawing on the unique knowledge of your visitors. Invite visitors to share information about *highly specific* questions. For example, "How has the habitat in your community changed over the years?"

Use specific questions about which visitors might have personal knowledge. Although thought-provoking questions — "What kinds of food might be good to eat in zero-gravity?" — are excellent teaching tools; in your exhibit, questions should be purely rhetorical, and not solicit written replies.

However, there are some questions, particularly historical and sociological questions, for which it is very effective to ask the public. For example, the *Victoria & Albert* museum asked, "The famous and the fashionable wore Ossie Clark designs. Were you one of them? Did you have an Ossie Clark outfit? Why did you buy it? What made it distinct? Do you have any special memories of wearing it? How do you feel now when you look back? Share your memories and any scanned photographs you might have by e-mailing..." Out of several thousand visitors to the online exhibit, 25 visitors sent replies (none with photos). The museum received several excellent and informative replies, which they posted online (Durbin, 2004).

Inform visitors that their submissions may be published, and become property of your museum.

Community Features

Consider the option to create an online community. Although the traditional science center tends to practice top-down pedagogy, with visitors rarely interacting; the Internet offers the opportunity to create virtual communities where visitors can correspond with other visitors. These can be implemented using a range of technologies, including blogs, forums and guest books, e-mail lists, page comments, ratings, and voting. Generally speaking, we discourage chat rooms and voting because they offer little value and are prone to abuse. The community technologies offer various opportunities.

- *Blogs:* A frequently updated Web log (blog) incrementally adds short postings over a period of time, arranging the postings in a reverse chronology. An expert creates the content, and the public has a chance to post comments. This works well in exhibits based on adventures, such as dispatches from a trip around the world, or a research project that will have ongoing "discoveries." A similar approach is an *ask-the-scientist* feature, but that tends to receive mostly homework questions from students.

- *Forums:* The late 1990s "guest book" evolved into the forum where visitors can leave comments (postings) on a variety of topics or start their own topics. Forums are tricky because they only work for a narrow range of visitors and a narrow range of topics. Too few participants and it is not interesting; too many and it is overwhelming. Shy participants fail to ignite interesting dialog; ornery participants quickly mire the conversation with belligerence (Ross, 2003). There is also always the risk of abuse and free-for-all off-topic postings. The value of a forum is only in topics where the visitors are potentially more expert than the curators. Avoid any other use of visitor input otherwise. For layman commentary, redirect comments to an off-line e-mail address.

- *E-mail lists:* If your science center is going to be revisiting a topic or theme, consider establishing a newsletter to keep the public informed of new developments. Also, as an extension to the teacher's resources, an e-mail discussion group will help teachers using the exhibit in their classroom share experiences and ideas with their virtual colleagues at other schools.

- *Page comments:* The option to post comments has become a popular feature on many online magazines and blogs. We have not found these useful in science exhibits. Similarly, we discourage real-time chat rooms because of the risk of abuse.

- *Ratings and voting:* Many Web sites seek to gauge public opinion by conducting a poll; but we suggest that you do not. An online poll, unless by invitation, is prone to abuse, misrepresentation, bias and sampling errors. A science museum should not encourage statistically invalid surveys. However, it can be quite successful to tally ratings for an individual visitor. For example, a visitor can tally their opinion about greenhouse gasses, and compare their opinion with averages or conclusions from experts. Likewise, a visitor could calculate their body-mass-index, and compare it with a national average. One method of prompting visitors to ask themselves questions is described in the next technique.

Figure 9.

A. Analysis of a famous map. The visitor analyses each piece of evidence individually and notes their opinion. Exhibit keeps a running tally of their opinions and displays the average on a dial atop every page. (WebExhibits)

Evaluating Evidence

Consider the need to explain a multi-faceted problem and have visitors evaluate the evidence, and see how a decision must look at a range of evidence. For example: Is an ancient map actually a fake? Did a suspect commit a crime?

Encourage visitors to evaluate information and make judgments about something by measuring it against a standard, determining a criteria for judging merits or ideas, prioritizing options, recognizing errors in reasoning, and verifying arguments and hypotheses through reality testing.

Encourage visitors to analyze information. Involve the visitors in recognizing patterns of organization; classifying objects into categories based on common attributes; identifying assumptions (suppositions and beliefs) that underlie positions; identifying central ideas in text, data, or creations; differentiating main ideas from supporting information; and finding sequences or order in organized information. Visitors should connect information, by comparing and contrasting objects or events, building arguments, conclusions, or inferences, and providing support for their assumptions (Hanna, Glowacki-Dudka, & Conceicao-Runlee, 2000).

If exploring unusual defect rates in a production line of a factory, introduce the history of the factory and the local suppliers, the nature of the quality control, and the limitations of different kinds of evidence. Explain to the visitor that there are several degrees of certainty:

- *Very certain:* There is abundant, compelling evidence... and only one reasonable interpretation.

- *Probably sure:* The evidence is very strong, and you can presume an opinion... but another interpretation might be possible. This can also be called "presumptive evidence."

- *It proves nothing:* The evidence is consistent with the supplier causing defects... We are speculating, this permissive evidence cannot prove anything.

Present the visitor with a series of pages, on which they are directed to note their opinion about each piece of evidence, and note how certain they are about their conclusion. For each piece of evidence, explain the basis for the evidence, and provide an unbiased explanation of what the evidence means. Alongside every page, show the following options, under a heading, "What is your analysis?" The visitor can click radio buttons.

This evidence suggests the supplier caused the defects: Yes; Not sure; No; Undecided.

How conclusive is this evidence? Very certain; Somewhat sure; It proves nothing; Undecided.

As the visitor clicks through the pages, a dial on the top of the page dynamically adjusts. The dial, labeled "your conclusion" has a range of positions, from "caused" to "undecided" to "didn't cause."

On a "your conclusion" page, total the pro and con votes. Some votes are "very certain," some are "somewhat sure" and some are "none." Add the votes, and calculate the overall determination. Implementation: A sample equation based on arctangent is listed in the appendix.

Allowing visitors to evaluate the evidence themselves creates an atmosphere of discovery. At the end of the exhibit, also tell visitors what the "experts" have concluded. In any good story, the experts often have widely differing opinions.

A related version of this technique is to present the visitor with a series of hypotheses, and the evidence to evaluate them. For example, what disease does a patient suffer from? What caused the Space Shuttle Columbia to explode in 2003? Why did the dinosaurs become extinct? For every hypothesis, present the visitor with evidence to review, and keep track of their replies. After evaluating every hypothesis, the exhibit tells them which hypothesis they personally think is most probable.

Conclusion

Over the last two decades, many notions for displaying data and information have passed through phases of theory, exploration, abstraction, and demonstration. Prior to the 1990s, ideas about interaction came from computer science academics using advanced computers (for the time).

In the mid-1990s, various interactive tools were developed, usually based on CD or kiosks, which had the data throughput necessary for rich media. Development was dominated by students, idealists, architects, failed artists, technological visionaries and brash young entrepreneurs (traditional designers, still reeling from the desktop publishing revolution a decade earlier, were not early adopters). The possibilities of this "new media" were embraced particularly quickly by the educational and cultural world, and many generously funded projects began. But these projects were often afflicted by immature software, and hardware that was not yet good enough for visitors to spend long periods of time learning in front of their computer monitors. Early productions were often clunky by today's standards, with inconsistent navigation (circles, squares, left-handed, right-handed, top, bottom, etc.), and lengthy introductory sequences. After failing to live up to expectations, both in terms of financial return and educational impact, much of educational world seemed to have lost its enthusiasm in multimedia and interactivity by the late 1990s, and a disproportionate amount of innovation was funded by relatively few organizations, including PBS, National Geographic, and the Smithsonian.

The turn of the century and the growing dot-com bubble introduced the Web into the lives of everyday people. Web technologies were riddled with flaws, but they were good enough to be quickly embraced by millions of users. Web design and development quickly became sought-after professions, and interactive design matured, with an explosion of projects in the commercial sector. From a design standpoint, interactive projects began looking very modern by 1999, with navigation bars generally in the same place as today, and logos placed in the upper left, and conventions becoming conventional. Vector graphics popularized by Macromedia Flash allowed designers to make animated and interactive content, propelling the educational possibilities, with fast-loading sites that included music and voiceovers, as well as animated interfaces. By the time the Internet bubble burst in 2001-2, and many poor ideas were lost to the graveyard of failed technologies, interactivity techniques had evolved to a point of maturity. Interactivity is now a mature field, offering tremendous opportunities for e-learning in science.

Computers capable of real-time visual manipulation are now mainstream and common-place. The vast majority of online visitors to e-science centers now have browsers capable of advanced interactivity (in 2004, this was achieved mostly with Macromedia Flash and Javascript). Network connections and hardware allow ever-faster data transfers, and video will soon be a common component of the online experience.

These technologies have been researched and implemented in many industries, with rich R&D budgets, and successful techniques have emerged and evolved. This chapter has provided an overview of many successful interactivity techniques appropriate for creating interactive science Web sites.

It is not possible to predict the roles that computers and interactivity will have in the future, but several trends are apparent. In addition to general improvements in speed, and reductions in cost, displays will attain higher resolutions. Developers in the future may have to consider large workstations (a monitor), small mobile (a phone) sizes, and other form factors that may begin appearing in visitors' living rooms and throughout their lives.

Many opportunities still lay on the horizon. New metaphors and means of visualization are waiting to be discovered. The nature of the interaction between the visitor and the Internet may change as newer search technologies (the semantic Web) begin to link together information. Sometimes an exhibit will be used wholesale, but other times visitors will use only a portion of an exhibit — like an appliance. Also, mobile devices are going to be used more in education. Students may use handheld devices for tasks as varied as data collection in a field, observing sociological data or weather, simulating predator/prey ecosystems, or viewing reference materials.

References

Alexa Data Services. (2004). Traffic Rank. *Amazon.com*

Allen, M. (2002). *Michael Allen's guide to e-learning*. NY: John Wiley & Sons.

Baxley, R. (2002) *Making the Web work*. Indianapolis: New Riders.

Bier, E., Stone, M., Pier, K., Buxton, W., & DeRose, T. (1993). Toolglass and magic lenses: The see-through interface. In *Proceedings of the 20th annual Conference on Computer Graphics and Interactive Techniques* (pp. 73-80). New York: ACM Press.

Bort, E., & Wheatley, G. (2003). *Virtual knee surgery*. Living Children Multimedia Development. COSI Columbus. Columbus, OH: EdHeads.org.

Bourge, D. (2001). *Physics for game developers*. CA: O'Reilly & Associates.

Bransford, J.D., Brown, A.L., & Cocking, R.R. (Eds.). (2000). *How people learn: Brain, mind, experience and school* (expanded ed.). Washington, DC: National Academy Press.

Card, S.K., Mackinlay, J.D., & Shneiderman, B. (1999). *Information visualization in information visualization: Using vision to think*. San Francisco: Morgan-Kaufmann.

Card, S.K., Pirolli, P., & Mackinlay, J.D. (1994). The cost-of-knowledge characteristic functions. In *Proceedings of CHI '94* (pp. 238-244).

Constantine, L., & Lockwood, L. (1999). *Software for use*. New York: ACM Press (Addison-Wesley).

Douma, M., & Henchman, M. (2000). Bringing the object to the viewer: Multimedia techniques for the scientific study of art. In D. Bearman & J. Trant (Eds.), *Museums and the Web 2000: Proceedings*. Toronto: Archives & Museum Informatics.

Douma, M. (2000). Lessons learned from WebExhibits.org: Practical suggestions for good design. In D. Bearman & J. Trant (Eds.), *Museums and the Web 2000: Proceedings*. Toronto: Archives & Museum Informatics.

Douma, M., & Henchman, M. (2000). Using multimedia techniques to present and interpret technical data from Art: Application to Bellini's Feast of the Gods. In J. Goupy & J.P. Mohen (Eds.), *Art et Chemie la Couleur* (pp. 213-217). Actes du Congre: CNRS.

Durbin, G. (2004). Learning from Amazon and eBay: User-generated material for museum Web sites. In D. Bearman & J. Trant (Eds.), *Museums and the Web 2004: Proceedings*. Toronto: Archives & Museum Informatics.

Feiner, S., & Beshers, C. (1990). Worlds within worlds: Metaphors for exploring n-dimensional virtual worlds. In *Proceedings of UIST '90* (pp. 76-83).

Fishkin, K., & Stone, M.C. (1995). Enhanced dynamic queries via movable filters. In *Proceedings of CHI '95* (pp. 415-420).

Furnas, G.W. (1997). Effective view navigation. In *Proceedings of CHI '97* (pp. 367-374).

Goldstein, M., Juanarena, C., Arend, M., &Bertram, T. (1997). CodeScope. In CD-ROM: *Leonardo da Vinci*. Seattle: Corbis.

Grove, T., Bristol, D., McFarland, M., & Meyers, T. (2003). For Educators: Teaching units & lesson plans: Politics & diplomacy. *Lewis & Clark: The National Bicentennial Exhibition*. Missouri Historical Society. Retrieved from *http://www.lewis andclarkexhibit.org/*

Hanna, D.E., Glowacki-Dudka, M., & Conceicao-Runlee, S. (2000). *147 practical tips for teaching online groups: Essentials of Web-based education*. Atwood.

Inselberg, A. (1997). Multidimensional detective. In *Proceedings of IEEE Information Visualization '97* (pp. 100-107).

Johnson, B. (2004). Beyond on-line collections: Putting objects to work. In D. Bearman & J. Trant (Eds.), *Museums and the Web 2004: Proceedings*. Toronto: Archives & Museum Informatics.

Lamping, J., & Rao, R. (1995). The hyperbolic browser: A focus+context technique for visualizing large hierarchies. In *Proceedings of CHI'95, ACM Conference on Human Factors in Computing Systems* (pp. 401-408). New York.

Lewis, L., Parsad, B., Carey, N., Bartfai, N., Farris, E., Smerdon, B., & Green, B. (1999). *Teacher quality: A report on the preparation and qualifications of public school teachers (NCES 1999-080)*. Washington, DC: National Center for Education Statistics.

McMillan, S. (2004, March/April). Visiting the virtual museum. *Communication Arts, 46*(1), 58.

Nielsen, J. (2000). *Designing Web usability*. Indianapolis: New Riders.

Norman, D.A. (1990). *The design of everyday things*. New York: Doubleday.

Norman, D.A. (1999). *The invisible computer: Why good products can fail, the personal computer is so complex, and information appliances are the solution*. Cambridge, MA: MIT Press.

North, C., Shneiderman, B., & Plaisant, C. (1996). User controlled overviews of an image library: A case study of the visible human. In *Proceedings of ACM Digital Libraries '96* (pp. 74-82).

Ross, A. (2003). What we can learn from organic online communities? How communities of practice seed themselves with conflict. In *Virtual Communities Conference*, London, June.

Songer, N.B., Lee, H.S., & Kam, R. (2002). Technology-rich inquiry science in urban Classrooms: What are the barriers to inquiry pedagogy? *Journal of Research in Science Teaching, 39*(2), 128-150.

Songer, N.B., Lee, H.S., & McDonald, S. (2003). Research towards an expanded understanding of inquiry science beyond one idealized standard. *Science Education, 87*(4), 490-516.

Sutcliffe, A. (2003). Multimedia user interface design. In J. Jacko & A. Sears (Eds.), *The human-computer interaction handbook: Fundamentals, evolving technologies and emerging applications* (pp. 245-262). NJ: Lawrence Erlbaum Associates.

Tan, M., Macdonald, J., Rhodes, G., Williams, B., Parker, K., Mulzer, G., Tarbell, J., Lettau, T., Hooge, J., Peters, K., Hirmes, D., Lifaros, Prudence, P., Kaluzhny, P., & Jokol, K. (2002). *Flash Math Creativity*. Birmingham, UK: Friends of ED.

Tufte, E. (1990). *Envisioning information*. Cheshire, CT: Graphics Press.

Tufte, E. (1997). *Visual explanations: Images and quantities, evidence and narrative*. Cheshire, CT: Graphics Press.

Tweedie, L., Spence, R., Dawkes, H., & Su, H. (1996). Externalizing abstract mathematical models. In *Proceedings of CHI'96* (pp. 406-412).

Watzman, S. (2003). Visual design principles for usable interfaces. In J. Jacko & A. Sears (Eds.), *The human-computer interaction handbook: Fundamentals, evolving technologies and emerging applications* (pp. 263-285). NJ: Lawrence Erlbaum Associates.

Wilson, B., & Lowry, M. (2000). Constructivist learning happens all the time on the Web. In E. Burge (Ed.), *Learning technologies: Reflective and strategic thinking, new directions for adult and continuing education. No. 88*. San Francisco: Jossey-Bass.

Wrigley, R. (2000, August). Information technology in education: Real perspective on classroom practice. In *Africa Connects*. Pretoria, South Africa.

Wrigley, R. (2001). The Odyssey Project: Creating a global community of classes following journeys around the world while engaged in project based learning. In *8th Annual I*EARN Conference*, September, Capetown, South Africa.

Zoomify technology and products white paper. (2003). Santa Cruz: Zoomify.

Appendix

Implementations for most of the techniques in this chapter are left to the programmers. We provide pseudocode for two techniques for which the algorithms are particularly challenging.

Creating an Image Map

1. Create the image map. Using Photoshop, create an image map using less than 91 indexed colors; down-sample by 4x; export as a "RAW" image (each pixel has a value 0-255 as a char).

2. Using perl, shift all the values to exclude special characters (newchar = char(ord(oldchar)+ 35) to begin with ascii char hash (35) and end with ascii char tidle (126); use perl to create code for creating an array in the language you are using, each element in the array being a row in the image map. For example,

   imagemap[4] = "88888888ssssssIIIIImmmXXXX}}}}}ww"

3.   Write display code. Using your preferred language, calculate the current mouse position on the image map, dividing by the amount you down-sampled (4). If the image is 400x400 pixels in size, and the user is pointing at pixel (120,20), that corresponds to imagemap(30,5), so the value of that pixel is the 30th char of the fifth element in the imagemap array (imagemap[5][30]). Display that value, for example, "m."

4.   Use your program to create a table of all values, and their corresponding captions. Store in an associative array. For example,

captions[ord("m")] = "Albert Einstein, German-American physicist who developed the Special and General Theories of Relativity"

5.   Update code to display the correct captions. These can be displayed in a floating balloon, or in a fixed-location legend.

Evaluating Evidence

Weighing pro and con evidence is like a balance, and a great technique is to use the inverse tangent, adjusted with the cube power.

Aggregate all the pro and con votes. Weight "very certain" as 1, "somewhat certain" as 0.13, and "none" as 0. Calculate the aggregate conclusion:

opinionPro is the weighted sum of the pro votes

opinionCon is the weighted sum of the con votes

m = minimum (1, maximum (opinionNeg, opinionCon))

y = 4 * arctan(opinionPro,opinionCon) / pi - 1

y = -m * ((y * y * y) + y * 2)/3

opinion = y

The value y will now vary from □ 1 to 1. If both variables are zero set opinion to zero. With -1 being definitely con, 0 being unsure, and 1 being definitely pro.

Chapter VIII

From the Physical to the Virtual:
Bringing Free-Choice
Science Education Online

Steven Allison-Bunnell, Educational Web Adventures, LLP, USA

David T. Schaller, Educational Web Adventures, LLP, USA

Abstract

This chapter proposes a series of strategies for recreating science center exhibits online. It argues that while physical and electronic exhibits share certain common features, electronic science interactives based on physical exhibits must be re-conceived in terms of the strengths of the electronic medium. Like a televised magic show, digital media allow any number of special effects that interfere with the immediacy and raw authenticity of an onsite physical demonstration. This interference is inherent in any mediated experience. Rather than trying to overcome it, we suggest alternate approaches that take online users deeper into the scientific concepts underlying the physical phenomena on exhibit in the physical galleries. We outline several strategies that we have successfully used to engage user's imaginations and emotions in online science activities, to foster motivation, and to provide an initial conceptual framework that supports the learning process.

Introduction: New Interactive Approaches for New Interactive Media

"Interactive" museum exhibits and "interactive" human-computer systems were both born in the late 1960s (Hein, 1990). In fact, the earliest usage of the term comes from computer science and not museums (Oxford English Dictionary). However, first the Exploratorium in San Francisco, and then its progeny, picked up the term and developed increasingly sophisticated physical interactive exhibits during the 1970s and 1980s. Many of the exhibits designed in this period are still canonical in science centers around the world. During this same time, interactive computing remained mostly a laboratory curiosity (Apple's Macintosh, the first personal computer with an interactive graphical user interface, was introduced in 1984). While many museums began experimenting with computer-based exhibits during the 1980s, few approached the sophistication and conceptual scale of their mechanical counterparts in the physical exhibit galleries. Many electronic interactives were variations on reference guides or trivia quizzes. By the time powerful desktop computers and sophisticated multimedia authoring tools finally enabled the production of high-quality electronic interactives in the 1990s, the pedagogical approach, design philosophy, and construction methods needed for physical science center exhibits had been articulated and formalized well enough for the Exploratorium to publish the first in its series of exhibit construction "cookbooks" (Bruman, 1991; Hein, 1990). Electronic exhibits, on the other hand, had no design principles or foundational best practices of their own. Few educational multimedia producers were (or are today) formally trained in computer science in general. Information architecture and interface design have only recently become specialties in their own right, and theories of virtual reality are still not widely applied to inform or justify design practices. Video games were (and often still are) seen as the educationally impoverished enemy in the same category as television, rather than as a model to be emulated (Prensky, 2001; Gee, 2003). But by the end of the 1990s, exhibit developers began to articulate best practices for electronic interactives based on watching visitors use them in the museum. These rubrics mostly focused on eliminating common hardware and software design flaws, rather than articulating an underlying approach to interaction design, or the process of conceiving the user's experience while working with the program (Gammon, 2000).

With the advent and explosive growth of the World Wide Web in the mid-1990s, usability came to the fore as a widespread concern in the commercial sector, and usability experts such as Jakob Nielsen have helped codify principles of sound design that make Web-based materials user-friendly (Nielsen, 2000). After nearly a decade of serious Web design experience, Web developers now know a lot about how people use Web sites and what makes them usable. Usability analysis focuses on efficiency and user success at pre-determined tasks such as information seeking for personal research or completing an electronic commerce transaction. As we recently observed, these metrics, while helpful in preventing or remedying basic design problems, do not provide much guidance in structuring free-choice learning experiences online (Schaller et al., 2004).

While the genre of hands-on museum exhibits has its own underlying paradigm based in constructivist learning and cognitive science, development of online interactives has largely been driven by trial and error and focused on technical issues such as the user's Web browser platform or working around the limitations of authoring tools and the user's Internet connection speed. Indeed, the new medium of online interactives has yet to consolidate a unique vision and paradigm about its distinctive strengths, weaknesses, and uses. If a paradigm for online interactives has been in play, it has most likely employed the same assumptions as those underlying physical interactives. While a natural and valuable starting point, this will only take us so far given the profound differences between physical and electronic exhibits.

In an effort to encourage visitation of their physical facilities, museums of all kinds (especially art museums) have tended to heavily promote their physical exhibits online by creating virtual tours of existing galleries. And because science museums in particular know how to build successful physical interactives, the first thought in constructing online interactives might be to transfer existing physical exhibitry or on-site computer programs to the Web. This is now more technically feasible than ever; current multimedia tools such as Macromedia Flash or Shockwave now allow high-quality animation, user interfaces with sophisticated functionality, and fairly complex simulation engines. The popularity of video games with simulated 3-D virtual spaces makes recreating even the exhibit galleries themselves as a literal virtual museum a strong temptation. Duplicating physical exhibits online is reasonable enough from an institutional point of view in terms of leveraging existing expertise and resources. Indeed, digital versions of some exhibits, such as the illustrations of optical illusions pioneered by the Exploratorium, can give the user a quite similar experience to the physical exhibit (Exploratorium, 1998).

From the standpoint of learning, however, a literal transfer is not likely to be effective for the majority of existing science center exhibits. Some of the reasons for this are obvious and others more subtle. For example, in the physical space, the user manipulates a mechanism and sees or feels the direct chain of cause and effect. Multimedia demonstrations of these sorts of physical phenomena are easy to produce technically, and, on the surface, appear to duplicate the content of the onsite interactives. But *virtual* demonstrations of *physical* processes, unless they show the user something that cannot be seen in the physical setting, do not make sense in cyberspace and will not effectively engage users. In general, simply recreating the exhibit gallery as a virtual space, either as an organizational principle for the site or as model to walk through, does not necessarily recreate the experience of being in the real museum gallery, nor does it necessarily help the user create or elaborate their own mental map of the exhibit content. Unless the virtual museum recreation shows us something we cannot otherwise see, like a reconstruction of an ancient archeological site, the sacrifice in ease of navigation is probably not worth the effort.

With these issues in mind, what new approaches can leverage the strengths of new media as effectively as onsite interactives utilize the physical environment? This chapter is an effort to outline the similarities and differences between physical and online interactives, and to roughly sketch out a conceptual framework for Web-based science interactives that, while not yet coherent enough to be called a paradigm, can help the medium stand on its own and continue to gain educational value and effectiveness. Given these differences, we propose several alternatives to the verbatim transplantation of physical exhibits to the Web. In our work as educational Web developers, we have experimented with various methods to engage Web users through imagination and emotion, rather than the cognitive hooks more traditionally relied on in informal and inquiry-based science education (Donovan et al., 1999).

We have been inspired by the work of several learning theorists and proponents of digital learning, including Howard Gardner (1991), Roger Schank (1992), Kieran Egan (1998), Mark Prensky (2001), and James Paul Gee (2003). We have adopted several of their concepts to break through the clutter of modern media and connect with users on their own terms, and we will describe how they can be used. Because Web technology continues to evolve rapidly, this chapter does not profile case studies that will only appear quaint in a few years. Neither will we dwell on the nuts and bolts of usability and interface design. Rather we will frame how we can think about Web-based interactives as a medium and as a learning experience. We hope that these concepts will be even more fruitful as the technology matures and user's access to our efforts expands.

Shared Assumptions, Strengths, and Weaknesses

We do not mean to imply that the need to articulate a distinct strategy for online science exhibits is a criticism of physical exhibits. Neither is it a call to replace physical exhibits with virtual ones. On the contrary, it is clear that the rapid embrace of the Internet by museums starting in the mid-1990s demonstrates some strong and deep affinities between the two arenas, which we will describe below. By pointing out these affinities, we are not ignoring the profound differences in form and function between working (either as a developer or a user) in physical and virtual spaces. However, for exhibit developers and informal educators more familiar with physical exhibits, the connections we outline here form a practical bridge to thinking about electronic interactives. Table 1 summarizes some of the salient shared characteristics of physical and Web interactive exhibits.

Table 1. Shared characteristics of on-site and Web interactive exhibits

Shared Characteristics of Physical & Web Interactive Exhibits
1. Constructivist model of learning and cognition.
2. Developers create a non-linear path through a space.
3. Heterogeneous representations carry the message.
4. Free-choice learning context of use.
5. Atmosphere of playful experimentation.
6. Most users and visitors would rather do than read.
7. Atomized information can loose invaluable context.

Constructivist Model of Learning and Cognition

Once controversial, most informal science educators now espouse some form of constructivist model of learning and teaching. In a nutshell, this model states that people only learn by actively creating, modifying, and elaborating their picture of how the world works, and that facts, or pieces of information, have no place (and are therefore not learned, or retained) outside the context of this picture (Donovan et al., 1999). George Hein (1998) has articulated this notion for learning in museums, and we have found it applicable online. Truth be told, our job as Web developers would be far easier if users would indeed unproblematically receive and understand the information we publish online as we intended it. However, along with the theoretical considerations that hold that this does not happen, formal and informal audience feedback during development and after deployment of numerous projects has disabused us of this assumption. In reality, Web learners actively construct their understanding of the material they see online based on their own prior knowledge and picture of the world. It is clear to us that online learning is only effective if it can successfully engage users on their own terms, and then motivate them to reconfigure their model of the world to incorporate that new material. Indeed, it has been argued that regardless of the teaching vehicle, learning has simply not taken place if this reorientation does *not* happen (Gee, 2003).

Developers Create a Non-Linear Path Through a Space

Both museums and Web sites not only allow, but also to some extent require, the visitor to choose a non-linear path through a space laid out by the developers of the exhibit. There may not be an infinite number of choices, but the user has a lot of control over where to focus his attention and the order to look at things. Exhibit developers have responded to such visitor behavior in various ways, sometimes by imposing linearity with a strong storyline, or in other cases by atomizing the exhibits to the point that many rich relationships between individual concepts are lost. As we will discuss further, the Web presents precisely this challenge as well, and there is no single perfect solution to how to offer clear orientation while still allowing meaningful choices. If anything, orientation and association are even harder in the virtual space, which is by nature abstract and multi-dimensional. It is for this reason that many Web designers fall back on literal organizational metaphors such as the floor plan of the physical building.

Heterogeneous Representations Carry the Message

"Multimedia" (moving and still images, text, and sounds) and "multi-modal" (physical manipulation along with reading, hearing, and looking) characterize science museum exhibits and Web interactives equally well. This is likely a major reason why museums understood and began colonizing the Web as early and as enthusiastically as they did. This richness is also likely a major contributor to the "Wow" factor of both media. That is, simply offering the user a novel array of stimuli creates a certain level of engagement.

While a good start, this cannot be relied upon online, where the environment itself is emotionally cold or neutral.

Free-Choice Learning Context of Use

Museum exhibits are built on the premise that visitors are in the museum mostly by choice, and that their behavior as free-choice learners is distinct from the compulsory classroom-learning environment. At the very least, it is well known that visitors vote with their feet and pass by museum content that disinterests them (Falk & Dierking, 2000). While much is made of the classroom value of online museum exhibits and other types of interactives, we strive to design online interactives as free-choice experiences that emphasize or support engagement, intrinsic satisfaction, personal meaning-making, and self-directed exploration. It is abundantly clear that Web surfers "vote with their mice," and will quickly leave any content behind that does not quickly and fully engage them.

Atmosphere of Playful Experimentation

In keeping with the assumption that physical and online exhibits are free-choice learning experiences, science museum galleries and their Web sites put a premium on having fun. Learning should follow from playful (but hopefully concerted) experimentation. Many successful museum interactives encourage creative play and do not assume a particular content-based outcome. Although Web usability experts eschew elements such as animation and music as blocking efficient access to information, we have found these things to be precisely what gives educational Web sites the fun factor that they need to engage free-choice learners. We have also seen significant differences between the information-seeking behaviors of expert and novice users in a subject domain, so it is clear that many of the choices in development strategy should be driven by an assessment of the audience (Schaller et al., 2004).

Most Users and Visitors Would Rather Do Than Read

It is a much-lamented truism that museum visitors of all ages are most likely to push the button or turn the crank before they read the label, if they ever do. Visitors may be lazy, or just responding to the overwhelmingly non-verbal nature of the museum environment, along with the strong expectation fostered by the museum itself that it is "hands-on" rather than didactic. In spite of an early feeling that the Web liberated designers from the tyranny of page sizes and counts imposed by paper-based publications, we now recognize that, except at the lowest level of detailed content, brevity of text is just as crucial, if not more so, to the success of Web exhibits. As we will elaborate below, a virtual experience should allow the user to do something they cannot easily do in a physical space.

Atomized Information Can Lose Invaluable Context

While many science museums have galleries organized around such themes as outer space, engineering, or human cognition, the basic physical principles many exhibits illustrate are often taken from their original context of the scientific research or technical problem-solving and displayed in relative isolation (Bradburne, 2000). Such exhibits run the risk of becoming mechanical factoids, lacking a framework to help the user to know why they should be interested in the phenomenon, or how to connect it to other phenomena and processes. This stems at least in part from the non-linear nature of museums and the Web as well as the limited amount of interpretation offered by short labels. This concern is even more acute on the Web, where a user may arrive at a page deep in a site via a search engine, or may bounce from page to page without a strong sense of how the pieces of the puzzle interlock. The good news is that the Web offers developers the chance to integrate fundamental principles with broader social and practical context.

This brief and non-comprehensive inventory of the affinities between physical and Web interactive science exhibits reveals both shared strengths and weaknesses. It means that while we may need some new tools and approaches for Web interactives, thinking about interactives on the Web does not mean forgetting everything we know about physical exhibits. It also means that we have a golden opportunity presented by a new medium to transcend some challenges that have been hard to deal with in the physical environment.

Challenges of Translating
Physical Exhibits to the Web

Those with deep experience with virtual reality and human-computer interactions might argue that the differences between physical exhibits and virtual reality mean that there is an insurmountable incommensurability between these environments, and that it is therefore futile to suggest moving any aspect of existing physical exhibits online. While we agree on the theoretical level, in practical terms a science center's physical exhibits represent a large investment of human capital and are a significant reservoir of expertise and content. While virtual exhibits cannot precisely embody physical exhibits, physical exhibits *do* embody the identity and vocabulary of the museum's educational framework. Rather than starting from scratch, it therefore seems useful to evaluate how some of the different types of exhibits in a physical science center might or might not be readily adapted to the Web. If in fact an exhibit might already be suited to an electronic version (perhaps even because the physical version is less physical than we thought!), we see no need to go out of our way to reframe it just for the sake of being different. If, on the other hand, a type of exhibit is fundamentally unsuited for life online, then it is useful to be aware of that before undertaking the conversion process. There are many ways to categorize exhibits, and this treatment is not meant to be definitive or exhaustive. It serves to highlight the issues encountered in thinking about online exhibits, and implies some

Table 2. Science museum exhibit types and their digital potential

Physical exhibit type	Electronic version potential
1. Object-centered	Authentic objects become part of reference encyclopedia.
2. Demonstration of a physical principle	Explain underlying principle instead of demonstrating surface phenomenon.
3. Illustration of function of human cognition	Can be equally effective.
4. Mechanical analogy for non-mechanical principle	Find new non-mechanical analogy.
5. Fun science or hardware-derived tricks	Create digitally native opportunities for creative play.
6. Showcase exotic technology	Explain underlying operation or encourage innovation.
7. Engage socially relevant topical issues	Strong opportunity to frame and enable community discussion.

of the strategies that we will develop further below. While above we highlighted the shared aspects of physical and virtual exhibit spaces, the following comparison of specific physical and virtual exhibits brings out several essential differences between the two. After this examination, we offer some general strategies for dealing with these differences. Table 2 lists several science museum exhibit types and their possible electronic equivalents.

Object-Centered

Museum: Although most science museums are not as collection-centered as natural history or art museums, many science centers and museums still exhibit objects of interest such as geology specimens, models of historically significant inventions like Leonardo's flying machine, or exotic hardware such as a space shuttle robot arm. Science museums also often create immersive environments where the visitor can assume a role like an astronaut or experience the shaking of an earthquake.

Web: If the object on exhibit is authentic, such as an *original* space suit or a *real* meteorite, a Web site obviously cannot have the immediacy and emotional power of seeing the real object in person (Pearce, 1992; Allison, 1995). Even as hard as the museum has to work to certify the authenticity of the objects it exhibits and to hide the mediation between the visitor and raw reality, a Web site can never pretend to display the object transparently free of mediation (Allison, 1995). While a museum visitor may feel transported to some other time and place, the Web user is always aware they are in cyberspace. That is not to say that objects should not be exhibited online. But the user's

Figure 1. This role-play game enlists users in Lewis & Clark's Corps of Discovery, where they can search the American West for plants, animals, and native peoples. Instead of relying on low-resolution images of the original artifacts to captivate users, the game provides a sense of exploration and discovery as well as historical context. (Courtesy Colonial Williamsburg Foundation, www.history.org/trips, demo at www.eduweb.com/portfolio/jeffwest)

purpose for looking at objects online will probably be different from the sense of awe and excitement engendered by seeing the Wright flyer or Apollo command module in person. Objects online can be framed by deeper interpretation of the object, transforming the Web site into a reference tool or an interpenetrating narrative that the museum cannot sustain. As we will discuss below, the sheer emotional impact of seeing an authentic object in the museum can be recovered on the Web by making the role-playing aspect of some museum exhibits more central to the activity than the physical exhibit can. While playing at being an astronaut or an archeologist in physical exhibit spaces is more often a design conceit than a learning strategy, online role playing games can offer the player substantive choices and subsequently present the consequences of those choices (Schank, 1992; Prensky, 2001; Gee, 2003).

Demonstration of a Physical Principle

Museum: A staple science museum exhibit is the demonstration of some basic physical principle using a mechanical device to instantiate the phenomenon in question. For instance, most science museums illustrate the Bernoulli effect with a fan that levitates a lightweight ball in the column of low-pressure air created by the blower (Exploratorium, 1995). Putting philosophical niceties about mediation and the construction of phenomena aside, this exhibit is as close as the museum comes to putting raw reality on display (Allison, 1995). The exhibit offers the user the chance to form hypotheses about the phenomenon and experiment to test them. However, written interpretation of these exhibits is usually brief, and without guidance from a staff explainer or a more knowledgeable member of the visitor group, most users do not spend enough time or lack the

Figure 2. This simulation lets users safely experiment with bridge types and seismic safety features to design an earthquake-safe bridge over Oakland Bay. (Courtesy California Alliance for Jobs, www.newbaybridge.org/classroom)

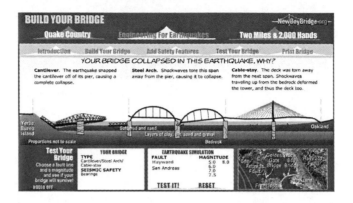

procedural or subject experience to carry out meaningful experiments that actually lead to an understanding of the underlying principle (Bradburne, 2000).

Web: As with authentic objects, an electronic version of something like the Bernoulli blower by itself lacks the immediacy and seemingly unmediated experience that makes these demonstrations compelling. Whereas in their galleries, museums may succeed in hiding their role in mediating the visitor's perception of reality, the Web can never claim to offer direct access to a physical experience (Allison, 1995). It obviously makes little sense to create an elaborate digital representation of the Bernoulli blower as a device. Rather, we need another way to frame the learning goal and the central task of the activity to focus on the *underlying principle* or the *inquiry process* rather than the *surface phenomenon*. One thing an electronic interactive *can* do that the physical exhibit cannot so easily do is to actually get at the underlying principle more clearly. By showing the user something that is not readily visible to the unaided senses, digital simulations can offer powerful insights into otherwise hidden cause and effect. For instance, a molecular-level model of how moving gasses and fluids create pressure differences could offer a more cogent *explanation* than putting one's hand in the stream of moving air coming from the blower. To get at the *process* of *investigating* the Bernoulli effect, a virtual wind tunnel could allow the user to design and test a wider range of airfoil shapes than could be done in the museum. Rather than emphasizing the underlying principle, the wind tunnel activity focuses on the practice of empirical inquiry. This may be more suitable to the free-choice learning arena in terms of a skill to be learned through practice rather than a body of content to be learned through didactic drilling. Electronic simulations are exceptionally good at creating safe places to explore things that are too dangerous or expensive to do in the real world, and as such open the door to demonstrating principles and processes that are off-limits inside the museum (Schank, 1992; Prensky, 2000).

Illustration of Function of Human Cognition

Museum: The Exploratorium was founded around exhibits that illustrate various aspects of human cognition, such as color perception, persistence of vision, and various optical illusions. These exhibits are extremely successful and popular because they often introduce a discrepant event, or an unexpected result, for the visitor to puzzle out. In principle this surprise motivates the visitor to observe closely and think about what they see (Rockley, 2001).

Web: This is easily the least problematic and the most obvious type of exhibit to directly translate to an electronic environment. Digital versions still convey the same message, perhaps even more clearly because of the opportunity for unlimited repetition and more focused experimentation. Optical illusions online can offer more information about the underlying phenomenon because they are not limited by label space. Different variations of perception are easy to display and juxtapose online, leading the user to a more complete understanding through repetition (Gee, 2003). And, as the Exploratorium site already does with its online exhibits, it can help the user follow a path between exhibits based on similar principles or perceptions by offering relevant links to other exhibits (Exploratorium, 1995, 1996).

Mechanical Analogy for Non-Mechanical Principle

Museum: Some of the most brilliant science museum exhibits use a mechanical device to create a physical *analogy* for a more complex, perhaps not directly visible, principle. For example, the normal distribution, used in statistics to characterize a group with random variation, is illustrated by a probability board consisting of an inclined or vertical grid of pegs. The balls bouncing through the grid stack up in the characteristic bell-shape of the normal distribution. Unlike the Bernoulli blower, the normal distribution itself is a non-physical mathematical concept. What creates the "Aha!" moment for museum visitors is the way the mechanical exhibit makes this abstract idea concrete. The probability board is particularly effective because it helps show how the curve arises from a series of independent yet similar events: all balls go in at the top, and if the board is kept stationary, the curve always ends up in the same place in the grid.

Web: Watching virtual balls roll down a virtual probability board would most certainly become rather boring after the initial novelty of the animation passes. But if we recognize that the power of the physical probability board is its power as an *analogy* that renders an invisible principle visible, that strategy is excellent for an online demonstration. The challenge is to find an analogy more suitable for the virtual environment that connects to a different aspect of the user's lived experience (Gee, 2003). For instance, people not well versed in probability often come to believe that coincidences such as having the same birthday as someone else, or meeting an old friend in an unexpected place, are not random. So perhaps an online demonstration of probability would incorporate calculators showing just how often seemingly rare but everyday events might happen, and how quickly the chances of a coincidence occurring increase with large numbers of concurrent events. In other cases, an online interactive may actually be more successful if it uses a *more* abstract representation that allows a greater deal of user manipulation to experiment with and build-up and understanding of the principle. For example, science educators have seen that mechanics can be better taught using very simple visual representations generated by a specialized computer language that allows the user to build up a descriptive, rather than algebraic, vocabulary about the cause and effect of motion (Gee, 2003). Along with providing the experience itself, these approaches give users a richer set of tools to understand their experience. The physical museum typically emphasizes the former while attempting to provide the latter through labels.

Fun Science or Hardware-Derived Tricks

Museum: Many science museum exhibits, such as giant soap bubble hoops, are in essence ingenious tricks that rely on exploiting some physical phenomenon with clever engineering. These encourage visitors to play and experiment within the parameters of the exhibit (i.e., "How big a bubble can I make before it pops?") without always overtly trying to teach specific content ("Surface tension governs bubble size"). These are successful at engaging the visitor's inquisitiveness and willingness to experiment and help to convey the playful nature of the scientific enterprise.

Web: Electronic interactives can capture this same spirit of play, inquiry, curiosity, and experimentation but within the electronic domain. It has been argued that this is what good video games have done for the past ten years, but few educators have taken them seriously until recently (Gee, 2003; Prensky, 2000). Making music, movies, or art with the help of the computer are ways to enable that sort of creative play online. The results may in fact be far more abstract than a physical exhibit, such as exploring or creating a fractal image, which exists nowhere but in the computer. The activity can, however, create a similar feeling and sense of engagement and exploration. And the more unique the online experience is, the greater the sense of awe and wonder will be.

Showcase Exotic Technology

Museum: Even before they were hands-on, early museums of science and industry such as the Chicago Museum of Science and Industry, the Science Museum in London, and the Deutsches Museum in Munich, were cathedrals of technology. They offered visitors the opportunity to see and learn about exotic or impressive technology that the visitor did not ordinarily see every day (Friedman, 1996). The amount and nature of the interpretation of the underlying functionality or principles of the equipment were highly variable. Of course, what counts as impressive and exotic is a constantly moving target as technology advances. As high technology has become more ubiquitous in the developed world, some exhibits, such as many using lasers or television equipment, have shed their technical presentations and become quite play-oriented. They may or may not offer the chance to observe and form an understanding of cause and effect or the science behind the device.

Web: Electronic versions can't replicate the cool factor of a "real" laser or other latest gadgetry. In a society surrounded by high technology, the original motivations behind these exhibits need to be assessed and a new rationale developed for online interactives. Are they about explaining the hardware? A simulation and "how it works" interactive might be appropriate if the functionality of the device is at the center. As with "fun tricks" above, if play and curiosity are the primary goal, then the answer might be to create another venue for playing with electronic objects that may not be physical things. If the goal is to celebrate and foster the inventive spirit that lead to the innovative hardware showcased in the museum, then a role-playing game that encourages tinkering and problem solving in a virtual domain could be effective.

Engage Socially Relevant Topical Issues

Museum: Museums of all genres have felt both internal and external pressure for their exhibits to have greater social relevance and to engage current issues (Davis & Gurian, 2003). Science museums have undertaken exhibits that venture beyond basic phenomena into the realm of social policy and decision-making. Subjects such as risk assessment, global climate change, and genetic engineering have been addressed. In recognition that these exhibits are topical and subject to becoming dated, most have been developed as lighter-weight traveling exhibits and many rely more on images and text than objects or some of the more traditional interactive exhibits about basic phenomena. Still, physical exhibits of any type are difficult to change as information and societal perceptions and policies change.

Web: The ability to construct Web sites that can be quickly and easily updated, perhaps even automatically drawing content from databases, means that the Web is an ideal place for science museums to contribute to public understanding and discussion of science and technology policy and the role of science and technology in society. With Web-based forums and message boards, science museums can create online communities around these issues that can continue as long as the subject is relevant. Some issues, such as climate change, would clearly benefit from simulations that help the user to understand the underlying science better. Issues such as environmental conservation or public health concerns can be brought alive by role-playing scenarios that are more about the social landscape than technical details. These experiences allow the user to discover the challenges and complexities of trying to formulate public policy that satisfies a diverse array of constituencies.

Table 3. Strategies for effective digital interactives

Challenge	Strategy
Overcome the emotional coldness of the computer.	Find new ways to make the affective connection that is lost in virtual space. • Apply Egan's Kinds of Understanding • Rediscover the Power of Narrative
Electronic media cannot offer direct access to authentic objects or physical phenomena.	Rethink what counts as "authentic" to make up for the heavy mediation of the electronic environment. • Rely on the relationship of trust with your visitor • Replace real phenomena with real data • Simulations offer authentic cause and effect
Very large and small temporal and spatial scales are hard to grasp.	Take advantage of virtual reality to cross boundaries of time and space.
Web users lack intrinsic motivation to engage didactic virtual exhibits.	Maintain the creativity and atmosphere of playful experimentation that physical exhibits establish.
Atomized content is disconnected from problems in the subject domain.	Take the opportunity to add context back to the content.

Strategies for Web-Based Science Exhibits

The opportunities and challenges of bringing these exhibit categories online suggest several general strategies for developing online science interactives that transcend exhibit type. Since no Web development effort is a one-size-fits-all proposition, we offer these strategies less as a mandatory checklist than a suite of tools, some of which will be more appropriate than others for any given project. Table 3 summarizes these strategies.

Find New Ways to Make the Affective Connection that is Lost in Virtual Space

Recent educational research has persuasively overturned Bloom's division between cognitive and affective domains (Falk & Dierking, 2000; Hein, 1998; Healy, 1994). Learning simply can't occur without an affective dimension. As the aphorism suggests, "People forget what you said, people forget what you did, but people will never forget

Table 4. Egan's kinds of cognitive understanding (Egan, 1998)

Kind of Understanding	Age	Concerns	Examples
Somatic	Birth to three	Body abilities	Walking Bodily functions
Mythic	Three to eight	Binary opposites (good/evil; survival/destruction)	Fantasy stories & play Rhyming & word play
Romantic	Eight to fifteen	Limits of reality & experience Heroes Idealism	Guinness Book of World Records Sports and pop stars Carl Sagan's Cosmos American Girl dolls
Philosophic	Fifteen to twenty	Systems & schema	Evolution Marxism Conspiracy theories
Ironic	Twenty & up	Playful appreciation of multiple perspectives	Post-modernism Multiculturalism

how you made them feel." Satisfying the user's information needs is not enough. Engaging the user on an affective level is critical, but how?

Tap into Basic Kinds of Understanding

Educational theorist Kieran Egan has outlined five ways, or "kinds of understanding," that can serve as effective (and *a*ffective) avenues to learning. Each is associated with particular stages of development from childhood to adulthood, but even those associated with childhood usually remain important filters throughout adulthood (Egan, 1998). Table 4 summarizes Egan's kinds of understanding and the role they play in shaping our view of how the world works.

According to Egan, the first three phases occur fairly naturally in human development, while the last two require substantial guidance and support from one's social milieu to achieve. They are by no means the only way for human beings to perceive and understand the world, but have come to be regarded in the Western tradition as valuable cognitive tools that define a high level of education. Simulations of physical and ecological processes are an obvious match for Egan's philosophic understanding, but that approach must be taken cautiously, since much of a science center's target audience has not developed that set of cognitive tools. Many youth and adults interpret the world

Figure 3. This ecosystem management simulation frames the activity in romantic terms, characterizing the goal (accurately but with dramatic flair) as a struggle against an invasive exotic grass. The intro animation also employs narrative to establish the scenario and further evoke in the user an affective response to the content. (Courtesy the JASON Foundation for Education, www.jason.org)

primarily through mythic and romantic lenses, focusing on individual characteristics (biggest, oldest, strongest) and actions (greedy, noble, heroic) rather than underlying traits and large-scale systems. While some systems are sufficiently simple and transparent to be meaningful to these audiences, the simulation should still present strong affective hooks that connect with their existing preoccupations and mental frameworks. For example, an ecosystem restoration interactive might have a simulation engine under the hood, but the content and presentation would deliberately engage romantic understanding through emphases on heroic or extreme aspects of the system and its inhabitants. This should not be construed as simply a veneer — the core content and learning goals must be aligned accordingly for a coherent and appropriate learning experience. Egan's schema has helped us with many subjects to find the kernel that our target audience will naturally find compelling.

Rediscover the Power of Narrative

Perhaps the most recognized method to hook the emotions is narrative. For millennia, stories have been our primary tool for embodying and transmitting knowledge. Information wrapped up in a dramatic, emotionally rich story ensures that the critical meanings endure through many retellings. However, due to visitor's non-linear paths through exhibits, museums at best have employed story concepts rather than actual narratives onsite. But the online environment offers more possibilities to exploit the value of storytelling about science.

Narrative can be employed in two complementary ways: introductory movies and structured goal-based scenarios. Introductory movies for online activities can introduce the subject in a dramatic fashion (for example, framing the story of reclaiming forest land from an invasive exotic weed as a campy horror movie). Intro movies can also provide a conceptual orientation that is crucial for the learning goals to be achieved. This has proven to be more effective than orientation text panels in exhibits (which visitors often ignore), since most users will view a short movie sequence as an introductory gateway to an online activity. Using an intro movie as an advance organizer is distinctly different from the all-too-familiar and much-maligned Web site splash page Flash animations that serve only to show off a designer's flair (Schaller et al., 2004).

Equally valuable is designing the entire activity around a narrative. This strategy maintains and extends the emotional connections and conceptual frameworks presented in the introductory movie. The narrative brings context and humanity to the activity, often in the form of a Goal Based Scenario (the familiar structure of many video games) which assigns the player a role to play, presents a context, or back-story, and then challenges the user with a mission that will conclude with a clear sense of success or failure (Schank, 1992).

Even storytelling can fail to engage if the audience does not identify with the characters, for it is the characters that channel the emotional content of the story. Such identification can take center stage when the user *becomes* a character in the story, although this also makes it especially critical that the story is clearly framed affectively, giving the user a strong sense of how to *feel* about the challenge, the obstacles and the outcome. Assuming an identity (through an avatar or first-person perspective) is a powerful way to further invest users in the story and the content. Learning, notes James Paul Gee (2003), requires the learner to "take on an identity as a scientific thinker, problem solver, or doer," gain new powers, and ultimately "sense new powers in themselves." As a genre of online interactives, role-playing stories can bridge the gap between novices and a scientific domain. For example, a role-playing game might tackle the subject of pollution by granting the user superpowers to see and remedy problems. As users explore, they can practice their new identity as an ecologist, reflecting on what skills and talents that identity requires that they already possess as well as the superpowers the role grants them. As the player develops skill in problem solving, the superpowers that initially drew them into the game as an exciting hook can fade into the background.

Rethink What Counts as "Authentic" to Make Up for the Heavy Mediation of the Electronic Environment

The notion of authenticity revolves around two connected ideas: something that is trustworthy, and something that is genuine or original. Both aspects take social work to establish and maintain. As discussed above, virtual media present a fundamental problem for authentic representation. As the "virtual" in virtual reality reminds us, VR always mediates the user's experience of the "real" object. Like museum exhibits, the more "real" an object is online, the more mediation has gone into it in the form of preparing and rendering models, animations, and etcetera. Thus authenticity in virtual reality is problematic both in terms of trustworthiness and genuineness.

Rely on the Relationship of Trust with Your Visitor

As graphic design tools have increased our ability to manipulate original images or generate realistic synthetic images to show things that did not happen, establishing the credibility and authenticity of images and media on the Web is critical (Bearman & Trant, 1998; Brower, 1998). As respected cultural institutions, museums can certify that the images they display are accurate or authentic by banking on their authority as reposi- tories of real knowledge about the world (Allison, 1995; Bearman & Trant, 1998). Web visitors will tend to trust the accuracy of material they find on a museum Web site. However, given the inherent mediation of the virtual environment, an online exhibit will always be a reproduction of the object or phenomenon represented. Thus a key aspect of authenticity in the museum gallery can never be transferred online.

Replace Real Phenomena with Real Data

The other aspect of authenticity, the genuineness of the experience, requires a much more dramatic alternative to cope with the fact that we cannot give users direct access to genuine phenomena online. The strategy we propose is to replace real phenomena with real data. That is, a major goal of informal science museum education is to provide visitors with experiences that in some way model the conventional description of the scientific method: observation, hypothesis formation, experimentation, revision of conclusions. Interaction with basic physical phenomena is the in-gallery context for visitors to try out these practices (Hein, 1998). While electronic media can offer user-friendly structures that embody the inquiry process, the domain of inquiry needs to be different. One solution is to turn from real phenomena to real data generated by working scientists as the basis for inquiry. Just a few examples include remote sensing images, astronomical observations and calculations, and animal tracking records. Many science centers offer

Figure 4. "The Bat Profiler" activity lets users analyze actual field data (such as these bat call sonograms) through a user-friendly fictional PDA device. (Courtesy the JASON Foundation for Education, www.jason.org)

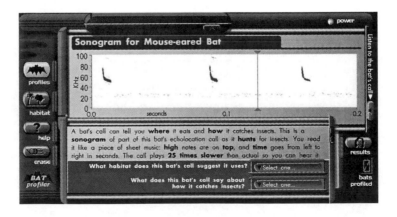

on-site displays of real-time data feeds such as weather observations and local earthquake activity. These data streams could easily be delivered online with added interpretation and activities for exploring and understanding them.

Instead of claiming to offer unmediated access to the world, this strategy offers access to real data generated and used by real scientists. Instead of focusing on cognition, as the physical exhibit does, this is an authentic engagement with another part of the scientific food chain: the interpretation of data. Non-technically trained users are connected to and learn to make sense out of material that is at the heart of the modern scientific enterprise. The challenge is to find or create a visualization that appropriately represents and explains the underlying patterns in the data. Such visualizations, or "front-ends" for real data may not look like the tools that practicing scientists have trained themselves to use. Among other things, most quantitative description needs to be replaced with qualitative displays that can convey an impressionistic feeling about the overall pattern and allow for basic comparisons. Whatever the solution, these new tools make the most of multimedia technologies to allow users to view and manipulate data in ways impossible without the computer.

Use Simulation to Offer Authentic Cause and Effect

The key feature of a simulation is that it contains a model that is sophisticated enough to respond to a range of user choices and generate multiple outcomes. A true simulation relies on a computational or logical decision engine to process user input and produce a result, often with random variations. Such an engine allows the player to truly explore combinations of inputs and compare outcomes to various runs of the simulation. It also allows the user to iteratively alter the system, using as inputs the last output of the model. This is why a computation engine of some sort is required; a fixed branching tree pointing to pre-produced outcomes cannot build on its own results and as such cannot demonstrate emergent properties. In a sense, a computationally driven simulation offers an authenticity of cause and effect. While every model simplifies the real world based on certain assumptions, the results follow from the choices the user makes and the nature of the system, rather than being more narrowly determined by the combinations the developer of the interactive chooses to include. Ideally, a good simulation allows some measure of open-ended exploration and discovery.

Along with digitally modeling the principle underlying the superficial phenomenon illustrated in a physical interactive exhibit, simulations have the potential to model situations or processes that are too expensive or dangerous to recreate on the exhibit floor. Or, as described below, it may cross scales of space and time that are otherwise hard to grasp. One of the greatest virtues of a simulation is that it creates a safe atmosphere for failure, whether it is the physical danger of failure or the social cost of failure. In the virtual world, the primary barrier to trying again is not the stigma attached to failure, but the ease of use of the interface and structure of the activity. Because it is safe to fail, and hopefully easy to recover in a simulation, players are encouraged to try again, and to experiment in taking risks that they would not so willingly try in real life (Schank, 1992; Gee, 2003).

Take Advantage of Virtual Reality to Cross Boundaries of Time and Space

One of the great challenges of science is grasping scales outside of regular human experience. The vast sweep of geological time and interstellar distance, and the infinitesimally small time and physical scales of molecular interactions and nuclear reactions all literally boggle our minds. Science museum exhibits frequently create analogies that calibrate these scales to something we can more readily comprehend. For example, various physical installations show the relative distances across the solar system scaled to fit a portion of the museum building or even an entire town (ScienCenter, 2001). Online, we are free to move across these scales of time and space with a facility that we don't have in the physical gallery, although in the confines of the computer screen this can be disorienting as well. Virtual exhibits can take advantage of this in the same way that Geographic Information Systems allow the user to change mapping scales to see different patterns at different levels of scale and to render the otherwise invisible visible.

Maintain the Creativity and Atmosphere of Playful Experimentation that Physical Exhibits Establish

The Web is on its way to becoming the world's largest library; it also has potential to become the world's communal playground for conversation, collaboration, and experimentation. Science centers can draw on their tradition of innovation to help lead this effort. This strategy perhaps even goes without saying, given the track record science

Figure 5. Playful experimentation and minimal text allows users to focus on the challenge of building a fish that can survive on the coral reef. Going beyond this implementation, a multi-player version would facilitate social interaction as well, creating a more motivating and playful experience. (Courtesy The John G. Shedd Aquarium, www.sheddaquarium.org/sea/)

centers have for creating quirky, creative exhibits. However, a few suggestions are still in order:

a.   Resist the urge to be too didactic online just because the medium allows you to have more words.

b.   Keep thinking in terms of activities (things to manipulate) and products (things that result from the manipulation), such as a geometric pattern, electronic music, or a movie. These products may be different, but serve the same purpose, as the creative results of physical exhibits.

c.   Don't neglect the social nature of the Web. Although most people use the Web by themselves, they do not have to be alone online. Publishing user creations for other visitors to see creates a pride in ownership of the creators and motivates others to join in. As the technology advances, multi-player games will become another means of connecting visitors in playful experimentation in real time. Multi-player games will also help bring the social nature of the scientific enterprise to the fore, something that most science museum exhibits, with their emphasis on cognition as the cornerstone of science, don't often do.

Take the Opportunity to Add Context Back to the Content

As we discussed above, both the Web and the exhibit hall can both feel like a collection of factoids rather than a unified whole. This is less the fault of the designer, and more a function of the user's behavior in either virtual or physical space. Here are some options for helping give a body of online content more coherence.

a.   Create linkages between content or activities that might often be isolated in the exhibit hall. Don't forget what the *Web* actually is: an interconnected set of related nodes. As a virtual science museum grows, it should have a mechanism for cross-referencing the content online in ways that help users see the relationships between different exhibits or activities. Amazon.com has pioneered this "What's related" or "You might also like" approach to electronic commerce. The Exploratorium has already done this to some extent with its categorization of online exhibits (Exploratorium, 1995).

b.   Role-playing games offer users scaffolding by setting expectations and laying out what's possible in the virtual world (Schank, 1992). The key elements of what Roger Schank calls a "Goal Based Scenario" include a problem-solving domain or world to inhabit; a task, or problem to solve; and a role, or identity to assume while solving the problem. Whereas a typical physical science center exhibit may pose a problem for the visitor to solve (What's causing the observed phenomenon?), the exhibit rarely explores the implications of the setting for solving the problem (What are the challenges and opportunities of working in the field or in the laboratory?) and the identity of the visitor (Are you solving the problem as a citizen, a scientist, or a policy-maker?). These added elements of setting and identity provide much-needed context, both in terms of adding motivation when there may not be any, and setting the parameters of the capabilities, expectations, and limitations of the problem.

Even if you do not want to create a complex role playing game in the mold of a commercial video game, using the basics of a Goal Based Scenario helps give structure and sequencing to the inquiry process that can be absent or very difficult to offer in the exhibit hall. It also helps the developer parse out the informational content for "just in time" delivery woven into an activity, rather than front-loading an otherwise engaging interactive with an overly didactic introduction.

Conclusion: Keep the Audience at the Center when Re-Conceiving Online Exhibits

In this chapter we explored the relationships between physical and electronic exhibits. While there are significant differences, development efforts in both areas can draw on shared expertise and practice. In enumerating several physical science museum exhibit types and what aspects of them might be successful online, we saw that fundamental differences between the physical and virtual spaces do require many science museum exhibits to be reworked for electronic versions. To facilitate this transition, we sketched out several strategies for creating affective engagement with online exhibits. While these outlines may seem to imply that moving physical exhibits to the Web will be as easy as going down the list and assessing an exhibit's suitability for the Web, and then choosing a new Web-oriented presentation strategy, we want to strongly urge that any program to develop exhibits on the Web should take place in a user-centered, top-down development process.

Along with considerable expertise in designing exhibits, science museums have also developed extensive experience with audience research and evaluation. In fact, almost all of the tools we have worked on for evaluating informal learning on the Web are derived from museum evaluation methodologies (Diamond, 1999; Burrough et al., 2003). Therefore the idea, while perhaps newer to commercial developers, is in principle well understood by science museums. But even as they *study* their visitors, science museums still struggle to take their visitors seriously as equal partners in the learning enterprise (Bradburne, 2000). Furthermore, in working with museum clients, we have seen that the temptation to dive in with an exciting implementation for an activity remains strong. We have found that following a structured information architecture process that focuses on the audience helps value the audience's needs and interests, enables the project to meet its fundamental goals, and leads to a smoother production phase (Educational Web Adventures, 2003). Doing so will also ensure that the Web exhibits truly take advantage of the strengths and fully account for the weaknesses of the Web as a medium. Prior to production, the first three most important stages of our information architecture process are as follow.

Key Formative Steps in Information Architecture for Online Interactives

Identify Needs and Characteristics of the Primary Target Audience

For a long time, science center exhibits appealed to the elusive "general public," expecting that visitors of all ages and levels of subject knowledge would get something out of an interactive exhibit. Because Web surfers expect highly personalized content, the most successful online projects target their audience more precisely, especially when the audience doesn't even have the common experience of stepping foot in the physical building. Age, subject domain expertise, motivations, and goals are all factors in the online visitor's experience (Schaller et al., 2002; Schaller et al., 2004; Haley Goldman & Schaller, 2004) and must be considered at the outset of project planning. In particular, graphic design strongly influences what age groups will be attracted to an activity, and even younger schoolchildren have a very well developed sense of whether something is for them or "for little kids" (Gilutz & Nielsen, 2002). As described above, Egan's Kinds of Understanding help us understand some of the fundamental perspectives of our audiences without knowing them individually.

Although it can be possible to layer information, and to expect different ages and levels of expertise to be engaged by a single online interactive, doing so means more explicitly accounting for those audience segments. As we have discussed elsewhere (Schaller et al., 2003), activities such as simulations have a great deal of conceptual power, but also may need to be framed to engage users with a kind of understanding shared by a wider range of the audience. While a simulation may be perfectly suited to explicate deep, complex connections between surface phenomena, in Egan's system such a method appeals mainly to Philosophic understanding, which most people do not attain until the teen years, if at all. Therefore to be successful, a simulation aimed at middle school students or adults without extensive formal education would need to be cast in more Romantic terms to connect to their primary kind of understanding.

State Learning Goals and Objectives in Terms of Successful Use of the Activity

Whether they focus on content knowledge, conceptual leaps, behavioral change, skill development, or other types of learning, the goals and objectives for an activity should clearly imagine what constitutes successful use, and what kinds of outcomes the developers hope for. While free-choice learning is highly idiosyncratic, the likelihood of accomplishing the project goals are far higher when these outcomes are visualized clearly so they can inform project development. Along with rethinking the nature of a physical exhibit as it migrates online, the underlying learning objectives for that exhibit may need to be reconsidered. Some goals, such as creating curiosity and encouraging inquiry about the world, can easily be retained. Other goals tied more closely to the cognitive and kinesthetic experiences offered by the physical exhibit space must be recast. While more abstract representations of principles may seem more didactic, they may in fact illustrate the underlying principle more clearly.

Define Central Concept or "Big Idea" that Serves as the Focus of the Entire Learning Experience

Museum exhibit consultant Beverly Serrell advocates framing a "Big Idea" for each exhibit (Serrell, 1996). The Big Idea distills the conceptual approach and subject addressed into a single declarative sentence. Choosing and clarifying the Big Idea may be a simple matter or more arduous if there are competing ideas under discussion, but it is essential to arrive at a single Big Idea for a coherent and meaningful final activity. Trying to combine competing topics into one activity will generally result in a convoluted and flawed activity that confuses users and leads to frustration at best and misinterpretation at worst. Having a Big Idea also helps promote the activity within and without the organization when brief explanations of various online exhibits are needed.

Define Structure and Function Before Visual Design

Good information architecture always determines how an interactive will work before deciding what it will look like. The look and feel of an interactive should follow from all the previous steps. Good visual design for interactives supports the functionality as well as the aesthetic sensibilities of the target audience.

In sum, we may have come perilously close to stating the obvious with regard to the similarities and differences between physical and online science center exhibits. Of course they are different. Of course they require different vocabularies and design paradigms. Of course you can't expect to recreate a physical space verbatim. Those reactions are to be expected for readers who have already assimilated Internet culture and understand its nascent vocabulary as well as anyone can at this stage in its evolution. However, we have seen enough reconstructions of physical spaces and interactives online to be confident that not every science center has fully grappled with the issues we raise. We do not intend this chapter as an indictment of those efforts. Rather, it is an invitation to view the next generation of online science exhibits in the same pioneering spirit of innovation of presentation, playfulness of approach, and clarity of intent that Frank Oppenheimer and his heirs brought to the Exploratorium and the science center movement it inspired. As we continue to differentiate further between the nature, purpose, and function of online versus physical exhibits, we expect that the two genres of exhibit will further diverge while retaining key family resemblances. Even as distinct vocabularies and design approaches emerge, we expect them to remain anchored in the affinities we have outlined. While we may still wish for much in the way of improved authoring tools, higher bandwidth, and greater user sophistication, we are confident that after close to a decade of growth and evolution, informal science education online offered by science centers around the world is ready to realize the full potential of the medium and the creativity of its professionals. We can only look forward to the results with great anticipation.

References

Allison, S. (1995). *Transplanting a rain forest: Natural history research and public exhibition at the Smithsonian Institution, 1960-1975.* Unpublished Dissertation. Ithaca, NY: Cornell University.

Bearman, D., & Trant, J. (1998). Authenticity of digital resources. *D-Lib Magazine.* Retrieved April 28, 2004, *http://www.dlib.org/dlib/june98/06bearman.html*

Bradburne, J. (2000). Tracing our routes: Museological strategies for the 21st century. In B. Schiele & E. Koster (Eds.), *Science centers for this century.* Sainte-Foy, Quebec: Editions MultiMondes.

Brower, K. (1998). Photography in the age of falsification. *Atlantic Monthly, 281*(5), 92-111.

Bruman, R. (1991). *Exploratorium cookbook I: A construction manual for Exploratorium exhibits* (rev. ed.). San Francisco: Exploratorium Press.

Burrough, L., Schaller, D., Beaumont, L., & Cannon, E. (2003). A rolling evaluation gathers no moss. In D. Bearman & J. Trant (Eds.), *Museums and the Web 2003: Selected papers from an international conference.* Pittsburgh, PA: Archives and Museum Informatics.

Davis, J., & Gurian, E. (2003). Timeliness: A discussion for museums. *Curator, 46*(4), 353-61.

Diamond, J. (1999). *Practical evaluation guide: Tools for museums and other informal educational settings.* Walnut Creek, CA: Altamira Press.

Donovan, M.S., Bransford, J.D., & Pelligrino, J.W. (Eds.). (1999). *How people learn: Bridging research and practice.* Washington, DC: National Academy Press.

Educational Web Adventures. (2003). Interactive design for online learning activities. Unpublished white paper. Retrieved April 28, 2004, from *http://www.eduweb.com/ Eduweb_Design_Process.pdf*

Egan, K. (1998). *The educated mind: How cognitive tools shape our understanding.* Chicago: University of Chicago Press.

Exploratorium. (1995). Exh*ibit and phenomena cross-reference: Balancing ball.* Retrieved April 28, 2004, from *http://www.exploratorium.edu/xref/exhibits/ balancing_ball.html*

Exploratorium. (1996). *Depth spinner.* Retrieved April 28, 2004, from *http:// www.exploratorium.edu/exhibits/depth_spinner/index.html*

Exploratorium. (1998). *Online exhibits: Seeing.* Retrieved April 28, 2004, from *http:// www.exploratorium.edu/exhibits/f_exhibits.html*

Falk, J.H., & L.D. Dierking. (2000). *Learning from museums: Visitor experiences and the making of meaning.* Walnut Creek, CA: Altamira Press.

Friedman, A. (1996, March/April). The evolution of science and technology museums. *Informal Science Review*, 14-17.

Gammon, B. (2001). *Assessing learning in museum environments: A practical guide for museum evaluators.* Unpublished Report. London: Science Museum.

Gardner, H. (1991). *The unschooled mind.* New York: Basic Books.

Gee, J.P. (2003). *What video games have to teach us about learning and literacy.* New York: Palgrave Macmillan.

Gilutz, S., & Nielsen, J. (2002). *Usability of websites for children: 70 design guidelines.* Fremont: Nielsen Norman Group.

Haley Goldman, K., & Schaller, D.T. (2004). Exploring motivational factors and visitor satisfaction in on-line museum visits. In D. Bearman & J. Trant (Eds.), *Museums and the Web 2004: Selected papers from an international conference.* Toronto: Archives and Museum Informatics.

Healy, J.M. (1994). *Your child's growing mind: A practical guide to brain development and learning from birth to adolescence.* New York: Doubleday.

Hein, G. (1998). *Learning in the Museum.* London: Routledge.

Hein, H. (1990). *The Exploratorium: The museum as laboratory.* Washington: Smithsonian Institution Press.

Jonassen, D., Peck, K., & Wilson, B. (1999). *Learning with technology: A constructivist approach.* Upper Saddle River, NJ: Merrill.

Nielsen, J. (2000). *Designing Web usability.* Indianapolis: New Riders Publishing.

Oxford English Dictionary. (2nd ed.). (n.d.). Definition of "interactive."

Pearce, S. (1992). *Museums, objects, and collections: A cultural study.* Washington: Smithsonian Institution Press.

Prensky, M. (2001). *Digital game-based learning.* New York: McGraw-Hill.

Rockley, M.G. (2001). *Hypothesis-based learning.* Retrieved April 28, 2004, from *http:/ /solomon.bond.okstate.edu/thinkchem97/*

Schaller, D., & Allison-Bunnell, S. (2003). Practicing what we teach: How learning theory can guide development of online educational activities. In D. Bearman & J. Trant (Eds.), *Museums and the Web 2003: Selected papers from an international conference.* Pittsburgh, PA: Archives and Museum Informatics.

Schaller, D., Allison-Bunnell, S., Borun, M., & Chambers, M. (2002). How do you like to learn? Comparing user preferences and visit length of educational Web sites. In D. Bearman & J. Trant (Eds.), *Museums and the Web 2002: Selected papers from an international conference.* Pittsburgh, PA: Archives and Museum Informatics.

Schaller, D., Allison-Bunnell, S., Chow, A., Marty, & Heo, M. (2004). To flash or not to flash? Usability and user engagement in flash vs. HTML. In D. Bearman & J. Trant

(Eds.), *Museums and the Web 2004: Selected papers from an international conference.* Toronto: Archives and Museum Informatics.

Schank, R. (1992). *Goal-based scenarios.* Technical Report #36. Chicago: Northwestern University Institute for the Learning Sciences.

ScienCenter. (2001). *The Sagan planet walk.* Ithaca, New York. Retrieved on April 28, 2004, from *http://www.sciencenter.org/saganpw/*

Serrell, B. (1996). *Exhibit labels: An interpretive approach.* Walnut Creek, CA: Altamira Press, Sage Publications.

Chapter IX

Storytelling-Based Edutainment Applications

Anja Hoffmann, ZGDV e.V. - Computer Graphics Center, Darmstadt, Germany

Stefan Göbel, ZGDV e.V. - Computer Graphics Center, Darmstadt, Germany

Oliver Schneider, ZGDV e.V. - Computer Graphics Center, Darmstadt, Germany

Ido Iurgel, ZGDV, ZGDV e.V. - Computer Graphics Center, Darmstadt, Germany

Abstract

Within this chapter, the authors — all members of the Digital Storytelling group at ZGDV Darmstadt e.V. — provide an overview of the potential of storytelling-based edutainment applications and approaches for narrative learning applications. This covers not only online applications, but also off-line edutainment components, as well as hybrid scenarios combining both types. The chapter is structured into five parts. At the beginning, a global scenario of edutainment applications for museums is introduced and key issues concerning the establishment of edutainment applications and the level of interactivity for online applications are highlighted. These open and relevant issues are discussed within a technology-oriented, state-of-the art analysis concentrating on the authoring process, storytelling aspects, dramaturgy and learning issues. Based on this brief STAR analysis, storytelling methods and concepts, as well as a technical platform for the establishment of storytelling-based edutainment applications, are described. The strengths and weaknesses of these approaches are discussed within the context of the edutainment projects, art-E-fact and DinoHunter Senckenberg. Finally, the major results are summarized in a short conclusion and further research and application-driven trends (context: museums) are pointed out.

Global Scenario of Edutainment Applications for Museums

The multifaceted scenario of edutainment applications for museums includes key players and user groups, as well as major components and aspects, for both online applications within the museums and Web-based scenarios using the museum (Web site) as document archive or knowledge pool, enabling teachers to enhance lessons through multimedia content provided by the museum's archives and collections (see Figure 1).

Some general questions address various needs and aspects of the different user groups involved in these museum scenarios:

- How to enter content into the exhibition and make it available via interactive artefacts?

- How to visualize (scientific) background information?

- How to build valuable exhibitions with learning effects?

- Which learning methods are appropriate?

- How is learning interconnected with Gaming/Fun?

- How much technology is appropriate?

- What are the benefits of combining museums with the Web?

- How to measure success of artefacts & exhibitions?

- How to finance artefacts and exhibitions?

- Which business models are appropriate?

Whereas the two first questions concern the authoring process and its outcome (as input) for run-time systems, such as interactive artefacts or terminal applications, the subsequent questions are more general in nature, encompassing learning and method-

Figure 1. Multifaceted scenario of edutainment applications for museums

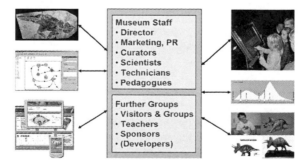

Figure 2. Levels of interactivity within edutainment applications for museums

Step 1: Website	Step 2: Interactive Tour	Step 3: Virtual Museum	Step 4: Virtual Exhibition	Step 5: Combine Visitors (M+W)
Levine Museum, Charlotte, NC	*Online Gallery, Deutsche Bank*	*Guggenheim, Bilbao*	*Sfmoma, CA*	*Lux – Markus Bader*

ological-didactic aspects, as well as marketing- and business-oriented issues. Within this chapter, we concentrate on the more technical aspects behind these questions and analyse the benefits of storytelling-based approaches in order to answer these questions and to make edutainment applications and museum exhibitions more valuable in a broader sense.

Referring to online edutainment applications and "Web sites with learning components" from the technical point of view, the most interesting question concerns the level of interactivity. Göbel & Sauer (2003) provided a short overview of those levels in their workshop, *Combine the Exhibition and Your Web Visitors. Integrated Concepts of Interactive Digital Media for Museums*, presented at Museums and the Web 2003 in Charlotte, NC:

> Whereas plain Web sites providing text, images or further media, such as audio and video clips, are very popular (low-cost production), enhanced technological methods and concepts increase the level of interactivity for users and, subsequently, the level of experience. Examples of this include interactive tours or virtual museums providing (more or less detailed) 3-D models of the real museum, enabling users to wander around and interact with elements, objects, and artefacts. Peculiarities of these scenarios are guided tours, such as the Interactive Tour provided by Deutsche Bank in Germany. Hereby, an isometric chat room is used to improve interactivity and overall experience of the users. Some virtual museums provide interaction metaphors and experience is focused on interactive media — others consciously avoid interactivity with objects/ artefacts and interaction is limited to navigation metaphors. Virtual exhibitions provide Web-based artefacts and users have a direct experience through the Web site — similar to the experience with "real/physical" artefacts within museums. Finally, Markus Bader's "Lux" provides an example of a hybrid system combining (visitors of) the virtual (Web) with the real (museum), enabling a bidirectional experience (Bader, 2004).

In summary, an appropriate slogan might be "the more interactivity, the more experience, the better the learning effect." Apart from the interactivity aspect, another current trend concerns stories and storytelling issues as new media for both knowledge transmission and interactive experience and the learning environment. An example of this trend is provided by Educational Web Adventures (Eduweb, 2004): "Eduweb's mission is to create exciting and effective learning experiences that hit the sweet spot where learning theory, Web technology, and fun meet." Underlying methods and concepts for such storytelling-based edutainment applications and learning environments are discussed in other sections of this chapter; other aspects, such as *Museums as Information Archives and Knowledge Pools* for teachers, pupils and interested people (Museum of Tolerance, 2004) or *Organizing Data Collections and Making Them Accessible through the Web* (Vernon, 2004), are not the focus of this chapter.

Current Research

The following paragraph gives an overview of current research in interactive storytelling. Many disciplines affect this field and it is largely influenced by traditional forms of storytelling, such as literature and movies. For the creation of these complex applications, it is helpful to look at methods and structures coming from, for example, script writing. This state-of-the art survey concentrates on

- general aspects of storytelling and their application for interactive digital storytelling,

- the authoring process,

- dramaturgy in museums, and

- learning issues.

From our point of view, the most challenging issues encompass the provision of "appropriate" authoring environments for the different user groups and the integration or harmonization of interactive storytelling with learning issues.

Interactive Digital Storytelling

Although stories are widely used in game-based learning applications (for example, Chemicus, Physicus, Klett, 2004), there is still a lack of appropriate integration of story and instruction: "The instructional design was generally concentrated in an isolated instructional space that existed independently of the story arc" (Noah, 2004). The result is that the learning-supportive characteristics of stories are limited because of the interruption of immersion and engagement. To avoid this limitation, our approach aims at a full symbiosis of learning and story content.

Possible definitions for story models as the basis for story instances and story development are:

- Definition: "A Component, which integrates Structure, Content, Context and Development" (Mellon & Webb, 1997);

Figure 3. Suspense curve based on story models

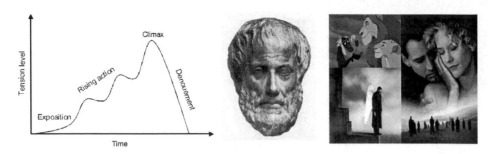

- Story Models represent narrations in an abstract way to underline the structure of the story;
- Story Models represent frameworks/templates for story instances.

Most story models are finally based on the simple dramaturgic arc model of Aristotle for telling linear stories: "Exposition," "Rising Action to Climax," and "Denouement." Other examples of widespread story models are provided by Syd Field (1988), who extended the Aristotle Model with regard to its usage for film scripts – hereby, script pages (= film minutes) are used for temporal structuring. The difference between Hollywood films and European film productions in terms of the usage of story models for films is very interesting. To form your own opinion, please compare the Hollywood film "City of Angels," directed by Brad Silberling (1998), and the European film "Wings of Desire," directed by Wim Wenders (1987). Tobias (1999) provides 20 master plots; the Russian formalist Vladimir Propp (1998) analysed hundreds of Russian fairy tales and extracted 15 morphological functions/components appearing in all these stories. Further, Propp defined characters (Dramatis Personae) representing rules within the stories, for example, an enemy, a hero, a magic agent (helper) or a princess (prize/award). Our integrated storytelling concepts and first reference examples, such as GEIST, developed at ZGDV Darmstadt, are primarily based on Propp's story model and morphological functions.

The application domains for Interactive Storytelling projects range from pure entertainment scenarios (e.g., Façade; OZ ; Mateas, 1997; Mateas & Stern, 2002) to marketing applications (e.g., interactive kiosk systems) or systems with therapeutic purposes (e.g., Carmen's Bright IDEAS, Marsella, 2003). All of the above-mentioned projects have one question in common: How can the narrative structure be combined with interaction? User interaction means an interruption of an ongoing story flow. The challenge is to design the story somewhere between emergent and predefined.

One of the major results of our comprehensive research in the area of Interactive Storytelling is the realization that the characteristics of stories foster the design of engaging and motivating learning environments for several different reasons:

1. **Cultural Tradition:** Stories are fundamental to culture and human understanding. They have familiar structures, which are recognizable and can easily be under-

stood. In human tradition, stories were a means for information transmission and knowledge acquisition, for example, within families and cultural communities. Today, kids are growing up with fairy tales (and moral education), learning words with story books and learning about several topics ranging from history to biology through TV shows, such as the famous French series "Il était une fois..." ("Once upon a time...") (1979). Unfortunately, as a means for the education of adults, storytelling is being widely lost.

Springer, Kajder, and Borst Brazas (2004) summarize the pedagogical dimensions of storytelling as follows:

Stories are:

a.   **Humanistic:** A culturally rich and venerated practice, global in relevance; encourages people to value their experiences, both imaginary and real, and it puts us in touch with ourselves and others. Stories communicate values.

b.   **Cross-disciplinary:** Stories apply to many K-12 subjects, including language arts, history, social studies, and humanities.

c.   **Cross-cultural:** Narrative structures cut across cultural and geographic spaces and unite oral, written, and technological literacies.

d.   **Multi-sensory, multi-modal:** They have visual, auditory, kinaesthetic properties.

e.   **Constructivist:** Storytelling is user-centred (learner); tales are created out of an individual's knowledge and experience.

f.   **Learning-directed:** "We learn in narrative structures and think in terms of stories."

2.   **Emotion and Immersion:** Another fact is that stories are structured in a suspenseful way and foster emotional engagement. Experiencing a good story (e.g., within films or novels) can cause total immersion in the imaginary world for the recipient, forgetting time and space. The research results in the area of Affective Computing show the considerable effects of emotional user interfaces (Picard, 1997). Studies in neuronal sciences point out the importance of emotional engagement for learning efforts and motivation (Spitzer, 2002).

3.   **Support of Basic Functionalities:** We can find essential functionalities for learning environments, such as focusing the learner's attention, provision of information and feedback about the learner's efforts (Gagne, Briggs, & Wagner, 1992). In addition, stories are not limited to certain topics. That means that any area of interest can be told in a narrative way. Furthermore, virtual worlds promote a deeper and active understanding.

4.   **Core Functions of Cognition:** Indeed, according to R. Shank (1995), stories constitute nothing less than the main building block of intelligence, memory, creativity, learning, and cognition in general. Educating with stories employs a most appropriate learning method, because it respects the way the mind truly works. According to Shank, we adapt to new situations and solve problems by

recurring to already available stories, rearranging and recombining them in an attempt to cope with new challenges ("Case Based Reasoning").

These statements totally cover current trends within the "Museums and Web" community: Everybody talks about "stories" and "storytelling," but, as the means of providing a background story about history, the biography of an author or circumstances of an artist creating some piece of art, without taking into account underlying storytelling concepts (e.g., story models) and theory. This fact motivates our daily work in developing methods and concepts for a storytelling platform as a basis for the wide range of edutainment applications based on storytelling fundamentals, taking into account the different needs of various people involved with the authoring process and global scenario of storytelling-based edutainment applications – with this paper focusing on edutainment applications for museums.

Authoring Process and Authoring Environment

Authoring of interactive stories is an iterative process, conducted by an interdisciplinary team consisting of designers, technicians, content providers and other participating disciplines. Figure 4 describes the multi-step authoring process with the three major phases *brainstorming*, *preparation* and *fine-tuning*.

All the engines for interactive storytelling are worthless without content. Therefore, content has to be created and be put into the interactive storytelling environment. There are lots of authoring environments for interactive presentation available, such as Macromedia Director and Flash or Blender (Macromedia Director MX, 2004; Flash MX, 2004; Blender 2004), but they do not provide any help in structuring a story or suspense-rich storytelling. For screenplay scripts the authoring software Dramatica (Dramatica, 2004) can be very helpful, but it does not offer any possibility to write an interactive story.

Figure 4. Authoring process

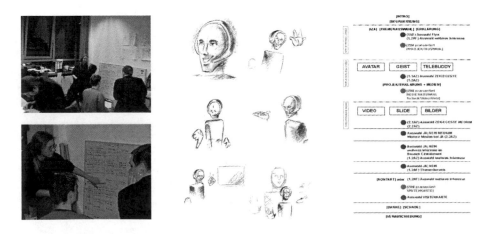

One of the main problems is that most of the software applications — including computer games - are only action-driven and most of the current stories are without interaction (e.g., films and books). If we just look back a few years, we can see that interactive storytelling had once been very common. Stories were relayed to the recipients by narrators. Because this had been a life process, the recipients could interact with the storyteller by asking questions or just being interested (or not). Today, parents often tell their children fairy tales before they sleep in a similar manner. Before any narrator can tell a story, he needs to know the story. And because of the possible interactions, he needs to know more about the story and its world than he can narrate on his own. This is the main concept of our Authoring Environment for Interactive Storytelling.

As mentioned above, there are three groups: The author, the narrator and the recipient. The author creates the story and explains it to the narrator. The narrator tells the story to the recipients. At the end, everybody has his own story world in mind. But if everybody has done a good job, the recipient's story world is somewhat the same as the author's story world. The better the author has explained the story to the narrator, the better he in turn is able to narrate the author's thoughts. Hence, it's easier for them if they each know how the other works and what to do to achieve a good result.

In the case of Interactive Storytelling, the role of the narrator is generated by the storytelling environment. It "knows" about story structure, suspense, immersion and how to narrate. But, of course, it does not know anything about the author's content. With our authoring environment, we primarily address regular book or film authors, because they are used to writing suspenseful stories. As much as possible, we give them the environment they are used to, but push them smoothly along toward interaction. Hence, we divided the authoring process into three parts: Brainstorming, Preparation and Tuning.

Within the brainstorming process, the author grasps the first ideas for his narrative. The main aspects are what should be told and how should it be told. Therefore, an abstract with about five lines is written down. Moreover, the author creates the characters of the story. Creating the main characters is hard work, because it includes complete CVs and everything about each character's life, behaviours, and more. The story's world is of great importance: In which era do the characters live, what are the circumstances, in which part of the world (or universe) do they live? Lots of sources have to be worked through for a believable and immersive story. Additionally, a first structuring of the narrative takes place during the brainstorming process: When and why should what happen, and which result should the story have? Normally, a concise version of the beginning and the end of the story are written down. The most important points of the story will be defined (in common stories, they are known as plot points). After that, the ideas for the missing scenes are noted and ordered, so that the journey through the story is full of suspense.

Now having an idea about the story, the preparation for the realization begins. The author has to choose the story model. He has got to decide which kinds of interactions should be possible (interaction metaphors). Furthermore, the modalities and media used are of huge importance for the following work. After all preparations are done, the story and its world can be created. This process depends a lot on the decisions taken before. This is the time when the storyteller gets prepared by the author. Hence, the whole storytelling

Figure 5. Authoring Environment Keating, Authoring Environment Cyranus

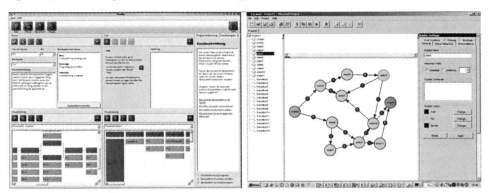

run-time environment and its helpers are fed the author's content. Similar to other media, this can't be done by the author alone. He needs a helping team as, for example, the director and cameraman for a film or the printer for a book. So, depending on the kind of presentation, a team of creative people will fulfil the author's imagined world.

To create these worlds, we use available software as much as possible. For example, for creating and animating characters and the surrounding world for Virtual/Augmented/ Mixed Reality projects, 3-D software, such as Maya (Alias Maya, 2004), Blender (Blender, 2004) or 3D Studio Max (Discreet 3D Studio Max, 2004), is used. The data is exported as VRML files, which the render engines can use for presenting the world. However, there is no authoring software for our interactive storytelling run-time environment. So, we created Keating with the help of authors, designers and programmers.

Keating is an authoring environment for structuring stories. With Keating, stories can be edited, combined with some media and verified. It helps during the brainstorming process and presents the decisions taken there for finishing a suspenseful story structure. Therefore, it gives different and flexible views of the story model, the story structure and the content. Via "drag and drop," structure can be changed or new content can be included. At each step of the creative process, the StoryEngine can be started to view and verify the work that has been completed thus far (Schneider, Braun, & Habinger, 2003; Schneider, 2002).

We have tested this tool in some projects and the way it works seems to fit an author's needs. Expansions have been undertaken with Cyranus (cf., Iurgel, 2004), which helps with authoring believable interactive dialogues. For the future, more work will be done to include an interactive world in the authoring software and to make it understand real storybooks for a much easier start in creating interactive narratives.

Dramaturgy in Museums

Museums are not only archives for cultural heritage and places for education and information, but also places for cultural development. Digital media and interactive exhibits within the museum's environment are of significant value to increase attractiveness and competitive advantage. In the same way, the presence of a museum on the Internet is of great importance, for example, for art galleries or science centres.

The ordinary presentation of exhibits, such as Web sites full of images and explanations, doesn't seem sufficient any more. Besides information, visitors of (virtual) museums and science centres are seeking entertainment, playful education and convincing experiences. Examples like "One Wright Way" of The Franklin Institute Science Museum demonstrate how collections could become classrooms with educational activities. There, students can make their own "Flight Forecasts" and learn more about the flight pioneers, the Wright brothers (Elinich, 2004). Being part of the story contributes to an active occupation with the topic. Springer, Kajder, & Borst Brazas (2004) also aim for the application of digital storytelling to make personal connections to visual art and museum artefacts in the National Gallery of Art in Chicago. In an interactive CD-ROM application, contemporary witnesses become the protagonists of a story: One example is a film that shows youngsters riding kickboards (a Swiss invention and therefore shown at Musée Suisse) and talking about the vehicles and their social context (Kraemer & Jaggi, 2003).

The examples show that stories are a powerful means to impart knowledge, especially for museums and science centres. However, none of them are interactive in such a way that the user might influence the story's flow and still experience a consistent narration. Therefore, we aim for interactive storytelling environments in order to improve individual learning efforts.

Learning Issues

E-learning is one of the current buzzwords in our information society and simultaneously represents a key area of EU Frameworks or national action lines for science and education. Recently, a number of approaches, such as notebook university Darmstadt (Notebook, 2004) or various school subject-oriented projects initiated and funded by the German Ministry for Science and Education (Neue Medien, 2004), have been initiated and carried out. On the other hand, in addition to those R&D projects, various learning applications, platforms and products, such as learnexact from Giunti (Learnexact, 2004), have been launched. Hereby, the amazing and exciting fact is that learning concepts are mostly limited to

- the usage of multimedia to express and explain content or,

- the usage of constructivist learning concepts, which could be freely interpreted, and

- the realization of learning applications as hypertext-oriented courses.

Another interesting issue concerns the usage (and benefit?) of 3-D learning environments or virtual characters as tutors guiding the user through learning applications. Examples of this are role games, such as the famous Final Fantasy Series (Final Fantasy, 2004) or Tomb Raider (Tomb Raider, 2004) and Doom (Doom, 2004), using the so-called

Figure 6. Learning software: 3-D and string of pearls

 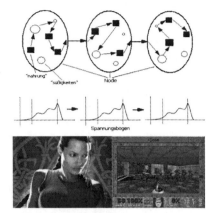

"String of Pearls" technique for sub-linear narration (see right half of *Figure 6*). Further on, *Figure 6* presents "Ritter Rost" as an example of interactive learning software for languages, "Mathica" (Klett-Heureka, 2004), proposed for pupils (age >= 10 years) using behaviouristic and cognitivistic learning methods, and "Der Manager im Handelsbetrieb" (Dekra, 2004), proposed for trainees in the commodities market economy using constructivistic learning methods.

Based on these approaches and learning examples, we invested research effort into the development of attractive, interactive and narrative learning software combining storytelling techniques with learning methods. Our first concepts providing a three-level concept are described in the following sections.

Storytelling Concepts for Edutainment Applications

Based on this state-of-the-art analysis, storytelling methods and concepts, as well as a technical platform for the establishment of storytelling-based edutainment applications, are described.

The global aim of our approaches developed by the interdisciplinary Digital Storytelling group at ZGDV Darmstadt is to combine different approaches from the fields of fairy tales, theatre, film or game-based learning and to establish narrative environments:

- For information & knowledge transmission

- For learning, training & education

- To increase immersion through suspense and suspenseful stories

Thus, taken altogether, some kind of new user interface paradigm in the form of "narrative user interfaces" has been introduced.

From a technical point of view, the basis for our storytelling-based edutainment applications is a storytelling platform providing a content layer (for story models, media, document archives, museum collections, etc.), an authoring environment with various editors, such as a story editor, scene editor, character or interaction editor, and a run-time system consisting of a story engine as control unit, as well as a scene engine, character engines and a rendering platform (scalable from mobile devices to simple Web sites or workstation screens up to complex physical set-ups for interactive artefacts within a museum).

Storytelling Run-Time Environment

As already indicated above, in any interactive storytelling application, the balancing act is between the degree of freedom (emergent stories that evolve from the parametric description of actions and interactions) and a predefined story plot. It is even more challenging for the design of interactive learning environments: the learner's autonomy must be balanced with the storyline and an instructional goal (Collins, 1996; Jonassen, 1989). How we deal with these problems will be explained in our approaches to an interactive storytelling environment. Then, we will present our integrated concept for educational applications.

For creating interactive storytelling projects, we use both a run-time and an authoring environment developed at the Digital Storytelling group at ZGDV Darmstadt. In order to follow a user-centred development approach, the user's needs directly influence the

Figure 7. Narration environment

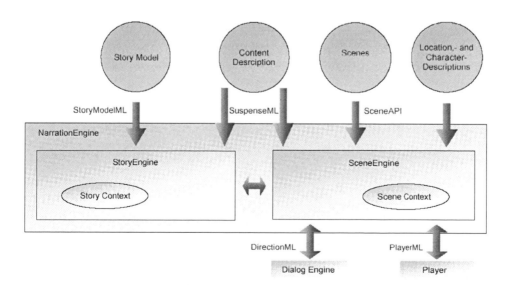

design of the run-time applications, as well. For example, the API used is very close to instructions used by directors on a film set. So both the run-time environment and the authoring environment influence the development process and vice versa. Authoring does not begin with creating an application for authors, but with building the whole environment, including the API, the MLs and the overall structure, with their needs in mind. The storytelling run-time environment consists of several modules for narration, scene and — for VR, AR and MR applications — character controlling and user behaviour interpretation. Additionally, a set of mark-up languages and scripts, as well as content databases, is used.

StoryEngine

The StoryEngine takes care of the narration of the overall story. Hence, it knows about the story structure and has implemented algorithms for creating suspense within an immersive form of storytelling. Therefore, it uses a story model provided by the StoryModelML. We have just developed the descriptions for some types of story-like presentations as fairy tales, story-driven education and business presentations. Additionally, some information is needed about suspense. This is done by the SuspenseML, which describes the storyline and how to combine it with the possible scenes. With this information, the StoryEngine processes the content interactively with regard to the user's preferences. All the user's interactions influence the kind of storytelling but, of course, not the story itself. Unlike common computer games, the story will always come to an end, so the user will experience satisfaction about the presented content.

SceneEngine

The SceneEngine is responsible for setting the scene and controlling the interaction with the user. A scene is the presentation of some content at a certain time and place. It gets a context description from the StoryEngine, so the SceneEngine controls which part of story should be presented at the moment. To have the same interpretation, it uses the SuspenseML, as well. Additionally, it loads a description about the actual place (stage) and the SceneScript. For SceneScript, we currently use an extension to the scripting language Phyton, which is well-known by game developers. It handles the content itself, the form of its presentation and the manner of interactions. For controlling the following modules, we specified the DirectionML. This is a non-blocking, asynchronous protocol with distributed systems and tasks in mind. Again, it has been developed for usage, which can be controlled by authors who are not programmers.

Integrated Learning Concepts

Our approach to an integrated concept for storytelling-based education purposes consists of three levels: Story Level, Knowledge Level and Learning Level. At each level, specific information must be described. In the following paragraph, we will explain what

Figure 8. Learning systems – Linear structures versus story-based 3-level concept

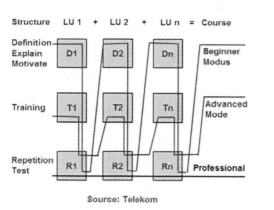
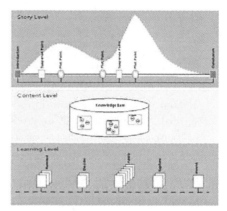

kind of information the author has to include in the design. Figure 8 contrasts our concept with traditional linear structures of existing learning software, such as a learning platform provided by Telekom, providing different levels for beginners, advanced and professional users, but very predefined structures.

Story Level

As described above, an interesting story with suspenseful structure has to be designed on the story level. As is already known from experiences from (non-linear) script writing and literature (McKee, 1997; Seger, 1990) we can differentiate between two approaches: story-driven approach (top-down) and character-driven approach (bottom-up) (Spierling, Grasbon, Braun, & Iurgel, 2002).

Starting from the top, the author defines the whole storyline, including beginning and ending of the story, as well as plot points where the story turns, under consideration of dramaturgical aspects. With knowledge of the whole story, the author can work out scenes, characters, interactions and dialogue in detail. Using the character-driven approach, the story evolves from a precise description of the (main) characters. In this case, authoring for interactive storytelling means a detailed definition of parameters and rules to control the character's behaviour. Practically speaking, a combination of both authoring methods is useful and must be supported by the authoring environment. During run-time, the data on the Story Level is processed by the StoryEngine and SceneEngine.

Knowledge Level

The knowledge base consists of modular fragments of information. It contains the information, which should be transferred and understood by the learner. Actually, the information is not part of the story, but rather serves as input for the author to develop an appropriate and consistent story. Therefore, content-related fragments should be designed for reuse and modification for other application domains. For museums, it is important to archive content for different presentation media, such as Web sites, onsite information or brochures.

Learning Level

At the learning level, the author decides about the education goal of the application. The design of the story depends on a decision as to whether the goal is to initialize general interest in a topic or achieve deeper knowledge. In any case, the learning part shouldn't undermine the power of the story (suspense, engagement, immersion). Educational parts must be integrated in the storyline. An exemplary situation shall explain a possibility of how it can be integrated: The learner needs to apply already acquired knowledge, but is obviously not able. During the following scenes, the necessary information should be presented in an alternative way so that he has another chance to succeed. In current (learning) games, the learner risks getting stuck in a similar situation.

Consequently, the learning model has to present how the user will proceed in his learning process — similar to a story model that presents the procession of the narrative.

Reference Examples

The strengths and weaknesses of these approaches are discussed in the context of the edutainment projects GEIST (2004), art-E-fact (2004) and DinoHunter Senckenberg.

Our approach to storytelling-based education and edutainment applications is currently realized in several projects:

- GEIST is an Augmented Reality system to experience historical coherences in the urban environment with interactive storytelling, funded by the Federal Ministry of Education and Research (BMBF)
- art-E-fact is an EU-funded project providing a generic platform for art (for both the creation and presentation of art)
- DinoHunter is an integrated concept in the wide range of museum applications and is partially implemented in the Senckenberg Museum in Frankfurt (DinoSim Senckenberg and DinoExplorer Senckenberg).

Hereby, the different projects address both edutainment applications within museums as tourist sites, as well as scenarios on the Web for museum Web sites, online courses and education or virtual science centres.

Figure 9. Impressions of the GEIST project

GEIST: Storytelling System for Experiencing History

The GEIST project (GEIST, 2004) represents a mobile outdoor Augmented Reality system to experience historical coherences in the urban environment with interactive storytelling. Hence, by telling a story, GEIST motivates the user to go sightseeing in Heidelberg while also introducing the user to the events that took place during the Thirty Years War. The users are supported to learn playfully, which awakens their natural interest to learn. Therefore, GEIST combines interactive storytelling with AR technology. The so-called "Magic Equipment" has been designed as user-centred input-/output equipment. It consists of AR glasses headsets for presenting the story. For tracking the user's position and view, a combination of GPS (Global Positioning System), a tracker and a video tracking system was developed. It provides the possibility to track the user's location and direction very accurately. Foremost, this equipment is used for displaying the AR information.

For the GEIST project, we have placed virtual stages around the city of Heidelberg. The story will be presented upon these stages. The information concerning which stage the user is on at the moment is provided by the tracking system, which exerts a huge influence on the kind of presentation from a narrative point of view. In combination with the other input equipment like magic maps and pointers, the GEIST system can always adjust the flow of the story presentation to the user's needs and keep it interesting. A magic book offers further historical information about the location, the people and the time. Thereby, users are able to answer questions which may arise independently, satisfying their developing thirst for knowledge right away.

Altogether, GEIST, as one representative of storytelling-based edutainment approaches, covers story models, learning and gaming aspects, Virtual Reality/Augmented Reality technology, location-based services, as well as virtual and physical props. These features show the great variety of interactive storytelling and storytelling-based

Figure 10. art-E-fact scenario and components of the platform

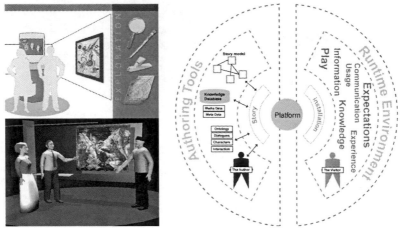

edutainment applications with a lot of different underlying research disciplines, such as computer graphics, interaction and communication design, but also history, pedagogy or artificial intelligence (concerning the dialogue modelling between the virtual characters and the user or among virtual characters).

art-E-fact: Interactive Edutainment Platform

art-E-fact (art-E-fact, 2004) is an EU-funded project for interactive storytelling in Mixed Reality. The aim of the project is to offer the user an engaging way to understand and experience art and art history from a philosophical perspective. The scenario is designed for a museum site or an exhibition hall. As the focus of the installation, a work of art (e.g., Byzantine icons) will be presented.

A group of virtual characters is situated close to the painting. When the visitor enters the installation, the narration begins and the characters begin a lively discussion. During the ongoing story, the user has the possibility to interact, for example, by text input via keyboard. At this moment, the visitor becomes part of the discussion group and can express his opinion, ask questions or change topics. He can also choose to enjoy the story passively. Then, one of the characters will take over the role of a non-expert to allow the visitor a certain degree of identification with the character. In any case, the narration touches on different areas of interest without massive interruption of the story line (Iurgel, 2002; Spierling & Iurgel, 2003).

Additionally, the visitor can use physical props to interact with the painting, for example, using a sponge for replacing layers or a magnifying glass for zooming in/out. Therefore, a video tracking system for gesture recognition is implemented.

A planned Web-based application will offer the same conversational interactions and adequate replacements for the interaction possibilities in Mixed Reality.

The art-E-fact approach aims at the transmission of information within an interactive dialogue with virtual characters and multi-modal interaction possibilities. Each character is representing a particular perspective on the work of art. The main topic of the story can be influenced by the user without departing from a consistent storyline. During the narration, the user will be prompted by the characters from time to time, for example, to give an answer. If he doesn't react, they will answer the question during the ongoing conversation.

art-E-fact is an example of the purposeful use of emotionally involving and personality-rich virtual humans in edutainment. Here, the information is transmitted in a natural way, involving emotion, dialogue and social aspects. Indeed, one of the core future issues for such virtual character-based educational applications will certainly be the guided establishment of social bounds for the human student with the virtual characters, since the importance of affective relations with teachers and comrades is a well-established phenomenon in learning. Accordingly, a learner should be able to establish a kind of "friendship" with the virtual companions, and feel affection and trust towards the virtual teacher (cf., Bandura, 1997; Iurgel, 2003).

DinoHunter: Edutainment Applications for Museums

The global aim of DinoHunter is to develop integrated concepts for mobile edutainment applications and knowledge environments. Typical examples of this are interactive scenarios for museums, theme parks or various kinds of exhibits and trade fairs. From the technical point of view, DinoHunter combines computer graphics technology with interactive storytelling, user interface and user interaction concepts, such as Kids Innovation or momuna (mobile museum navigator) (Göbel & Sauer, 2003).

From a global perspective, DinoHunter provides integrated concepts for the wide range of interactive museum (or any other edutainment) applications. The basic principle is to combine computer graphics technology, such as 3-D Rendering, Virtual and Augmented Reality or multi-modal interfaces (speech recognition, video recognition, gestures, etc.), with interactive storytelling approaches established in the field of film, theatre or fairy tales and further user interface and user interaction concepts. With the support of mobile devices, location-based services and pedagogic aspects (learning models and concepts), DinoHunter transforms the museum into an interactive learning and gaming environment.

Hereby, the setting of DinoHunter takes into account the needs and knowledge of various users and user groups involved in the multi-faceted domain of museum applications (see Figure 11): individual visitors, families or school classes as visitor groups, museum guides, scientific, administrative or marketing staff at the museum, or all the different user groups visiting virtual museums via the museum's Web site.

Apart from a comprehensive DinoHunter platform providing tools, methods and concepts for all these different user groups, additional case studies and pre-defined templates help to support:

Figure 11. Global DinoHunter scenario for museums

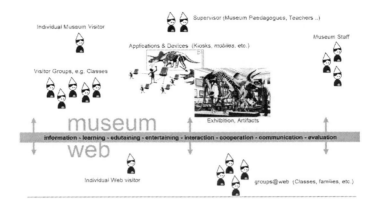

- Museum staff to archive library data and artefacts and make them available within digital museum applications.

- Administrative staff to monitor user behaviour and the success of individual artefacts or parts of exhibitions by measuring the retention period of visitors at special exhibits.

- Scientific staff to get a visual feedback of their research, providing 3-D reconstructions of dinosaurs or visualizing appearance and behaviour (such as walking). This also includes a rapid prototype environment as part of the authoring environment.

- Museum educators to enter digital media, didactic methods and learning models or any hints leading the visitor to "the most important" artefact.

- Teachers in preparing (and post-processing) the museum visit of a school class.

- Kids/pupils interacting and communicating among one another or sending messages in order to solve a group-based task associated with a museum's rally or game.

Figure 12. DinoSim Senckenberg

Figure 13. DinoSim Senckenberg: Modelling process

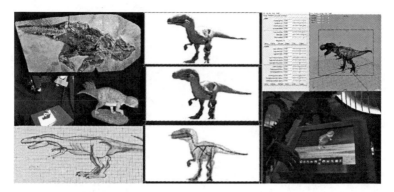

- Marketing people to combine the content layer (artefacts, exhibits and digital media or even access to further repositories) with the museum's shop or the Web site or event calendars.

DinoSim Senckenberg

One example of the successful combination of onsite exhibits and online services is the DinoSim project for the Senckenberg Natural History Museum (Senckenberg, 2004). On the occasion of its re-opening in November 2003, Senckenberg had a great interest in improving their exhibition by integrating multimedia systems. In addition to a general visitor information and navigation system, two kiosk terminals around the most exciting exhibits (skeletons of T-Rex and Diplodocus) are enhanced by the DinoSim application, which provides a 3-D real-time simulation and animation of dinosaurs. The primary goal of DinoSim is to visualize different scientific-related theories about appearance, movements and behaviour of dinosaurs. Visitors can freely navigate around the dinosaurs within a 3-D environment and can take pictures from their own view (point). These pictures are then sent by e-mail to the visitor's e-mail address, where users can then take advantage of the T-Online Fotoservice (T-Online Fotoservice, 2004) and have T-shirts, cups or bags printed with their individual dinosaur. The visitor will get his unique souvenir and, for the museum, the application is part of its comprehensive customer relationship management.

Figure 13 shows DinoSim's multi-phase modelling process, starting with real fossils, 3-D reconstructions by the palaeontologists, the extraction of geometry, appearance and skeleton, and the animation with Maya software, resulting in a touch screen application on terminals in front of the T-Rex and Diplodocus in the dinosaur hall of the Senckenberg Museum.

Figure 14. DinoExplorer Senckenberg

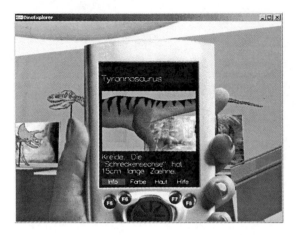

DinoExplorer Senckenberg

DinoExplorer represents a game-oriented application that is available for download on the Senckenberg Web site. Within a hide-and-seek style game the user can explore the virtual Senckenberg and its exhibits with the task of finding a particular animal (Leptictidium). As the user succeeds in finding the Leptictidium, further functionality is available on a virtual mobile device. For example, users can experience different layers for the skeleton, inner organs, muscles or possible appearances (colour, structure) of skins. Further along, the breathing of dinosaurs is presented via 3-D animations.

Figure 14 shows a snapshot of the DinoExplorer Senckenberg hide-and-seek game with an emulated PDA providing augmented information about the appearance of dinosaurs.

Detailed descriptions about the global DinoHunter scenario, as well as DinoSim, DinoExplorer and further applications out of the DinoHunter series, are provided in various publications of the Museums and Web and ICHIM conference series (Sauer et al., 2004; Göbel & Sauer, 2003a; Sauer & Göbel, 2003b).

A similar approach to DinoSim and DinoExplorer is provided by the Canadian Museum of Nature (Canadian Museum, 2004), underlining the usage of 3-D computer graphics technology to visualize geometry, appearance and behaviour of dinosaurs. Concerning the question, "What are the uses and benefits of 3-D imaging?," they list both exhibition enhancement by virtually displayed artefacts and specimens, as well as benefits in education by enhanced and animated data, for the provision of dynamic learning experiences.

Conclusion

Finally, the major results are summarized in a short conclusion and further research and application-driven trends (context: museums) are pointed out.

Within the wide range of museum applications and scenarios, this chapter describes methods and concepts for the establishment of storytelling-based edutainment applications. Hereby, the global aim is to combine traditional learning methods and concepts with dramaturgic and narrative elements in order to improve knowledge transmission, to increase the user immersion and finally to produce positive learning effects. This is realized by a storytelling platform, providing a content layer, an authoring environment with various editors for different user groups (such as storytellers, museum pedagogues and scientists or teachers), a complex run-time system with different story engines and a set of player components ranging from online Web sites for simple PC workstations up to complex physical set-ups and installations for interactive exhibits within museums. With regard to the integration of storytelling methods and components, such as story models or story engines and learning methods, a 3-level concept is introduced, providing a story level, knowledge level and learning level. The practical use and usage of these concepts are discussed in the context of the reference examples, GEIST, art-E-fact (EU-funded project concentrating on character-based conversations and Mixed Reality) and the two DinoHunter applications, DinoSim and DinoExplorer, developed and realized for the Senckenberg Museum in Frankfurt, Germany. Regarding DinoHunter, first user studies have shown the great benefit of storytelling and game-based approaches enhanced by interactive Virtual and Augmented Reality technology: Especially young visitors — used to computer games and new media — are fascinated by interactive applications and like this form of playing and learning.

From the research-oriented point of view, further effort will be invested into the field of integration and harmonization of various story models providing suspense and dramaturgy with learning models providing learning methods, content and media. Later on, both the authoring and run-time environment of the storytelling platform for edutainment applications will be improved in relation to learning aspects, as well as personalization and individualization. On the other hand, application-driven issues and obstacles affect the integration of content management systems with edutainment applications or the necessity (especially for museums) to find appropriate business models financing digital media, (interactive) museum Web sites, high-end installations or hardware and software for storytelling-based edutainment applications in general.

References

Alias Maya. (2004). Retrieved June 16, 2004, from *http://www.alias.com/eng/products-services/maya/*

art-E-fact. (2004). *Generic platform for interactive storytelling in mixed reality.* EU-funded project (IST-2001-37924). Retrieved March 15, 2004, from *http://www.art-e-fact.org*

Bader, M. (2004). Retrieved June 16, 2004, from *http://www.natural-reality.de/*

Bandura, A. (1977). *Social learning theory.* Englewood Cliffs, NJ: Prentice-Hall.*Blender.* (2004). Retrieved June 16, 2004 from *http://www.blender3d.com/*

Campbell, J. (1973). *The hero with a thousand faces.* Princeton, NJ: Princeton University Press.

Canadian Museum. (2004). *Canadian Museum of Nature, Canada, 3D Imaging Center*. Retrieved March 15, 2004, from *http://www.nature.ca/3D/*

Collins, A. (1996). Design issues for learning environments. In S. Vosniadou, E.D. Corte, R. Glaser, & H. Mandl (Eds.), *International perspectives on the design of technology-supported learning environments* (pp. 347-362). Mahway, NJ: Lawrence Erlbaum Associates.

Crawford, C. (n.d.). *Interactivizing stories*. Retrieved March 15, 2004, from *http://www.erasmatazz.com/library/Lilan/interactivizing.html*

Crawford, C. (n.d.). *Plot versus interactivity*. Retrieved March 15, 2004, from *http://www.erasmatazz.com/library/Lilan/plot.html*

Csikszentmihalyi, M., & Csikszentmihalyi, I.S. (1997). *Optimal experience*. Cambridge, UK: Cambridge University Press.

Dekra Akademie GmbH. (2004). *Der Manager im Handelsbetrieb*. Retrieved March 15, 2004, from *http://www.dekra-akademie.de/*

Discreet 3D Studio Max. (2004). Retrieved June 16, 2004, from *http://www.3dmax.com/*

Doom. (2004). Retrieved March 15, 2004, from *http://www.doom3.com/*

Dramatica. (2004). Retrieved June 16, 2004, from *http://www.dramatica.com/Eduweb*

Educational Web Adventures. (n.d.). Retrieved March 15, 2004, from *http://www.eduweb.com/*

Elinich, K. (2004). One Wright way: From collections to classrooms. In D. Bearman, & J. Trant (Eds.), *Museums and the Web 2004: Proceedings*. Toronto: Archives & Museum Informatics.

Feix, A., Hoffmann, A., Osswald, K., & Sauer, S. (2003). DinoHunter: Collaborative learning experience in museums with interactive storytelling and kids innovation. In S. Göbel, N. Braun, U. Spierling, J. Dechau, & H. Diener (Eds.), *Proceedings of TIDSE 2003* (pp. 388-393). Darmstadt.

Field, S. (1988). *The screenwriter's workbook*. New York: Dell Publishing Company

Final Fantasy. (2004). Final Fantasy Series. Retrieved March 15, 2004, from *http://www.ffonline.com/*

Gagne, R.M., Briggs, L.J., & Wager, W.W. (1992). *Principles of instructional design*. Fort Worth: HBJ College Publishers.

GEIST. (2004). Project (01IRA12B GEIST) funded by the German Ministry of Science and Education (BMBF). Retrieved March 15, 2004, from *http://www.tourgeist.com*

Göbel, S., & Sauer, S. (2003a). Combine the exhibition and your Web-visitors. Integrated concepts of interactive digital media for museums. In *Workshop at Museums and the Web 2003*. Charlotte, NC.

Göbel, S., & Sauer, S. (2003b). DinoHunter: Game based learning experience in museums. In *Proceedings of ICHIM'03*, École du Louvre, Paris. CD-ROM. Archives & Museum Informatics.

Il était une fois... L'homme, France/Japan. (1978). Retrieved March 15, 2004, from *http://www.generiquestele.com/details_id-20_n-Il,etait,une,fois,l,homme.htm*

Iurgel, I. (2002). Emotional interaction in a hybrid conversation group. In *PRICAI-02. Workshop on Lifelike Animated Agents*. Tokyo, Japan. JSPS.

Iurgel, I. (2003). Virtual actors in interactivated storytelling. In *IVA 2003* (pp. 254-258).

Iurgel, I. (2004). Narrative dialogues for educational installations. In *NILE 2004* (in press).

Jonassen, D.H. (1989). *Hypertext/hypermedia*. Englewood Cliffs, NJ: Educational Technology Publications.

Klett-Heureka. (2004). Retrieved March 15, 2004, from *http://www.klett-verlag.de/heureka/*

Kraemer, H., & Jaggi, K. (2003). *Virtual Transfer Musee Suisse*. Unpublished CD-ROM. Retrieved from *http://www.musee-suisse.com*

Learnexact. (2004). Retrieved March 15, 2004, from *http://www.learnexact.com/*

Macromedia Director. (2004). Retrieved June 16, 2004, from *http://www.macromedia.com/software/director/*

Macromedia Flash. (2004). Retrieved June 16, 2004, from *http://www.macromedia.com/software/flash/*

Marsella, S.C. (2003). Interactive pedagogical drama: Carmen's bright IDEAS assessed. In *Intelligent virtual agents: Lecture notes in artificial intelligence (LNAI 2792)* (pp. 1-4). Berlin: Springer.

Mateas, M. (1997). *An oz-centric review of interactive drama and believable agents*. Technical Report. Pittsburgh PA: School of Computer Science, Carnegie Mellon University.

Mateas, M., & Stern, A. (2003). Integrating plot, character and natural language processing in the interactive drama façade. In S. Göbel, N. Braun, U. Spierling, J. Dechau, & H. Diener (Eds.), *Proceedings of TIDSE 2003* (pp. 139-151). Darmstadt.

McKee, R. (1997). *Story: Substance, structure, style and the principles of screenwriting*. Regan Books.

Museum of Tolerance. (2004). *Teachers' guide Web site*. Museum of Tolerance, New York. Retrieved March 15, 2004, from *http://teachers.museumof tolerance.com/*

Noah, D. (2003). *An analysis of narrative-based educational software*. Retrieved March 15, 2004, from *http://naturalhistory.uga.edu/narrative_paper.htm*

Notebook. (2004). *Notebook University Darmstadt*. Retrieved March 15, 2004, from *http://www.nu.tu-darmstadt.de/*

Picard, R.W. (1997). *Affective computing*. Cambridge, MA: MIT Press.

Propp, V. (1998). *Morphology of the folktale*. Austin: University of Texas Press.

Sauer, S., & Göbel, S. (2003). Focus your young visitors: Kids Innovation, Fundamental changes in digital edutainment. In D. Bearman & J. Trant (Eds.), *Museums and the Web 2003: Selected papers from an international conference* (pp. 131-141). Toronto: Archives and Museums Informatics.

Sauer, S., Osswald, K., Göbel, S., Feix, A., & Zumack, R. (2004). Edutainment environments. A field report on DinoHunter: Technologies, methods and evaluation results. In D. Bearman & J. Trant (Eds.), *Museums and the Web 2004: Selected papers from an international conference* (pp. 165-172). Toronto: Archives and Museums Informatics.

Schell, J. (2002). Understanding entertainment: Story and gameplay are one. In J.A. Jacko,& A. Sears (Eds.), *The human-computer interaction handbook: Fundamentals, evolving technologies and emerging applications.* Mahwah, NJ: Lawrence Erlbaum Associates.

Schneider, O. (2002). Storyworld creation: Authoring for interactive storytelling. In V. Skala (Ed.), *Journal of WSCG (Vol. 10, No. 2). International Conference in Central Europe on Computer Graphics, Visualization and Computer Vision 2002* (pp. pp. 405-412). Plzen: University of West Bohemia.

Schneider, O., & Braun, N. (2003). Content presentation in augmented spaces by the narration of interactive scenes. In *Proceedings of First Research Workshop on Augmented Virtual Reality (AVIR).* Genf, Switzerland.

Schneider, O., Braun, N., & Habinger, G. (2003). Storylining suspense: An authoring environment for structuring non-linear interactive narratives. In V. Skala (Ed.), *Journal of WSCG (Vol 11, No. 3). International Conference in Central Europe on Computer Graphics, Visualization and Computer Vision 2002* (pp. 411-417). Plzen: University of West Bohemia.

Seger, L. (1990). *Creating unforgettable characters.* New York: Henry Holt & Company.

Senckenberg. (2004). Retrieved June 16, 2004, from *http://www.senckenberg.de/*

Shank, R., & Cleary, C. (1995). *Engines for education.* Mahwah, NJ: Lawrence Erlbaum.

Spierling, U., & Iurgel, I. (2003). Just talking about art. In Virtual Storytelling. *Proceedings Second International Conference, ICVS 2003* (pp. 189-197). (LNCS 2897). Toulouse, France.

Spierling, U., Grasbon, D., Braun, N., & Iurgel, I. (2002). Setting the scene: Playing digital director in interactive storytelling and creation. *Computers & Graphics, 26*(1), 31-44.

Spitzer, M. (2002). *Lernen.* Gehirnforschung und die Schule des Lebens. Spektrum Akademischer Verlag, Heidelberg.

Springer, J., Kajder, K., & Borst Brazas, J. (2004). Digital storytelling at the National Gallery of Art. In D. Bearman & J. Trant (Eds.), *Museums and the Web 2004: Proceedings.* Toronto: Archives & Museum Informatics.

T-Online Fotoservice. (2004). Retrieved June 16, 2004, from *http://service.t-online.de/ c/00/01/36/1360.html*

Tobias, R.B.(1999). *20 Masterplots.* Frankfurt am Main: Zweitausendeins.

Vernon. (2004). Vernon Systems. Auckland, New Zealand. Retrieved March 15, 2004, from *http://www.vernonsystems.com/*

Vogler, C. (1998). *A writer's journey.* Studio City, CA: Wiese Productions.

Chapter X

Revolutionizing Information Architectures within Learning-Focused Web Sites

Ramesh Srinivasan, Harvard University, USA

Abstract

This chapter points to the potential new information architectures hold in the design of virtual science centers. Science centers are treated as education-focused institutions and the argument is made that that extending the power of the science center as an educational platform warrants an answer to the question of how to share knowledge across the community of visitors without physical co-assembly. Two approaches toward information design are discussed: community-driven ontologies and social information filtering agents. These approaches are introduced within the context of two pieces of previous research and hold great potential when applied to the Web environment of the science center.

Introduction

Virtual environments have begun to proliferate as complements to physical centers of learning and exhibition. The presence of these spaces has profoundly impacted the ability of an individual to learn, reflect, comment, and engage with the cultural, artistic, scientific, or educational material these institutions offer without physically visiting them. Literature abounds discussing the implications of the Internet on art museums and classrooms, and less so with non-formal science centers interested in finding ways to make science more understandable to the public.

The museum setting is more observational, and reflective, but not integrally related to the experience of learning in the way of the science center. However, they are connected by the important discussion between designers, curators and technologists. The question these parties together are focused on is how to bring new information technologies to the museum or science center and vice versa.

Already, work is underway to create and design new technological devices for these institutions, embedded within the "brick and mortar" of the building, or made portable for a visitor to wear. Fascinatingly, the museum or science center visitor can no longer be defined as the person who walks into the door of the building, but instead could be he or she who accesses the appropriate Web site.

The key question surrounding all these issues is one of interaction design: how can interactive technology be deployed to facilitate dialogue between Web visitors and physical visitors, curators and visitors, lecturers across the world with both types of visitors, and etcetera. Dialogue, learning, observation, reflection – in union these terms point to a new paradigm that is not individualistic, but instead community-focused. Thus, my design research has focused on the mechanisms by which technology can enable and sustain community, and how distance can be bridged to create community around the objects of the museum or science center without the need for physical co-presence. Hence, the term **virtual** science center or **virtual** museum emerges.

There are lessons to be learnt from the classroom and museum that can be applied to the science center, but there are also fundamental differences. The experience of interacting with a science center is integrally learning-based, though within an environment of observation that suggests the typology of a museum. One does not "learn" a painting or sculpture with the same pedagogical process as he or she does about electricity or Einstein. This chapter is focused on mechanisms by which the design of virtual science centers can advance the experience of learning. Learning can be understood as a construction, a shared activity of discovery and dialogue between the individual participants (Papert, 1980). Thus, rather than understanding content as created by a single source, extending the power of the science center as an educational platform warrants an answer to the question of how to share knowledge across the community of visitors without physical co-assembly.

Introducing Virtual Science Centers and the Question of Information Architecture

Virtual science centers are a new typology that has emerged to focus specifically on the education of science to a general public. In a sense they are a hybrid of the traditional museum and the educational environment. While their goals are to provide virtual visitors with an exposure to objects and knowledge emanating from the physical site (i.e., the science museum that administers the sites), they also maintain a focus on how to educate visitors. Additionally, science centers tend to have a much more hands-on explorative approach inspired by the first major science center, San Francisco's Exploratorium. It is even more critical then that the virtual component of the science center also maintains the interactive and responsive edge that set science centers apart as exciting institutions to visit.

> (Science centers) are non-formal educational institutions of relatively recent origins. They have been established primarily for the purpose of popularizing science and technology to the public and students, thereby contributing to the enhancement of science literacy levels. Initially, the tasks were performed by science museums which were, and generally still continue to be, repositories of scientific artifacts. (Tan et al., 2002)

Visits to virtual science centers have increased dramatically over the last ten years (Tan et al., 2002), and they have begun to be understood as a phenomenon that warrant dedicated study and analysis. At the same time, noteworthy research conducted on digital museums and e-learning remains relevant for virtual science centers.

Discussion of virtual environments for museums, education, and science centers has begun to reveal itself in publications. One key question that has been largely unanswered in this literature, however, is related to the design of the information architecture of the virtual environment that is so important to the visitor's experience. In this brief chapter, my discussion will center on two assertions:

1. The argument is made for the strong benefits that focusing on a variety of relationships between individual pieces of content could provide in revealing the multiple threads of knowledge and perspective the science center holds.

2. Science centers would do well to provide visitors with the topics, exhibits, and ideas they may truly be interested in but yet to have been exposed to. This could potentially provide greater meaning to the experience of the virtual science center visit by revealing an interesting path to the disorienting deluge of information and objects the virtual science center makes available. The use of such techniques as social information filtering agents merit investigation within this scenario.

In the next section, a virtual science center (Boston Museum of Science) is briefly analyzed. This is followed by an introduction of previous research on community-generated exhibitions and my current doctoral research on the use of social information filtering techniques to disseminate cultural and educational material across a distributed set of Native American reservations. Finally, the chapter closes with some thoughts and recommendations of how to bridge the gaps between the status quo with the ongoing findings of my research.

Boston Museum of Science: Dissecting a State-of-the-Art Virtual Science Center

A pioneer of science education in the Boston area, the Museum of Science has established strong partnerships with academic, media, and corporate institutions in the area. There is strong overlap in particular with MIT's Media Laboratory and different science-related organizations at Harvard. Perusing through the Museum of Science's website (http://www.mos.org) reveals many intersections between these institutions, as many exhibitions have been created by local university faculty, students, or alumni. Exhibitions include IMAX movies, notices for local teachers, the life and theories of Einstein, Costumes of the Lord of the Rings, the History of Computing, and many others. The virtual component of this museum is notable, particularly because it is built as a complete analogue of the museum itself. Of course, this is not surprising, but because the museum is so strongly involved in so many efforts, the Web site at first appears as a large list without any uniform integration.

More specifically, as Figure 1 demonstrates, the Museum of Science site is set up with links to top attractions, a selection of a top Science news story (in the news section), calls to get involved, and administrative information. Navigating through the site, it is easy

Figure 1. A snapshot of the Boston Museum of Science's main Web page (http://www.mos.org)

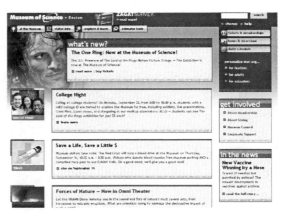

to be awestruck by the amount of incredibly rich content the Museum maintains, but completely disoriented by the journey. The Virtual Fishtank, created with the cooperation of the MIT Media Laboratory and a local educational technology firm, is an example of the powerful online exhibitions featured by this virtual science center. Originating from new ideas in the 1980s about the study of complex systems, where patterns originate from the synthesis of a set of simple parts, the virtual fishtank was created to provide virtual visitors to the Museum of Science with the ability to create their own simple element of a powerful exhibit.

> The virtual fishtank also builds on recent research in learning and education. Constructionist theories of learning are based on the idea that people learn with particular effectiveness when they are engaged in design and construction activities. . .In line with recent research, we believe that these design activities will offer rich opportunities for learning scientific concepts. (www.mos.org/exhibits/online_exhibits.html)

Thus, each visitor is given the ability to create their own fish and give it a simple condition-action rule (e.g., if you sense food nearby, follow it), and join existing schools. The ultimate effect is of viewing an exhibit of a set of these fish and their interrelationships over time.

Such rich concepts as the virtual fishtank are complemented by incredible exhibits that have a Web-life such as, "Ancient Egypt: Mummification." This exhibit allows visitors to view a 2,500 year old mummy and manipulate its 3-D model. It allows the visitor to learn about the process of excavating, the mythologies and gods associated with the practice, and prepares one for the journey to the afterlife.

What becomes clear when experiencing the Museum of Science site is a set of powerful ideas and experiences that are placed within an ad hoc information architecture. There are common imageries, appropriate age categories, conceptual ideas, and exhibit statuses around which more relational architectures can be at least experimented. Instead, the experience of interacting with the site involves wonderful discoveries and interactive exhibits, but that are only serendipitously accessed. Whether a specific exhibit is in "what's happening," a subcategory under "exhibits," or as a table under the main page is unclear. Another example of this can be found with the divisions between the Current Science and Technology section, the Virtual Exhibits section, and the lectures section. Within the virtual exhibits section, one exhibit featured is the "archaeology of the big dig," based around Boston's massive urban infrastructure project. While this is a very appropriate topic to base a virtual exhibit around, the site would be more powerful if the exhibit were integrated with content that would also be appropriate within the lectures and current technology sections. Information related to the current technology used in the big dig projects as well as lectures by construction engineers and architects could be integrated into the exhibit, and also placed within the other sub-sections. Essentially, the approach of providing the site visitor with greater associations between different pieces of content could add a previously unseen level of fluidity.

Across these discussions, it is clear that there are limited paths by which the rich content can be accessed, and the discovery of these paths sometimes is quite difficult. How can the Museum of Science begin to illuminate the intersecting images, ideas, actors, and stories of the content it places on its web site? This is a critical question that research into virtual science centers is still yet to answer.

Visitor-Driven Ontologies: Integrating a Resevoir of Experiences and Knowledge

The rich reservoir of content within the Museum of Science Web site is in need of an architecture as flexible as its visitors are varied. In this section, a possible solution is proposed to the problem of designing for the myriad of mechanisms and architectures by which knowledge within the science center's Web site should be organized.

The architecture of community-designed ontology can serve as an effective structure to describe an interwoven set of knowledge objects. I invoke my previous research, Village Voice, for this purpose. Designed for a community of distributed Somali refugees (of a civil war) in the Boston area, Village Voice is a Web portal designed to integrate the community through a digital medium that could provide voice for common issues the community faced collectively (Srinivasan, 2002).

Village Voice is built upon the premise that storytelling is fundamental to the sharing of experience. In cultures throughout the world, story exists to serve a range of purposes from teaching a moral, contemplating divinity, or preserving history. Stories are clearly one of the many ways in which we, as humans, present who we are to others (McAdams, 1993).

The Somali refugee community in the Boston area is concentrated amongst a few pockets in Jamaica Plain, Roxbury, Revere, and Charlestown. The population of this group has expanded over the last five years, from about 3,000 to 5,000 (Srinivasan, 2002). Refugees span a variety of ages, however, because of the mercurial nature of some of the programs that brought Somalis to Boston, a number of families have been broken up in the process. Refugees today are victims of a civil war that has torn apart these families and decimated a once thriving culture. This community has dramatically expanded over the last five years due to the civil war in Somalia, but remains fragmented with little identification of the common issues and experiences that could unite it.

Additionally, according to many community members, there is a desire to archive their experiences as they face new challenges in the United States. They wish to find a means to tell stories to their community, as well as to incoming refugees and others outside of the community. Traditionally, story has been orally transmitted in Somali culture, so the use of a medium that records and retells story is new to them.

Thus, there is a reservoir of knowledge and experience across community members being blocked from communication and in danger of being forgotten. An information system could potentially create a safe, communicative archive that would assist the community in identifying the common specific stories related to common contemporary issues while providing a window into their diaspora authored and designed by the community itself.

With this understanding, I introduced myself to the community as a graduate student interested in using video to document the experiences of people in the community with the purpose of creating a community artifact that could serve as a growing archive of their shared issues, challenges, and experiences as recent immigrants to Boston. Story creators were identified cutting across gaps of generation and gender, and a number of simple video stories using basic documentary format were created.

These stories were employed to stimulate the design of a representation, or ontology, to illustrate the intersecting issues of the community. The goal was to engage the community in the reflective process of creating an ontology that could articulate the relationships between relevant community issues. As issues in the community would change, the community could redesign this representation through future ontology design meetings. This representation would be more than a static structure with which to represent community knowledge: when continuously populated with their stories, ontology becomes a dynamic structure that is used by members to model the evolution of their community. Thus, ontology becomes a mechanism by which the stories and artifacts of a community can be represented and exhibited in the landscape of a multimedia system. In short, it could provide a powerful architecture for the distributed and wide-ranging content that characterized the Somali refugee diaspora within the Boston area.

Artificial Intelligence pioneers and experts on knowledge representation have focused on architectures to represent abstract knowledge in the form of stories (Schank, 1990), common sense (Minsky, 1986), and formal relationships of temporality, space, and hierarchy (Guha & Lenat, 1990). More specifically to the domain of learning, culture, and exhibition, there are a number of techniques of dynamic ontology modeling focused on the representation of pieces of content within a system of cultural heritage. For example, The Dublin Core metadata project (www.dublincore.org/projects) has attempted on a large scale to create a new interoperable set of standards to tie distributed content together by worldwide creators. The successes of this group are truly admirable, and serve as a first solution in the quest to create cross-reference-able, deeper repositories of knowledge. Another interesting example is a prototype system developed by Hunter and Newmarch (1999) to extend metadata models on top of the Dublin Core successes to audiovisual content present within the State Library of Queensland, Australia. Thus, the influential models which build on top of the descriptive RDF (resource description framework) and structural XML include:

1. The CIDOC metadata model is focused on the integration and interchange of media within the diverse setting of objects of cultural heritage. As opposed to broader standards in existence (RDF, XML), its ontological approach involves a restriction to the underlying semantics of database schemata and document structures within objects of cultural heritage.

2. Funded by the Harmony project, the ABC Model was developed very much to provide interoperability between different metadata ontologies, and enable communities to begin to develop their own descriptive ontologies. It has become the

emerging standard to resolve some of the issues raised in the RDF/XML discussion, and uses formal logical categories such as situation, temporality, and object oriented relationships (Lagoze & Hunter, 2001).

Village Voice has taken a much less formal mode in its approach toward the content of the Somali community. Specifically, it builds on the findings of Concept Mapping pioneered by Novak and Canas (Novak, 1998). Concept Maps are learner-created knowledge models where students dissect a topic into its constituent conceptual pieces. An example of such a map is shown in Figure 2.

In a similar spirit, over several sessions Somali community members met to view the submitted video pieces and develop a concept map of their own. This involved a great deal of discussion and an understanding of the community topics that would emerge in a video and their relationships to each other. At the end of several initial meetings involving close to one hundred community members, an initial ontology was created, as shown in Figure 3.

The iterative ontology drives the Village Voice media system's architecture by enabling community members to browse through different video submissions based on their relationship to the topic/s queried. This interface utilized a dynamic collage to show the relationships between different pieces of content to one another or selected ontology nodes.

Evaluations of this architecture have demonstrated greater system usage, submission of content, and general engagement with other community members and issues within the architecture defined by the community itself. This was compared to standard indexing techniques of keyword as a control variable and finally to the ad hoc information design so prevalent in cultural, and educational Web systems (Srinivasan, 2002). There is much more to the Village Voice story, but the key finding was that the diversity of content and knowledge within a cultural Web site can be better integrated when the authors themselves are given power over the relationships and representations within the

Figure 2. An example of a concept map

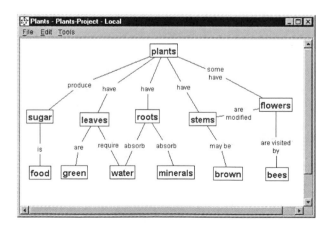

Figure 3. The collective Somali Community Ontology

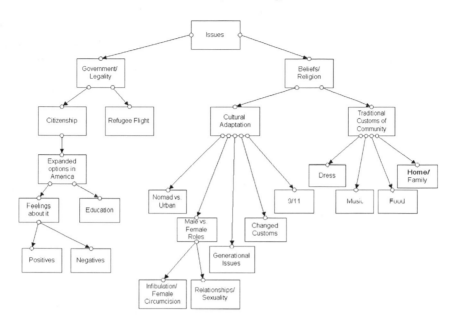

system. The result is a Web artifact that creates a truer representation of its knowledge network – the narratives of a community.

Applying the ideas of Village Voice to virtual science centers reveals the potential to create multiple concept map-type ontologies to describe different mechanisms of experiencing knowledge within the site. For example, one concept map could be just

Figure 4. A snapshot of Village Voice's interface – Each thumbnail is a different video story within the system. The relative brightness of different pieces indicates the level of correspondence between the story and the ontology node/s selected.

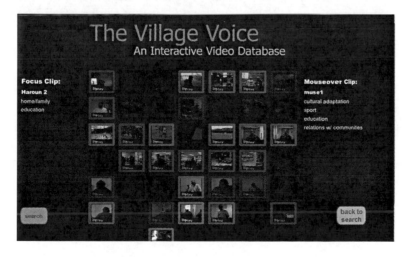

focused on major historical figures or science gurus featured across the exhibitions of the museums. Another mapping could be temporally organized, as the discoveries across exhibits could be experienced based on their epochs rather than the traditional thematic subject. Virtual science centers and their exhibits maintain a number of interrelations, some conceptual, some temporal, some related to a nation or part of the world, some related to a certain type of scientific approach, and much more. What is advocated from the discussion of Village Voice is the creation of multiple ontologies by the site architect by which the virtual science center content may be organized. This can be translated into a wider "web" of paths, trails, and navigational experiences for the site visitor. Ultimately, I believe, this approach can help generate a deeper learning experience.

Curators and designers of virtual science centers can begin to solicit multiple ontologies and tether the content of the site to these different architectures. The result could be a wonderfully diverse set of paths through which the exhilarating lessons of the science center could be provided.

Tribal Peace: Agents as Mediation for System Visitors

My current research involves extending the ideas of Village Voice to try to enable the virtual environment to begin to provide paths through the repository of knowledge that are particularly meaningful to the specific visitor. The idea of creating a set of ontology concept maps still involves a discrete set of knowledge architectures, metadata structures that engender separate but ultimately a limited set of experiences of the virtual science center. However, when the Web environment can make recommendations that integrate and compare the interests of all Web visitors then the possibility is raised to create paths of exploring the site that have not been pre-defined.

This idea has informed the development of Tribal Peace, the centerpiece of my ongoing research. Tribal Peace is a Web portal designed to integrate 19 Native American reservations distributed across a fairly large region within Southern California. Lacking interconnective infrastructure, these reservations have suffered from a lack of knowledge exchange across the critical educational, social, and political issues that tie them together (Shipek, 1986). Moving from a once centralized set of nations (Kumeyaay, Luiseno and Cupeno) to decentralized reservations without vibrant knowledge networks has created a situation in which educational and social problems have risen to the forefront.

Receiving a Hewlett-Packard Digital Village grant has now provided many reservations with basic Internet capabilities and computers, but the presence of an infrastructure is only one small step toward goals of exchange of knowledge and deeper dialogue and discourse. What is missing is a space that can enable the submissions of different reservation members to be shared, reflected upon, and framed in terms of collective priorities.

The situation described here is quite similar to the Somali setting of the previous section with the difference that there are deeper challenges of greater physical distance and historical fracture within the Native American communities (Shipek, 1986). Furthermore, there are concerns raised that a simple archive, while important, would not necessarily arm community members with critical information that they may truly be seeking. This question certainly intensifies within a situation where a database would populate with greater and greater amounts of content.

Within this context, the Tribal Peace system is in the process of creation. Serving as an archive of video, sound, and image-based content, the system enables community members to upload and browse content according the community-driven ontology model described in Village Voice. However, this system is also continuing to extend these ideas to focus on how to more deeply "guide" a community member within the information landscape of the system.

With this goal in mind, Tribal Peace is driven by the novel architecture of proactive intelligent agents normally employed in economically focused contexts. We define agents as technological entities that attempt to accomplish a task for a user based on an understanding of the user, the content of the task, and the environment in which it is situated (other agents, the information being expressed, etc.). Through an interaction with a respected elder hosting the Web system, reservation members will be guided through the Tribal Peace system as it makes recommendations on different media pieces to view and community topics to investigate. The algorithms empowering this engine will be based on the literature of social information filtering, which makes recommendations based on a set of evolving user profiles. These ideas have been explored in commercial setting (such as Amazon.com) and within a number of academic papers (Shardanand &

Figure 5 Tribal Peace's Manzanita tree inspired interface, with different selected community pieces retrieved based on a selected topic. Site visitors can also have pieces selected for them on the tree without querying a particular topic.

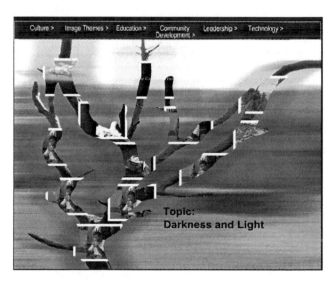

Maes, 1995). By recommending topics or media pieces the reservation member may not have been exposed but clearly would be interested in, Tribal Peace is designed to navigate a complex landscape of media through statistical algorithms that can make intelligent decisions related to otherwise deeply complex social, cultural, or political material. The potential is held to provide an ever-changing path through the knowledge of the Web environment to the visitor based on how the visitor's interests compare to other science center visitors. Today, the system is being created and used by reservation members throughout the San Diego County, and holds great promise as a new artifact of dialogue, archive, and communication across a previously dispersed set of reservations.

I believe that virtual science centers could benefit from the use of social information filtering agents. The virtual science center site could monitor all the exhibits, lectures, and pieces of information the visitor has accessed with various levels of interest. The level of interest a visitor has in a certain idea, individual, or time period contained within the site can be ascertained by such simple data as the amount of time online, number of times accessed, whether any comments were left, and etcetera. These pieces of data could be used to begin to allow the site to evolve into a more active educator of the site visitor, recommending new or old exhibits, lectures, ideas, time periods, or other Web sites to investigate. Essentially, I believe, agents can begin to serve as the teacher or virtual docent of the virtual science center, actively enabling the experience of exposing the site visitor to new information in which he or she is likely to be interested.

Providing paths that are unpredictable and previously not accessed yet meaningful to the Web visitor makes the experience of visiting the virtual science center improvisational and spontaneous. Different elements of knowledge can be suggested to the visitor that are individuated and ever-changing, thus adding greater flexibility than the multiple pre-created profiles featured within Village Voice's concept mapping metadata architecture. The potential is there for the Web visitor to come to the virtual science center without any idea of what he or she will be presented with, but reassured that the virtual science center would make appropriate recommendations.

Conclusion and Future Directions

The state-of-the-art virtual science centers provide visitors with an array of powerful exhibits. They provide strong interactive components that enable users to experience scientific dynamics. However, the paths, overall themes, and relationships between different exhibits and events are clouded and chaotic. There is an overload of information, exhibitions, and perspectives without a correspondingly flexible set of architectures by which these can be imbibed. In this chapter, two examples of existing Web-based environments have been presented as potential enablers of the learning experience from a repository of community knowledge.

The findings of Village Voice and Tribal Peace can begin to add to the palette of the information designer of the virtual science center. The two approaches (of social information filtering agents and concept map-driven ontologies) provide the ability to create a variety of paths that are all representative of the constituent. They serve as appropriate mechanisms of organizing and managing knowledge, by considering different narratives as "knowledge objects" (Srinivasan, 2004). Village Voice has revealed some of the potential of community or organization-created ontologies to reveal the intricacy of a landscape of informational artifacts. Tribal Peace continues to take these ideas further to bridge distance and historical divides within an architecture that has the potential to enable reservation members to access the knowledge and narratives that can satisfy their goals. The next step is for virtual science centers to begin to explore these approaches.

References

Guha, R. et al. (1990). *Building large knowledge based systems*. Reading, MA: Addison Wesley.

Hunter, J., & Newmarch, J. (1999). An indexing, browsing, search and retrieval system for audiovisual libraries. In *Proceedings of Research and Advanced Technology for Digital Libraries: Third European Conference, ECDL '99* (pp. 76-91). Heidelberg: Springer-Verlag.

Lagoze, C., & Hunter, J. (2001). The ABC ontology and model. *Journal of Digital Information, 2*(2).

McAdams, D. (1993). *The stories we live by*. New York: William Morrow & Company.

Minsky, M. (1986). *The society of mind*. New York: Simon & Schuster.

Novak, J. (1998). *The theory underlying concept maps and how to construct them*. Retrieved from *http://cmap.coginst.uwf.edu/info/*

Papert, S. (1980). *Mindstorms: Children, computers, and powerful ideas*. New York: Basic Books.

Schank, R. (1990). *Tell me a story*. New York: Charles Scribner's Sons

Shardanand, U., & Maes, P. (1995). Social information filtering: Algorithms for automating "word of mouth." In *Proceedings of CHI 1995* (pp. 210-217). Denver, CO.

Shipek, F. (1986). *Pushed into the rocks*. Lincoln, NE; London: University of Nebraska Press.

Srinivasan, R. (2002). *Village voice: Expressing narrative through community designed ontologies*. Master's thesis. Cambridge, MA: Massachusetts Institute of Technology.

Srinivasan, R. (2004). Knowledge architectures for cultural narratives. *Journal of Knowledge Management, 8*(4), 65-75.

Tan, W.H.L., Subramaniam, R., & Aggarwal, A. (2002). Virtual science centers: A new genre of learning in Web-based promotion of science education. In *Proceedings of the 36th Hawaii International Conference on System Sciences*.

Chapter XI

From Information Dissemination to Information Gathering:
Using Virtual Exhibits and Content Databases in E-Learning Centers

Joan C. Nordbotten, University of Bergen, Norway

Abstract

Concurrently, large multimedia databases, termed content databases, *are being created to store and manage digital representations of a museum's physical collections, such as scanned images, text documents and videos. These databases can also provide valuable information and data for use in the development of e-learning centers[1] for tasks ranging from presentation of information about the museum's educational resources to providing a full interactive learning experience for students and casual information seekers. There are at least three ways an e-learning center can support learning: first, by providing information on a given set of topics, second, by providing educational activities to reinforce learning, and third, by supporting information gathering. In this chapter, we will present and discuss how different e-learning center architectures support these different forms of learning.*

Background Assumptions
and Definitions

Figure 1 shows the three basic components of an e-learning center, the user/learner, the learning site and the (optional) content database.

The users of a museum e-learning center include anyone who 'discovers' the Web site. The users most commonly thought of during design of an e-learning center include teachers – who are expected to guide their students in their use of the site – and students – who are expected to use the site as a learning tool. However, since the site will also be available to the general public, it is important to consider the needs of interested information seekers. We can assume that these users are interested in the site topic (most likely they found the site through a search engine), but will not have other guidance in the use of the site than that which is given on the site. In order to assure effectiveness, it is important that the e-learning site is designed for these multiple user types. In the following, both the student and interested information seeker will be termed a *learner*.

A database can be generally defined as "*a logically coherent collection of related data, representing some aspect of the real world, that is designed, built, and populated for some purpose*" (Nordbotten, 2000-2004). In a museum[2] context, a *content database* contains digital representations — scanned images, text documents, videos, and etcetera — of physical objects in the museum's collections. In the following, we assume that a content database contains such representations and that the e-learning site can utilize them.

An *e-learning site* consists of a set of Web pages that present material intended to educate its users, the learners. A site's architecture can be classified according to the degree of user control over the material that is presented. A basic *virtual exhibit* presents topic material as a hyper-linked story that the viewer can navigate, much as one finds in traditional museum exhibits. An *interactive* site also presents its topic as a hyper-linked story but adds a variety of interactive activities aimed at increasing the learning effect for the learner. An interactive e-learning site is based on the idea of a "hands-on" physical exhibit, in which experiments can be run and questions answered. Typical for both of these architectures is that they are museum controlled in the sense that the user is presented with prepared material that is to be explored within a pre-defined site.

Increasingly, Internet users, both students and casual information seekers, search for information about a specific topic of individual interest. The topics of these searches vary widely and many, though well within the interest areas of the museum community, may not be well covered by existing virtual exhibits. Alternatively, if the topic is popular

Figure 1. E-learning center components

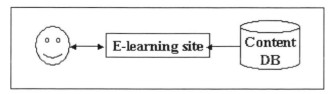

and/or part of a common interest area, it is likely that relevant information can be found at multiple sources. For example, someone interested in *space exploration* can find relevant information from numerous sources[3] including science museums, universities, and government research centers. A challenge for the museum community would be to provide multiple-site responses to a user request for information, that is, to support construction of a virtual "exhibit" on-demand (Nordbotten, 2001).

Current pedagogical practice emphasizes activating the learner and learning by doing, thus encouraging active exploration of available information and gathering of information to address his/her own interest area. We consider three learning models in order of increasing user initiative, which can be supported by different e-learning center architectures:

1. **Information browsing**, in which the learner explores material provided by a museum through a pre-defined virtual exhibit.

2. **Problem solving**, in which the learner interactively works through pre-defined, interactive activities.

3. **Information gathering**, in which the user searches for information on a self-defined topic, without concern for the source of the response set.

In the following, we will discuss e-learning site architectures and the information technology tools needed to support each of the learning models listed above. Each section will present the primary audience and anticipated use, discuss site structures through the presentation of examples, and finally present the strengths and the principal problems of the architecture type as a learning environment.

Note that it is beyond the scope of this chapter to give detailed information on the implementation tools mentioned. This information can be found in numerous books on Web site implementation, such as those by Lane and Williams (2002), Lowe and Prince, (2003) and Murray and Everett-Church (2003).

Information Dissemination: The Basic Virtual Exhibit

The basic virtual exhibit is one of the most common site architectures on the Web today and the one most frequently used for e-learning centers. The exhibit presents a topic as a hyperlinked story, patterned on that of a physical museum exhibit or textbook. As an e-learning site, the learner can have more information available and greater choice in its presentation sequence than that which is possible in a physical exhibit.

When using a virtual exhibit as an e-learning environment, the learner is expected to be a student or an interested information gatherer. He/she is assumed to be looking for information on the topic of the exhibit and willing to view, explore and read the material presented. The learner browses through the exhibit by following established links.

Ideally, both a navigation bar and site map are included so that the learner can change paths to explore alternative sections of the exhibit.

The basic virtual exhibit can be an excellent method for information dissemination, since the museum/center determines the material presented and guarantees its correctness and educational quality. A potential problem is that the exhibit is *fixed* in the sense that the material and the links used to determine the presentation sequence are pre-defined, thus reducing the possibility for the user to follow his/her own interests.

Architecture and Implementation

The typical exhibit architecture consists of a core *home* page with information about the exhibit and links to separate sub-topic sections. Topic units can be developed independently and then attached to the core forming a *starburst* or tree structure. The embedded links enable the choice of topic to be viewed, selection of additional in-depth explanatory material, and/or connection to "outside" (of the parent museum) information sources. Since the "lost in space" phenomenon is a frequent problem for viewers of even small sites, a navigation bar and/or site map should be included so that the viewer can re-orient him/herself (Conklin, 1987; Shneiderman, 1998).

The design of the basic virtual exhibit is analogous to the design of a traditional physical exhibit and is not unlike the structure of a textbook. Exhibit objects are selected, scanned, commented using textual and/or audio-visual material, and finally organized using links for presentation. For large exhibits, material can/should be stored in a content database to facilitate site maintenance. Smaller exhibits may not make use of the museum's content database. Rather, the site will be built using scanned objects that have been stored on separate file areas and given URLs that can be directly referenced using the html <img> tag. The current primary implementation language is HTML, frequently enhanced with FLASH™ animation, though other implementation tools may also be used.

There are two main architectures for virtual exhibits:

1. Single-tiered exhibits in which objects and information items are presented completely before continuing with the exhibit presentation.

2. Multi-tiered exhibits that include links within the main story that allow the viewer to select more detail about individual objects from an underlying information layer, possibly retrieved from the museum's content database.

The single-tiered exhibit *consists of a, possibly multi-threaded, story in which most links select the next part of the story. Depending on the topic and available object media, single-tiered exhibits generally use one of two metaphors:*

- The traditional *museum exhibit*, which relies on presentation of visual/ image material with accompanying text, or
- The *textbook*, which presents textual material with some illustrations.

Figure 2. Museum exhibit metaphor (from The Wright Experience™ ©2004)

THE FLYER IN THE AIR!

In the Air!
In November and December,
2003, The Wright Experience
made several test flights in the
1902 Glider and 1903 Flyer in
preparation for the Centennial of
Flight on December 17. See
video and pictures and read
descriptions of each flight!

An example of use of the traditional museum exhibit metaphor can be seen in *The Wright Experience — Preparing for Kitty Hawk* at http://www.wrightexperience.com/. This exhibit first presents the user with brief introductions to a series of videos documenting various activities from the planning, construction and execution of the centennial celebration of the first flight by the Wright brothers. Figure 2 shows a screen shot from the exhibit in which a photo is accompanied by a short text that gives the context and introduces the video from which the image was taken. In this example, each video clip contains the compete presentation of its sub-topic.

The San Francisco Exploratorium, at www.exploratorium.org/, has used a *textbook metaphor* for their presentation of a set of stories and myths about solar eclipses in an exhibit titled: *Solar Eclipse* at http://www.exploratorium.org/eclipse/index.html. Figure 3 gives a screen shot of part of a typical page from this exhibit. In the exhibit, each story is presented as a single-tiered exhibit on a series of linked pages.

The single-tiered exhibit architecture provides an excellent environment for information dissemination. For e-learning sites, this helps assure pedagogical and quality control of the story/message being presented and supports focusing of the presentation to a particular age and/or interest group. General familiarity with the presentation metaphor makes these sites "easy" to design and use. An added advantage is ease of implementation and deployment since well-known tools and technology can be used.

A common problem/mistake made in the implementation of the virtual exhibit is the inclusion of too much explanatory text for each visual object, such as one would expect to find in a traditional textbook. The Web is a very visual medium and studies of casual viewers indicate that they spend less than 30 seconds on a page (Nordbotten, 2000;

Figure 3. Single-tier: Textbook metaphor (reproduced with permission (c) Exploratorium, www.exploratorium.edu)

Nordbotten & Nordbotten, 2002; Shneiderman, 1989; Yamada, 1995). This is insufficient time to read lengthy text sections.

Multi-tiered exhibits aim to reduce the information overload that can be triggered by text quantity in the primary exhibit, by dividing the exhibit into multiple tiers, a primary tier for an image-based principle narrative with underlying layer(s) of detail that the interested/curious user can select from various points in the main exhibit. This support material can be accessed using different strategies, including through the use of:

- Active zones in the display that are used as links to pages with object enlargement or alternative media presentations.
- Zooming or a "magnifying glass" to highlight detail and/or provide more information.

The Natural History (Natuurhistorisch) Museum in Maastricht has used the multi-tiered approach in their *Virtual tour,* at http://www.nhmmaastricht.nl/engels/index.html, of the museum. The tour consists of a series of images that give the illusion of entering a room or passing into a further section of the current room. Wandering through the exhibit is accomplished by clicking on the highlighted rectangle, shown in the screen shot given in Figure 4a. In this tour, animated object names are used as links to a detailed presentation, as illustrated in *Figure 4b.* The detailed page appears "on top" of the primary page thus supporting orientation within the site.

Supportive detail can also be retrieved using *"magnifying glass"* technology, where a window is passed over an image and an enlarged version of the window is shown

Figure 4. A multi-tiered presentation of the Mosasaur dinosaur (reproduced with permission© Natural History Museum at Maastricht)

Figure 4a. Hot spots for detail and navigation

Figure 4b. Detailed presentation selected from Figure 4a

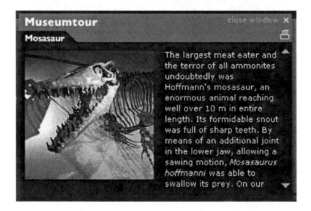

elsewhere on the user's screen. This technique is especially effective when the original image is content rich, as is the case in the screen shot from the *Ology/Marine Biology* exhibit at http://www.ology.amnh.org/marinebiology/workthesystem/mangrove.html, shown in Figure 5. In this case the primary image contains many creatures that are barely visible. The magnified version also gives the common and scientific name for the central object within the magnifying glass.

As learning tools, multi-tiered virtual exhibits transfer some control to the user so that he/she can choose to select added detail for those objects that are most interesting for him/her, thus helping to tailor the information presented to the information requirement of the viewer/learner. Multi-tiered exhibits support inclusion of a wealth of information that can support interested information seekers, while avoiding some of the information

Figure 5. Spiny lobster under a Mangrove tree (from the American Natural History Museum©)

overload problems for the learner. This is a useful strategy for addressing multiple audiences from school to university students to interested, casual information seekers.

The main problems of multi-tiered exhibits include the temptation to make increasing use of multi-media, such as Flash animations and "real" video, which require high bandwidth for seamless presentation systems. Though presentation software is often free, they still may not be available to all users. A growing problem is that as the quantity of secondary support material grows, users can get lost within the site, in the sense that they lose the flow of the overall story.

Both single and multi-tiered virtual exhibits can provide very good support for the learner and information gatherer needing an introduction and/or overview to a topic area as well as supporting those who wish more depth of information. Content and quality of the information provided are assured and maintained by the information provider. Increasingly, these sites are well advertised and easy to locate. And finally, the technology required to design and launch interesting and attractive exhibits, as well as to download them, is readily available to both the provider and information user.

As learning centers, virtual exhibits function well as a source of information but frequently do not include user activity aimed at reinforcing knowledge about the exhibit topic. Though suggestions for associated user activity may be included in a teacher area of the site, these are often off-line activities and may be difficult to find for the learner, particularly the casual information seeker.

At a certain level, fixed exhibits are a passive information source, well suited to information dissemination, but where user interaction is limited to selection of predetermined topic presentations. There is little support for reinforcement of the information content, other than to repeat a sub-topic presentation, often in its entirety.

A technical problem is that the information is commonly kept on a growing number of Web pages/presentations that may overlap and will become difficult to manage — update, extend and reuse.

Interactive Exhibits for Active Learning

The current leading pedagogical paradigm is "learning-by-doing" in which the student builds knowledge and expertise through active participation in development projects. Techniques to support learner activities abound and range from paper-and-pencil workbooks, through hands-on experimentation, to Web games and simulations.

An *interactive exhibit* is one that can adapt its presentations to user input and thereby support active learning through such interactive activities as *games* or *experiments* that are based on the thematic material of the parent exhibit. Interactive exhibits are similar in concept to the "hands-on" exhibits found in museums.

Architecture and Implementation

Dynamic Web programming tools[4] provide the implementation environment for construction of dynamic exhibits. *Figure 6* gives an overview of the system components necessary to support these exhibits. When the user activates an interactive exhibit, his/her browser sends a request and receives the exhibit through a Web server from the application server. The browser presents the exhibit with its audio-visual material and interactive scripts. During use, the client-side application controls the user interaction and for some exhibits the whole presentation, while the host machine's application server controls such server-side activities as access to a content database.

Exhibit scripts can be developed in Flash™, Java or some other programming language. Communication with the user is through his/her browser. When a client-side application is used, it is executed on the client machine, thus saving interaction time between the client and host machines at the cost of the initial load time. However, when material stored in a content database is required, a request must be sent from the client system to the host system's application server, which converts the request into one or more DB queries to the available databases, marked C_DB in Figure 6.

Figure 6. Interactive exhibit support

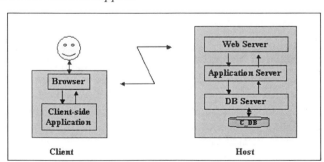

Figure 7. Quiz in a game setting (example from the Mint Museum©)

Games

Perhaps the easiest learning tool to add to an e-learning site is inclusion of a quiz on the material presented in the virtual exhibit. An interesting example can be found on the Ology site at http://www.ology.amnh.org/index.html of the American Museum of Natural History. Here both facts and quiz questions are placed on cards that can be "collected" and saved on the Museum site, that is, in a learner DB for future reference.

With some added effort from the developer, the quiz can be presented in the form of a game, encouraging learners of all ages to use the information acquired from the exhibit so that it can become useful knowledge. Typically the game is formed as a set of questions about topics that have been presented in the virtual exhibit. The system evaluates the answer(s) and forms a response commonly consisting of point assignment, a reinforcement answer, and a new question. This cycle of questions and answers repeats until the game completes. If questions are selected randomly, the game can be replayed with new questions repeatedly. The success of a game, as a learning tool, lies in its ability to engage the learner(s).

A good example (for all age groups) "teaches" the discovery and use of rubber for making balls through a Mesoamerican ball game, at http://www.ballgame.org/[5]. The Mesoamericans were the first to use rubber to make balls and to use a team ballgame for conflict resolution. This game uses animation and sound effects, as well as period costumes and background. The ball is first "batted" between the teams in a manner similar to volley ball. A question appears with three answer options, two of which are shown in Figure 7. Answering correctly gives the users' team a point. Sound effects clearly signal the quality of the answer, which is enhanced with a textual explanation. (To play, go to: http://www.ballgame.org/sub_section.asp?section=3&sub_section=2). All of the material in the above ballgame has been packaged in Flash™ and sent to the client as a client-side application. The trade-off is between download time (long) and a high frequency of interaction between the client and server machines for the question and answer sets.

*Figure 8. Particle acceleration — Simulation game (Reproduced with permission ©
CERN)*

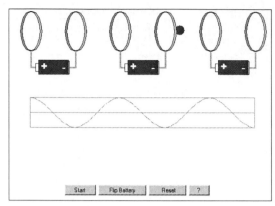

Virtual Experiments

Games can also be used to frame simulations that allow the user to explore some physical phenomena that might otherwise be inaccessible. The idea is to give the user an understanding through experimentation. An example is the "particle accelerator" at http://public.web.cern.ch/public/Content/Chapters/Education/OnlineResources/ LHCGame/LHCGame.html, developed by the Education Department of the European Organization for Nuclear Research, CERN. The goal of this simulation is to get the particle to accelerate by passing it through magnetic fields that are "charged" by flipping the batteries. Figure 8 shows a screen shot[6] as the positively charged particle is entering the second negatively charged loop from the right.

A simulation game can be packaged as a client-side application, as the one presented, or it can utilize data that has been stored in the museum databases. Data generated from the simulation can be saved onto the user's machine for local use. In more extensive simulations, both the status for the simulation and data generated "to-date" can be stored on the museum site. This would support long term and/or multiple player games or experiments.

Virtual Projects

Virtual projects are similar in concept to the physical "hands-on" activity that allows the museum visitor to develop/build something that can be taken home. In the virtual environment, the user is encouraged to solve a problem by using/collecting objects from the museum's content database(s). Frequently, support is provided for adding new objects, which could include notes taken and/or experiences, created during the project. The resulting material can then be stored on the museum site for future use and discussion, for example, in the classroom or with online friends or colleagues.

Interactive exhibits, when used as part of e-learning sites, can enhance learning by encouraging the learner and interested information seeker to become engaged in interactive activities that have been selected by his/her own interest, thereby reinforcing the information acquired from the exhibit. In providing interactive activities, the museum /science center can guarantee the correctness of the material presented, as well as its relevance to the theme of the parent exhibit. The technology and expertise for development of interactive exhibits are available and ability to access and use this type of site is increasing, in part through the introduction of IT into the school systems.

A problem with interactive exhibits is that it can be difficult to adjust the system questions, activities, and responses to the learner's level of expertise, potentially making the interaction too difficult or too trivial. To alleviate this, professional educators should be involved in the design of these sites.

While both the basic and interactive virtual exhibits can provide good support for learning, they are still fixed sites, in the sense that the learner is "bound" to the material provided in a single exhibit and to the sequences that have been implemented for topic presentation. The learner is invited to explore and use the site, but is only infrequently given the possibility to search for particular information on the site or in the museum's content databases. This limits the possibility for the learner to gather or focus on information about a topic of personal interest (or an assignment given by a teacher).

Information Gathering: Objective "On-Demand" Exhibits

Increasingly, students and general Internet users search for information about current and/or personal interests by using one or more of the Internet search engines, much as one would use an encyclopedia. Often, relevant information exists in one or several museum exhibits and/or content databases. The problem is to find and retrieve it.

To locate information on the Internet, the information seeker must:

1. Choose one or more search engines.
2. Formulate and submit a query statement of his/her information need.
3. Select potentially relevant sites from a large result list.
4. Evaluate selected sites for "real" relevance.
5. Review and extract (copy) information from the different sources.
6. Compile and integrate the information.

This can be a daunting task unless a real need to learn exists. Happily, a number of consortiums and research programs have been formed to find ways to ease information retrieval from the multitude of Internet sites.

Preparing a virtual exhibit for location and retrieval by the Internet search engines can be accomplished by including html <meta> tags in the site header that specify both the thematic content description and a set of search keywords. These will be found by the Internet search engines and used for indexing and retrieval. The advantage is that the learner/information gatherer will then be more likely to locate museum exhibit material relevant to his/her information need.

Internet search engines use "crawlers" that retrieve text from html pages to index a site. Currently, crawlers are not capable of extracting text or features from Flash™ or visual/audio exhibits for indexing. In addition, the crawlers cannot access the indexes of content databases. The result is that search engines are unable to locate video-based sites or information from content databases. The current work-a-round is to construct an html "envelope" containing a good description of the multimedia presentation to be used for indexing the site. In addition, a specific search facility for the database must be provided.

The above strategy only addresses the first two steps on the above search list. Accessing multiple content databases through their home pages and specific database search techniques still puts a high burden on the information gatherer in the form of multiple site and DB accesses with follow-up manual/local information consolidation. It must be a goal for the information provider community, in this case the museum community, to provide seamless interfaces to their multiple information resources.

Web Databases: Architecture and Implementation

A Web database is simply one that is accessible from the Internet. In a museum context, the content database must be assumed to contain multiple media objects, ranging from text documents through scanned images of physical objects to full multimedia presentations. Ideally, one would want to support user access to the media objects through a description of the type of information required without concern for the media type (text, scanned image, streamed media) that was used to represent the information.

Figure 9 presents an extension of the architectural environment given in Figure 6 that includes the components needed for multiple multimedia database access. The figure can be explained by following a request for information about a particular topic, for example

Figure 9. Accessing Web databases

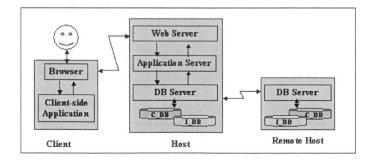

North Atlantic Whales[7]. The request is initiated when a user submits a query containing the three keywords, in the same way as using a Google™ search. The user's browser sends the request to the Web server for processing by the application server, which translates the input into one or more DB queries that are sent to relevant DB servers. The DB servers search for and retrieve relevant data, for example, multiple texts and videos of (some of) the 18 species of North Atlantic whales from each DB. Each DB server returns relevant content data to the application server, which packages it for return by the Web server and display by the browser to the user. If the information includes multimedia (Flash, audio, video…) presentations, appropriate media player(s) must be retrieved as part of the client-side application for presentation of the media data.

The DB server uses information from the DB indexes, labeled I_DB in the figure, to locate the actual data within the content database, labeled C_DB. If the search is to be preformed on multiple databases, then the indexes for all databases within the scope of the query need to be searched in order to locate those content databases that contain data relevant to the user query.

The data within an index database is derived from two primary sources:

- The metadata used to describe each object upon entry into the database and
- The terms and features in the object itself.

Table 1 gives an example of some of the index data that could be assigned to the story about Eclipses of the sun, illustrated in Figure 3. Note that the metadata elements used in this example have been selected from the 15 core elements in the Dublin Core Standard (Hillmann, 2003).

Object features are selected from the object itself. For text objects, the feature set consists of the most descriptive terms contained in the document. For visual objects — images, video — automatically extracted features are currently limited to color and texture distribution and possibly the primary shapes in the image. Semantic description of image objects is still a predominantly manual task (Hove, 2004).

There are at least three problems in using the above index to search for multimedia data:

1. The search terms given by the user in his/her request for information may not match exactly the index terms used to describe the DB content.
2. The terms in the controlled vocabulary may not match exactly the terms used by the creator of the text objects or the descriptive text for image objects.
3. The DB descriptive terms used for one content database are unlikely to match exactly those used for other content databases covering the same topic.

Therefore, in order to locate the most relevant information, some form of semantic matching must be made between the terms that the searcher uses to characterize his/her information requirement and the terms used by the cataloger to index the items in the database as well as the terms used between the cataloger and the creator of the media

Table 1. Example of index data for a specific text object

DC Metadata element	Value	Comment
Title	The Sun-Eating Dragon and other ways to think about an eclipse	
Creator	Noel Wanner	
Publisher	Exploratorium	
Identifier (location)	http://www.exploratorium.org/eclipse/dragon.html	
Source	Solar Eclipse http://www.exploratorium.org/eclipse/index.html	
Description	Myths and stories from Greek, Chinese and Indian cultures describing eclipses of the sun	Given in free/full text
Subject (keywords)	Eclipse, sun, dragon, myth, mythology …	From a controlled vocabulary or formal classification scheme
…		
Object features	**Location**	
Abandon	§2	These search terms may
Dragon	Title, §3, …	be weighted according to
Eclipse	Sub-title, Source, §2, 3, …	descriptive importance
Sun	Title, §1, 2, 3, …	
Sun-Eating	Title	
…		
Earth image	Lower right quadrant (LRQ) above Sun image	As seen in *Figure 2*
Sun image	LRQ, below Earth image	
…		

objects in the database. Information retrieval systems have long used thesauri as an aid for interpretation of user queries, thus addressing the 1st of the previous problems.

Current best practice proposals from research and the semantic Web community (Sowa, 2000) advise creation of a set of *domain ontologies* that all content database providers use to describe the objects in their database. This addresses the 3rd problem in the above list by standardizing the values chosen for the metadata elements. However, as seen in the example in Table 1, the metadata values may not match completely the index terms selected from the document, the 2nd problem in the above list.

Optimally, the content of the ontology must combine the information in the hierarchic structure and detail of a domain taxonomy, with the term relationships and descriptions in a thesaurus, and with the historical development of the common natural language terminology that information searchers and object creators can be expected to use. This is a tall order that has not yet been fulfilled, even though many years of research and development effort have been addressed to the synonym resolution problem in the context of both database integration (Elmagarmid et al., 1999) and information retrieval from both text and image databases (Baeza-Yates & Ribeiro-Neto, 1999).

Current Practice for Information Retrieval from Content Databases

A recent study of 100 museum Web sites, nominated for awards at the 2003 Museum and Web conference (http://www.archimuse.com/mw2003/) found that only 37 had included any form for database search facility on their sites (Peacock, 2004). Of these, 20 sites supported only keyword searches, while 17 provided some supplementary form for search support. The search support included:

- Keyword search, possibly with keyword selection from a given list of terms
- Browsing/selection from an index of terms or set of thumbnail images
- Selection via a category structure
- Linking to related and/or popular items.

Figure 10. A multi DB index – Hand constructed (reproduced with permission © The National Museum of Denmark)

Keyword search is known to be difficult for the information gatherer. Without a term list to choose from, the user is faced with an empty box into which search terms need to be entered. This assumes that the user has rather detailed knowledge about the database content and structure, knows which topics and themes are included in the database as well as the index terms used to describe the data. Often a labeled multiple-box form is given to the user who is then expected to be able to submit a title, creator name(s), and dates in addition to one or more descriptive keywords. Adding drop-down term selection lists can greatly ease the selection of appropriate keywords and fill-in options.

Searching an image database can be eased by the presentation of an index of thumbnail images, as shown in *Figure 10*. Here, each image is a link to an enlargement and possible completion of the thumbnail image with descriptive data about the object presented.

A problem with the *thumbnail index* is that the number of elements that can be presented (and thus indexed) are limited by the screen area. This makes it necessary to provide a *category index* for the collection. The user must first select a category for his/her area of interest and can then browse the thumbnails picking items of interest.

The example in *Figure 10* shows the index for the "mathematical chamber" of the Kings Kunstkammer, at http://www.kunstkammer.dk/MathematischeGB/gemach_ mathematischeGB.shtml, which held examples of instruments and inventions. There are actually nine chambers in the Kings Kunstkammer, at http://www.kunstkammer.dk, each with its own thumbnail index. Even so, it is a problem to display all of the 250 items in this relatively small content database. One way to expand the index scope would be to implement it as a hierarchy of related items or to link to "hidden" (not in the index) items. Items related by theme, materials, time, creator or any combinations of perspectives can be displayed with the selected item. An example can be found at the virtual Walker Art Center at http://collections.walkerart.org/item/object/7440.

Providing Access to Multiple Data Collections

Often information relevant for a user request can be located in the content databases of multiple museums. Given the widespread use of Google, the user may be unaware of (and basically uninterested in) the physical locations of the information. Creating seamless interfaces to multiple information sources is a current, highly active research and development area. There are a number of approaches:

1. Hand crafting a "traditional" exhibit of objects from multiple sources.
2. Constructing an integration system or layer "on-top" of a specific collection of participating databases, creating a federated database system.
3. Constructing a multi-database language that can access multiple independent databases, similar to the Google approach.

Hand crafted integrated databases are only realistic when the size of the data collections, in number of items to be integrated, is relatively small, for example less than 1,000. The strategy is to collect links to the items that one wants to include in the virtual, integrated collection and build an index with this information. *Figure 10* shows an example of this strategy in which the 250 digital copies of items have been collected (manually) from object descriptions from multiple museums in Denmark (Gundestrup & Wanning, 2004).

In the current implementation, each image links to an object presentation – enlarged image with historical data description – that has been stored on the exhibit site. The same index presentation technique could be used to link directly to the multiple source content databases, thereby supporting a seamless (for the viewer) interface to the current locations of the original collection. Unfortunately, this strategy for database integration, though conceptually easy, does not scale since it is limited by available manual labor.

A *federated database approach — portal integration* — requires creation of a generic description — metadata — that can describe all of the databases in the federation, as well as translations between the generic and local terminology as well as links to the participating database elements. Referring to the architecture given in Figure 9, the generic description would cover the data in each local I_DB.

It has been suggested that development and acceptance of metadata and ontological standards will allow creation of Web portals to multiple independent databases. (Sowa, 2000; Hyvönen et al., 2004). User queries for information would be directed to the portal, which would then have the linkage and translation data to direct the query on to relevant database systems. Two elements are required:

1.   An agreement on the metadata that must be supplied to describe the content of the local databases, and

2.   An agreement on the controlled vocabulary /ontology to be used for selection of metadata values.

In the museum world, there are at least three "standard" metadata structures in use: Dublin Core www.dublincore.org, Mpeg-7 http://archive.dstc.edu.au/mpeg7-ddl/, and CIDOC CRM http://cidoc.ics.forth.gr/. Each of these standards has been created/ proposed for the needs of a specific community: library, moving pictures, and cultural heritage institutions, respectively. Each defines a framework for the metadata to be collected when describing objects. The standards overlap, but are not identical. Their use creates a problem of searching metadata specified according to different standards.

Defining a general ontology, or controlled vocabulary, has proved very difficult, not least due to evolving language and cataloging cultures. To reduce the complexity and size of the task, current proposals are focused on development of domain specific ontologies.

Assuming that agreement can be achieved among a set of information providers as to the metadata framework and the ontologies to be used, a generic description or *integration schema* for the members of the federation can be constructed. Becoming a member of the federation then requires that the participating museum describe its content database using the metadata structure and ontology of the integration schema. A prototype for a portal or integrated schema for several of Finland's museum databases, MuseumFinland, is under construction, is located at http://museosuomi.cs.helsinki.fi/ and is described by Hyvönen et al. (2004). Once constructed, the generic integration schema can be used to form the base for development of a single interface for the users to provide a seamless search to multiple content databases, and thus reduce the information retrieval search task noted earlier.

The idea of constructing a *multi-database language*, similar to the Google approach, that can access multiple independent databases based on a user query, possibly with synonym definitions was first published almost 20 years ago (Litwin, 1986). The idea is that the user would express his/her information requirement and give relevant synonyms to the query search terms. The processing system would then expand the query and broadcast it to participating database systems. Since little integration work is required, the number of systems that could be accessed is much higher than in the federated approach, which is limited to the number of systems willing and able to join the federation.

A slight modification of this idea would be to give the query language processor access to a generic thesaurus that could map user terminology to the ontologies used for describing content databases. At this writing, this author is unaware of any prototypes attempting implementation of this idea.

Information Presentation

For each of the integration strategies mentioned, there is the problem of information presentation. Most prototypes and systems present query results in a list format of text and/or image links. The user is then required to link to each referenced system to determine its real relevance to his/her information need and then to collect and integrate the retrieved data.

Ideally, the information retrieved should be presented in context with the descriptive information presented as an exhibit story. For this to be possible, more research on story construction will be needed. A problem is that much of the metadata recorded today about museum objects is "administrative," in the sense that it describes the object context (creator, location, materials, etc.) rather than the semantic interpretation of the object and its relationships to other objects. If an *exhibit* presentation format is to be automatically developed, more semantic information about the objects in the content databases will have to be given. Then it will be possible to construct "real" *exhibits-on-demand*.

Summary and Research Issues

Museums have a vast store of material that can be made available on the Internet. Some is available through virtual exhibits and more is stored in content databases. Virtual exhibits are good vehicles for information dissemination, and when they contain interactive components, can be a good foundation for e-learning systems. However, if an information gatherer/learner is looking for information that is not adequately covered by existing exhibits, but is in existing content databases, it can be difficult to find.

A system to support information gathering should have the following components:

1. An interface structure similar to that found in current, familiar search engines, such as Google.com™. Query terms used to describe the information requirement should be translated into and /or supplemented by synonymous terms that have been used to describe the data. Current proposals to facilitate this translation include the construction of domain ontologies that combine the characteristics and data of taxonomies and thesauri.

2. The system should be able to search at three levels for information that match the relevance criteria, as determined from the interpretation of the user request:

   a. Through the museum's virtual exhibits.

   b. Within the museum's database(s).

   c. Through the combined material from multiple museums.

3. Relevant material should then be returned to the user/requester, either as a set of links to the information or, preferably, as a system constructed exhibit, that is, an exhibit-on-demand.

The advantages of such a system would be to give the user/learner control over selection of information about a topic of interest, expressed in his/her own terms. The system would then be able to interpret the query and utilize known information retrieval technology. However, before this type of system becomes a reality:

- Museums must either agree on the structure of the metadata and the ontologies that they will use to describe their data.

- Integration tools must be developed to mediate the differences between the set of metadata and ontologies in use.

- It will be necessary to extend the facilities of current indexing techniques to include feature extraction from multimedia databases.

- Methods for automating database integration and/or multi-database querying processors must be developed. And finally,

- Presentation techniques must be developed to provide more meaningful result presentation than the current list structures.

Acknowledgments

This work has been funded in part by the Norwegian National Research Council through project #148827/530. Sincere thanks are extended to the many students and colleagues who have participated in discussions and development of the ideas presented in this chapter while working within the research context of the Virtual Exhibit on Demand project, http://nordbotten.ifi.uib.no/VirtualMuseum/VMwebSite/VEDweb-site.htm.

References

Baeza-Yates, R., & Ribeiro-Neto, B. (1999). *Modern information retrieval.* Addison Wesley.

Conklin, J. (1987). Hypertext: An introduction and survey. *IEEE Computer, 20*(1), 17-41.

Elmagarmid, A, Rusinkiewicz, M., & Sheth, A. (Eds). (1999). *Management of heterogeneous and autonomous database systems.* San Francisco: Morgan Kaufmann.

Gundestrup, B., & Wanning, T. (2004). The King's Kunstkammer: Presenting systems of knowledge on the Web. In D. Bearman & J. Trant (Eds.), *Museums and the Web 2004: Proceedings* (pp. 79-87). Toronto: Archives & Museum Informatics.

Hillmann, D. (2003). *Using Dublin Core.* Retrieved from *http://dublincore.org/documents/usageguide*

Hove, L.J. (2004). *Improving image retrieval with a thesaurus for shapes.* Master's Thesis. Department of Information and Media Science, University of Bergen. Retrieved from *http://nordbotten.ifi.uib.no/VirtualMuseum/Publications/LarsJacob/Thesis.pdf*

Hyvönen, E., Junnila, M., Kettaula, S., Mäkelä, E., Samppa, S., Salminen, M., Syreeni, A., Valo, A., & Viljanen, K. (2004). Finnish Museums on the Semantic Web: The User's Perspective on MuseumFinland. In D. Bearman & J. Trant (Eds.), *Museums and the Web 2004: Proceedings* (pp. 21-32). Toronto: Archives & Museum Informatics.

Lane, D., & Williams, H.E. (2002). *Web database applications with PHP & MySQL.* O'Reilly & Associates.

Litwin, W., & Abdellatif, A. (1986). Multidatabase interoperability. *IEEE Computer, 19*(12), 10-18.

Lowe, D., & Prince, A. (2003). *Murach's ASP.NET Web programming with VB.NET.* Mike Murach & Associates

Murray, C., & Everett-Church, J. (2003). *Macromedia Flash Mx 2004 Game Programming.* Portland: Premier Press.

Nordbotten, J. (2000). Entering through the side door: A usage analysis of a Web presentation. In D. Bearman & J. Trant (Eds.), *Museums and the Web 2000: Proceedings* (pp. 145-151). Toronto: Archives & Museum Informatics.

Nordbotten, J. (2000-2004). *ADM: Advanced data management.* Retrieved from *http://nordbotten.ifi.uib.no/ADM/ADM_text/ADM-frame.htm*

Nordbotten, J. (2001). *Virtual exhibits on demand.* Retrieved from *http://nordbotten.ifi.uib.no/VirtualMuseum/projectDescription.htm*

Nordbotten, J., & Nordbotten, S. (2002). Evaluation of user search in a Web-database. In *Proceedings of the Hawaii International Conference on Systems Sciences, HICSS-35*. CD. IEEE Computer Society.

Peacock, P., Ellis, D., & Doolan, J. (2004). *Searching for meaning: Not just records*. In D. Bearman & J. Trant (Eds.), *Museums and the Web 2004: Proceedings* (pp. 11-20). Toronto: Archives & Museum Informatics.

Shneiderman, B. (1998). *Designing the user interface: Strategies for effective human-computer interaction* (3rd ed). Addison-Wesley.

Shneiderman, B. et al. (1989). Evaluating three museum installations of a hypertext system. *Journal of the American Society for Information Science, 40*(3), 172-182.

Sowa, J. (2000). *Knowledge representation: Logical, philosophical, and computational foundations*. Books Cole Publishing.

Yamada, S. et al. (1995). Development and evaluation of hypermedia for museum education: Validation of metrics. *ACM Transactions of Computer-Human Interaction, 2*(4), 284-307.

Exhibit Sites Referenced:

American Natural History Museum. (n.d.). *Ology/Marine Biology Exhibit*. Retrieved March 25, 2005, from *http://www.ology.amnh.org/marinebiology/workthesystem/mangrove.html*. Created by National Center for Science Literacy, Education & Technology at the American Museum of Natural History.

CERN. (2004). *The LHC game*. Retrieved March 25, 2005, from *http://public.web.cern.ch/public/Content/Chapters/Education/OnlineResources/LHCGame/LHCGame.html*

Exploratorium. (1998-99). *Solar Eclipse- Stories for the path of totality*. Noel Wanner. *The Sun-Eating Dragon*. Retrieved March 25, 2005, from *http://www.exploratorium.org/eclipse/dragon.html*. Produced in conjunction with AboveNet Communications, Inc.

International Conference on Museums and Web. (2003). Retrieved March 25, 2005, from *http://www.archimuse.com/mw2003*. Archives & Museum Informatics.

Mint Museum of Art. (n.d.). *The Mesoamerican ball game*. Retrieved March 25, 2005, from *http://www.ballgame.org/main.asp*. Produced in conjunction with Interactive Knowledge, Inc.

Museum Finland (2004). Retrieved March 25, 2005 from *http://museosuomi.ca.helsinki.fi*

National Museum of Denmark. (n.d.). *The King's Kunstkammer*. Retrieved March 25, 2005, from *http://www.kunstkammer.dk/GBindex.shtml*. Produced with the support of CulturNet Denmark.

Natuurhistorisch Museum Maastricht. (1995-2004). *The virtual tour of the museum*. Retrieved March 25, 2005, from *http://www.nhmmaastricht.nl/enge*

The Wright Experience™. (2004). *The Wright Experience in Flight*. Retrieved March 25, 2005, from *http://www.wrightexperience.com/edu/12_17_03/html/index.htm*. Produced in conjunction with Cognitive Applications, Inc.

Endnotes

[1]   The term *e-learning center* is used here to refer to a Web site developed with the intention to provide educational material to students as well as to information gatherers from the general public.

[2]   In this chapter, museums are organizations that collect objects and present them in context as information to their public. From a data management and information dissemination point of view, there is no difference between a science and an art museum/center.

[3]   Google.com found 32,500 links for the query "space exploration in science museums." The list included links to museums, as well as to research institutes, business and newspaper articles (retrieved February 2004).

[4]   Programming tools that enable functions to be embedded in a Web exhibit to support interaction with the user, receiving user input and possibly accessing one or more databases for construction of new response pages.

[5]   The ball game was developed by the Mint Museum of Art and Interactive Knowledge, Inc., and placed in an exhibit on Mesoamerican cultures.

[6]   CERN developed the version of the simulation shown in Figure 8 in 2000. The simulation presentation was updated and extended in 2004.

[7]   An example, developed in the virtual exhibits project (2004), can be found at *http://nordbotten.ifi.uib.no/VirtualMuseum/Prototypes/Osdal2/*

Chapter XII

Challenges in Virtual Environment Design:
An Architectural Approach to Virtual Spaces

Renata Piazzalunga,
Information Technology Research Institute, Brazil

Saulo Faria Almeida Barretto,
Information Technology Research Institute, Brazil

Abstract

In this chapter we will discuss some fundamental questions concerning creation and development of interfaces searching for the best way to promote interaction between the subject and information/interface. It starts from the fact that the fundamental and most revolutionary aspect introduced by the Internet is based on its sophisticated technological mechanisms that enhance substantially the concepts of space, time, perception, representation, limits, distance, presence, etc. Our everyday practices gain access to a new realm, cyberspace, which enables us to embrace multiple experiences where we exist in the propagation of our "Id." This condition represents a huge challenge, for example, the necessity to (re)design the image we have from the world in its physical and virtual spaces. We discuss the imagined trends related to the conception and development of virtual environments, addressing the issue of virtual environments in three levels of complexity: realized spaces, possible spaces and imagined spaces.

Introduction

As a result of the innovations and the potential inherent to this media, the expansion of the Internet in the 1990s launched a multitude of transformations in the ways of living of what is now referred to as the information society. Analysis of a social context requires us to analyze its technological model, which from a contemporary point of view means presupposing that both technology and society are part of the same system and, thus, should not be analyzed as isolated entities. As Castells puts it, scientific discovery, and consequently, the challenging of established paradigms is a process dependent on a series of imponderable factors, the result of which is intimately linked to how these factors interact (Castells, 2000).

The observation of the technological context in the information society shows significant changes in social organizations and in the fundamental principles related to conditions of existence, such as, for example, space, time, perception, representation and presence. All of them are altered in their fundamental aspects when faced with the fact that sophisticated technological resources, developed from a technology that is imbedded in this society, allow one to conceive of immersive environments, interactive environments, simulated spaces and so on. There seem to be three essential aspects to the characterization of this information society context: information technology (IT) as a technological paradigm, the dynamism with which processes evolve in this context, and the virtual dimension as a principle for events.

The significance of information technology as a paradigm for the social, economic and cultural models surpass the idea of a simple change in the productive system of society. IT is connected to a productive system which is in fact composed of a set of technological and computational resources that have an intangible good as an object: information. This means that we have moved from the implicit tangibility of the industrial society, a world with established limitations, to a fluid and inexhaustible world.

Up until recently, when we simply contributed to industrial society, we saw our physical abilities being reproduced by gadgets able to replace us in most mechanical tasks. Nowadays, we participate in simulations of our cognitive capacities by means of intelligent systems, in a dimension that transcends the physical space of objects and takes place in virtual space or cyberspace, as suggested by William Gibson's 1984 book *Neuromancer*. In this book, for the first time, the term cyberspace makes reference to "a physical and multidimensional representation of the abstract world of 'information.'" "A place you can visit with your mind, catapulted by technology, while your body is left behind," as Alex Antunes, responsible for the book's preface in its Portuguese version, adds (Gibson, 2003). Since then, not only the term cyberspace has expanded to a myriad of connotations, but also many other terms have been used to refer to it. Virtual environment, virtual space and virtual reality are some of the other terms employed by specialists when referring to cyberspace.

In this context, things basically take place in a virtual-digital dimension. Commercial relations, for example, become virtualized. Similarly, human actions become virtualized, in a way that the concept of presence, inherent to tangible bodies, is altered. Therefore, computerized intelligent systems simulate human cognitive capacities, from the simplest ones, like sharing information in public places, such as shopping malls, banks and supermarkets, to the most complex ones in highly specialized areas such as biotechnology and nanotechnology, among others.

The technological model for the information society introduces a condition never seen before in the history of civilization: a new spatial dimension, cyberspace, that not only becomes the parameter for the development of socio-cultural processes, but also opens up the possibility of exploiting new forms of experiencing and thinking within new models of experimenting in the most diverse fields of our social life.

In the Virtual Dimension: Virtualization, Space and Time

The *cyber* culture, the virtualization culture, reinvents time and space. Understanding aspects related to this reinvented time and space necessarily depends upon the virtual because it is the reference to this domain. In Pierre Lévy's *Becoming Virtual*, the author explores the virtual and contrasts it to other confusing concepts about this theme.

Pierre Lévy's perspective is significantly influenced by the Bergsonian philosophy and attests to the differences and congruencies between concepts like real, possible, actual and virtual. Let us first analyze the difference between virtual and possible. The virtual is not the possible, as the possible belongs to a pre-determined universe, already constituted, in which the only thing missing is realization. Therefore, the real is the complement to the possible because there is a correspondence between their elements and the only difference is that the possible lacks existence as a condition. An example that clarifies the relationship between the possible and the real is the transfer process of a photographic image on to paper. By means of an enlarger, the negative image is projected on to the photographic paper. In this process, the paper is sensitized and "impregnated" with the image from the negative. It so happens that the image is already constituted and pre-determined on the paper, however it has not been developed yet. When the image is developed on the paper by means of chemical baths, it comes into existence. The image has left the realm of the possible and has become real, in the perception of our human vision.

The next pair to be considered involves the virtual and the present. The virtual is not the opposite of the real, but of the actual. While the real *"resembles* the possible, the actual *responds* to the virtual" (Lévy, 1999). There are no correspondences between the actual and the virtual, as they are two different ways of being. In other words, the virtual is a field of indeterminateness or a "problematic complex" that, differently from the realization of the possible which does not imply a creative process, demands a creative resolution, established by means of actualization. Therefore, even though actualization is the solution given to a problem, due to its creative character it does not stop the dynamics of the problematic complex, but rather, feeds the problematic complex with the new qualities produced. As an example, let us consider Internet search engines. When we interact with these systems, each actualization opens a myriad of new relations that not only solve our problematic complex but also feed it.

Besides the actualization that takes place as a reaction of the virtual to the actual, virtualization can also be considered as a reaction, only that it works in the reverse direction to actualization, that is, from the actual to the virtual. While actualization defines the entity by means of a solution to the *problematic complex*, virtualization is where the entity finds its essential consistency. Virtualization is therefore a mutation of the identity of the object under consideration that, rather than being defined by its actuality, is defined by its virtuality, that is, the object is still within the problematic realm. Virtualization "consists in the transition from the actual to the virtual, an exponentiation of the entity under construction" (Lévy, 1999).

In the virtual, there is yet another important condition which is related to the "not there" (Serres, 1998). In this domain, the common sense that says in order to have an experience, one must *be* there is altered. Therefore, by means of the manifestation of a consciousness, it is possible, even *not being*, to assume an existence and to be an experience. As Gibson foresaw, it is possible to go with your mind and leave your body behind.

The "not there" emerges as a principle to virtualization processes. It implies the de-materialization and displacement of the virtualized object, that is, "A kind of clutch mechanism detaches them from conventional physical or geographical space and temporality of the clock or calendar" (Lévy, 1999). In this way, virtualization frees its objects from the inertia imposed by the matter. Moreover, it implies fundamental changes in terms of the condition of space itself, which becomes virtualized instead of materialized or concrete. This transition from concrete space to virtualized space is very close to what Merleau-Ponty describes in his *Phenomenology of Perception* as the transition from spatialized space to spatializing space: "In the first case, my body and things, their concrete relationships expressed in such terms as top and bottom, right and left, near and far, may appear to me as an irreducibly manifold variety, whereas in the second case I discover a single and indivisible ability to trace out space" (Merleau-Ponty, 1962)[1]. The space in the virtual domain would be similar to the spatializing space where, instead of being a possible which is realized, already pre-determined by the concreteness of the physical world, is seen as a field of potential actions to be actualized or virtualized, that is an entity that can withstand changes in nature.

This analysis can suggest that it is time, or more precisely, duration, and not the space itself that is the fundamental entity to virtual experiences. Let us analyze the following excerpt: "Duration is always the location and the environment of differences in kind; it is even their totality and multiplicity. There are no differences in kind except in duration — while space is nothing other than the location, the environment, the totality of differences in degree" (Deleuze, 1988)[2]. Thus, as in the virtual, space is a field of potential actions subject to differences of nature, because it can be actualized or virtualized. It is by means of duration that these differences make their occurrence possible. Unlike duration, space itself only withstands differences in degree, such as: open/closed, horizontal/vertical, and built/non-built.

Duration in the Bergsonian sense is the indistinct succession of the qualitative multiplicity of the states of consciousness that interpenetrate each other in constant and continuous change (Bergson, 2001). That is, it is through our changes of perception, of our internal differences and anxieties that time, or rather duration, reveals itself. In this sense, duration corresponds to how our consciousness habitually operates, that is, it is an integral part of our experience and is, therefore, related to the subjectivity of the human being. While we exist, the "field structured around potential actions" (Prado Jr., 2003) is in operation in our consciousness. In order to further our understanding of these concepts, let us take a look at *Phenomenology of Perception.* In the chapter on temporality, Merleau-Ponty holds that "we must understand time as the subject and the subject as time" (Merleau-Ponty, 1962)[3]. This image reinforces the fact that we are not in time, as suggested by the idea of chronological time, where time "passes" outside us, but rather that we are time because "it has meaning for us only because 'we are it'" (Merleau-Ponty, 1962)[4].

How Do These Concepts Influence the Design of Learning Environments?

In the case of virtual environments, the relationship between the subject and virtual space is established by means of the interface, which exerts influence on the levels of interaction established by these spaces with their users. When we relate the design of virtual environments, or more specifically of learning environments, to the set of complexities that define the features of virtual spaces, the intention is to verify the intrinsic characteristics that qualify these spaces and extract from them the elements that can guide the process of creating virtual environments. A field is opened up for new references to creation based on the principles that constitute these spaces. The moment we notice that one of the properties of virtual space is to be a field of potential actions and not a pre-determined field, as in the case of physical spaces; there is a new paradigm in the conception of design. That is, while we should consider certain actions and objectives specific to the environment under consideration, we should at the same time also consider plasticity and the transformation of the system according to each user's interactions.

Design cannot be a finished whole, closed to possible interferences created by users. The environment has to be conceived as a living system in constant transformation that takes place in time, that is, of the indistinct succession of the various states of consciousness that "inhabit" these environments. As mentioned above, the sense of existence in the virtual space is related not to physical presence, but to the manifestations of consciousness that can be identified there. There is, therefore, another important factor to be considered: virtual environments should be considered as places where the various consciousnesses that are surfing around can co-exist. These places should therefore allow the action and the perception of each user to be used as a constituent element and to be a constructive support to the space.

Thus, based on an approach that considers the dissonances and the congruencies existing between physical and virtual space, new concepts related to virtual learning environments can be explored. We believe that virtual environment design should be treated as an extension of physical reality. The information society brings the experiential plane to cyberspace and, thus, reveals a dimension of existence where it is no longer possible to think of a world that is not hybrid. We will first question the new educational scenarios resulting from this co-existence between physical and virtual spaces. We will then return to the discussion over the meaning of simulated worlds, where new forms of building experiences are introduced. Also, within this idea of new meanings, we will discuss the challenges to the construction of interfaces for learning environments.

New Scenarios of Experience: Simulation, Interaction, Immersion and Telepresence

Basically, four concepts or features, that also act as effects around which the experiencing condition in virtual space can be understood, define the conditions for the experiences constructed in virtual spaces: simulation, interaction, immersion and telepresence. It is always risky to present concepts because we could condition or restrict our understanding of the subject. In the fertile research field of exploring experiences in virtual space, running this risk is justifiable for two reasons: the first one has to do with the fact that these concepts, independently from the many others discussed by researchers in the area, are related to essential aspects, to the very nature of the virtual space; the second one is that, in this paper, these essential aspects are the object of our investigations.

The focus of the following definition is more conceptual and philosophical than technological. What is of interest to us in the definition of these concepts is exactly what they represent in terms of actions that lead to experiences in the virtual domain.

Simulation emerges as a formal representation in the virtual domain of occurrences from the physical world, such as actions, movements, expressions, places, and etcetera: the transference of forms, made material in the physical space, to the virtual. Simulation is the idealization in the virtual of a scenario for the exploration of experiences.

Interaction emerges as a communication between the physical world and the virtual, in the widest sense of the word communication, meaning to share something with someone. Hence, it is the reciprocity between actions from the physical to the virtual and from the virtual to the physical.

Immersion appears as the possibility to penetrate and move in the virtual dimension, characterizing a relationship between the directions that the body assumes and the changes in the configuration of the environment as seen by changing points of view. Lastly, the concept of telepresence, that allows one to be there without the presence of a tangible body. Telepresence allows one to occupy an existence virtually.

The conception of an existence in virtual space is therefore essentially connected to the alternation of these concepts that are translated into forms of actions in this space, producing a reality that is modelled according to our psychic states. Simulation, interaction, immersion and telepresence create a system of possible actions in virtual space and, consequently, other concepts such as virtual reality (Furness, 2001), CAVE or immersive environments become more real. A great project is created: one of a disembodied world of experiences. This context corroborates Merleau-Ponty's position:

> What counts for the orientation of the spectacle is not my body as it in fact is, as a thing in objective space, but as a system of possible actions, a virtual body with its phenomenal 'place' defined by its task and situation. My body is wherever there is something to be done. (Merleau-Ponty, 1962)[5]

These properties of virtual space create possibilities in which practices normally inscribed in a spatial domain and apparently dominated by our perception (the physical world), are extended to a new domain where we experience a multiplicity of experiences. Our existence is no longer restricted to the carnal body, we also exist as representations of our "I" immersed in cyberspace, experiencing a new reality.

We can, therefore, think of new scenarios based on the concept of spaces that become virtual. Experiences in virtual worlds can be simulated that in principle would only be possible in contexts of total materiality. Through technology, we can expand our experiences and create artificial worlds filled with meaning.

Simulated Realities and Augmented Realities

Virtualization not only changes the nature of space, but also extends the meaning of presence, that is, brings new ways of perceiving and experiencing an experience; and establishes a new field for experiences. It is fundamental that we understand how this experience takes place, as our tangible body is not present in this domain.

This field of experience can only be experienced because physical space and virtual space are interconnected. This interconnection, which allows us to travel between different spatial domains, extends the field of our sensorial experiences. This extension or amplification of our experiences to another spatial domain takes place by means of technological devices attached to our body (Figure 1), extending our sensorial and

Figure 1. Sterlac's Amplified Body – An interactive performance between the physiological control of the body and electronic modulation of human functions with the interface

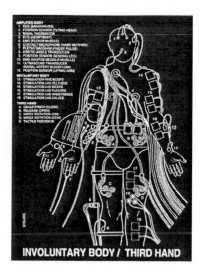

cognitive capacities. The sensory surfaces of our body are stimulated by means of interfaces, expanding our existence beyond the physical and concrete space. The fundamental principle for decoding the concept of experience in cyberspace lies in a change of perception of the role of the body in the world. That is, the body as suggested by Merleau-Ponty: "the body no longer conceived as an object of the world, but as our means of communication with it."[6]

Experience is not, therefore, conditional on the presence of the body, but is a consequence of this body's capacity to communicate with the world. Consequently, "[I]n the case of the normal subject, the body is available not only in real situations into which it is drawn. It can turn aside from the world, apply its activity to stimuli which affect its sensory surfaces, lend itself to experimentation, and generally speaking take its place in the realm of the potential" (the virtual) (Merleau-Ponty, 1962)[7].

In cyberspace, all the technology employed extends our body's communication with the virtual space and thus situates our experience within this domain. As William Mitchell suggests, "We are entering an era of electronically extended bodies living at intersection points of the physical and the virtual worlds" (Mitchell, 1995). Therefore, more than an anatomical body that moves within a medium, the body is also seen as associated with souls that either make things important or not for us. Moreover, the body takes into account our location in a particular context, where we have to deal with things and people and where we have to face all sorts of events.

There are two groups of devices used to establish communication between our body and the virtual space. The first group refers to output devices that allow the computer to be connected to human senses. Some examples are the optical devices known as HMDs (*Head-Mounted Display*), which are individual helmets used for stereoscopic vision, tactile devices or haptic interfaces, hearing devices, and etcetera...

Figure 2. Sterlac's Virtual Body: An interactive performance between the physical body and a system of interfaces (virtual body)

The second group refers to input devices that capture movements and actions of the user's body in order to produce a representation in virtual space. These devices "translate" the language used by the body to communicate into cyberspace. Some examples of input devices are the cinematic devices (*position trackers*; *eye-movement trackers*) that translate the movements performed by the body into computer language, voice and audio input devices, and etcetera...

The context produced by the information society creates the possibility of acting in another spatial domain and needs to be considered vis-à-vis the context in its entirety, that is, this new domain established in the virtual space should be analyzed in terms of its interaction or its co-existence with the physical space. "With cyberspace, a whole new space sprouts from the complexity of Earth's life: a new niche for a domain that is between two worlds" (Benedikt, 1991).

Hence, the co-existence of spaces in different domains points to a condition where physical space is not only a means of action but also fundamental to this level in which occurrences take place between two domains: the physical and the virtual.

This realm, known as augmented reality, is at the crossroads between the physical and the virtual world. It represents the synthesis and the interaction of the processes developed both in the physical and in the virtual planes. It is considered to be an important zone for the fostering of experiences related to the representational and perceptive processes, thus creating a new condition for the exploration of perceptive and representational experiences. The era of extended bodies adds the possibility for the user to live in a world of experiences produced via virtual reality, which incorporates aspects from the physical world as well as the virtual world (Figure 2).

From a technological point of view, the forms of interaction between the physical and the virtual spaces, more and more favour the exploration of full experiences in the context of augmented reality. We already have portable systems, which allow for mobile occurrences of augmented reality, such as the one developed by researchers from Columbia

Table 1. A summary of relationships in the virtual dimension

Concepts	Physical Dimension	Virtual Dimension
Space	Spacialized	Spacializing
Time	Linear, homogeneous space and time	Duration (non-linear, heterogeneous, multi-dimensional space-time)
Perception	Mainly visual	Mainly haptic
Presence	Tangible Body	Intangible Body – Telepresence (consciousness manifested)
Representation	From the physical world	Simulation
Communication	Emission/Reception of messages (person to person)	Interaction (with computerized systems)
Displacement	Local causation Objects with defined locations in space and time	Immersion (non-local causation; intertwining of space and time)

University, Battlefield Augmented Reality System (BARS). Hence, these experiences can take place independently of fixed infrastructures.

In reality, what this system signals is the fact that our experiencing could be ever more determined by experiences that take into account not only the domain of the physical space or the virtual space, but also the overlapping of these two domains in an augmented reality plane. Very much like cell phones which brought mobility to communication, wearable AR systems bring mobility to an experiencing that effectively incorporates data obtained from a physical dimension associated with data obtained from a virtual dimension. We can ultimately conceive of experiences based on the conditions developed in an augmented reality context. Furthermore, we can conceive of virtual science centers that would result from this interaction between the physical and the virtual. Consequently, a new scenario would be built, based on the construction of a world composed of inhabitable physical and virtual spaces.

We have thus far discussed a set of conceptual relationships and before going into directly related questions concerning challenges in building interfaces, we will summarize the concepts and relations presented until now, as shown in the Table 1.

Challenges in Designing Interfaces

With the technological expansion and modification of our individual sensations by means of interfaces, new ways of being in the space and making space are created. That is, *cyberception* occurs, as defined by Roy Ascott: the perception given by the enabled senses of perceiving in a domain devoid of material things. It is this extension, brought about by technology, of our cognition and our perception that suggests this perception.

This cyberception is also an essential underlay to representations. Experience or experiencing in the digital space generates a continuum of configurations and reconfigurations in this space. Thus, in this domain the perceptive domain is not restricted to the individual, but rather there is an expansion of each subject's consciousness in the context of cyberspace, in such a way that the subject's perceptions are imprinted there and become a constituent part of this space. Cyberspace is a dynamic and continuous flow that is updated at each interaction. A fundamental difference between the physical environment and the virtual environment is that, while in the former the form assumed is absolutely indifferent to the subject, in the latter the form may assume qualities derived from the interactions between subject and space.

The dynamic of cyberspace allows for metamorphoses of space depending on the perceptive subject. Cyberspace allows the possibility of extending the forms of spatial representation, as well as, and most importantly, the possibility of extending the sensorial processes. Galofaro (1999) points out that cyberspace allows one to reshape reality. To do so, the author holds that it is necessary to leave behind our one dimensional point of view, focused on visual experiences, and recognize that it is being replaced by a new perception fostered fundamentally by tactile and emotional experiences, which have a strong realistic connotation, based on corporeal sensations. The tactile dimension of cyberspace is reached by means of interactivity.

In cyberspace, the subject is at the same time a constituent part of this space and constitutes his/herself as he/she interacts with it. In this context, space and the subject could be inseparable and the representation could be evidence of the truth in saying that the world is within me and I am entirely outside of myself. In the virtual space, constituent subject and the constituted subject become one. In this condition, we become participants of the simulation of our own cognitive and perceptive capacities by means of intelligent systems. We exist as representations of our "Id" immersed in cyberspace and experiencing a new reality. What we see here is the "Id" as evidence of Merleau-Ponty's tacit *cogito*, that is, as evidence of "an experience of me by myself." This sort of existence propitiated by cyberspace makes the giving form to this "experience of me by myself" possible, where the "Id" is a field, an experience in the face of new possibilities of situations in a new environment, where new layers of significance are both created and represented.

If we accept this, it becomes clear that the interaction between the subject and the environment ought to be a fundamental element in the definition of the configuration of these spaces. Moreover, in cyberspace, we can think of adaptive and individualized environments, that is, environments that "respond" to the particular characteristics of each user. Thus, the great challenge in designing interfaces is to provide environments that, rather than being defined beforehand, are intelligent systems and that are being configured during the processes of interaction with each user.

On the other hand, the more open the systems are to user interactions (systems designed to be built during the processes of interaction with each user), the bigger will be the possibilities of their (the systems') adaptability. The feasibility of conceiving an interface with a higher or lower adaptability depends on, among other things, the technological resources available. In the specific case of instructional design, the system's adaptability seems to be a very significant resource in maximizing both the motivation and collaboration of users, making the learning process much more effective.

Virtual museums of science function as repositories of scientific knowledge but it is our observation that interfaces still need to be improved. Although content quality is taken into account, on most occasions, browsing the pages is unattractive due to their inability to make us identify with the space in a way that a virtual museum of science deserves.

In making an analogy with the space of a real, physical museum of science, a good example could be the *Ciudad de las Artes y las Ciencias de Valencia*, in Spain. Designed by the architect Santiago Calatrava, this museum is an educational center for the promotion of science and technology. When we visit this museum, we become involved in the different atmospheres of the conceived spaces. Equally important as the contents, the configurations of spaces in the specific atmospheres help not only our perception but also our assimilation of information. Virtual museums of science in our opinion should have such an appeal, too. That is to say, spaces where information and environment are conceived as an integrated unit with a language specifically designed according to the qualities required by these kinds of spaces.

The elements that make up the scenarios presented here must be suitably taken into account, according to what is to be implemented and the function desired. We will now discuss three different stages of complexity in the construction of interfaces that vary according to conceptual and technological possibilities: realized spaces, possible spaces and imagined spaces. In this way, what is intended is to show how the concepts presented in this article can be explored with greater or lesser degrees of complexity according to the purposes and possibilities of each project.

Interfaces: Realized Spaces

The discussion about interfaces as realized spaces can be illustrated by three examples: one related to an interface devoted to educational aims and the others to typical examples of contents designed for a virtual museum of science. To insert content into environments may be seen as a suggestion to create what we call a specific atmosphere to the conceived space.

Frame of Mind

Frame of Mind[8] is an environment aimed at promoting learning via Internet, a Learning Management System (LMS). The system was configured to serve the specific pedagogical potential of the Internet. Special emphasis has been given to the design of the learning environment, considering the adaptation of the content to cyberspace. In the case of the Frame of Mind, firstly, we have tried to understand the role of an instructional interface and, then, explore the specific cyberspace mechanisms in dealing with the interaction between user and environment. This brought the integrated solution (Goodbrand, 1997) of the different layers that usually compose the development of an instructional design.

From the analysis of other distance education systems; we can say that, in general, the computational problems receive all the attention from their developers. However, the design of the graphic interface is rarely treated with the same attention as the computational modelling. The interface emerges as a mere product of the computational modelling proposal. The lack of an adequate methodology to conceptualize the graphic interface not only shatters the possibility of forming a cohesive project unit and, consequently, of achieving integrated solutions, but also creates a gap between the level of the computational solutions and the graphic interface proposal, in which, most of the time, the former is well elaborated while the latter is poorly elaborated.

In reality, this reveals that the interfaces are not considered as such, but simply as screens or panels that allow the user to activate a program. Apparently, very few researchers realize what we consider to be the real role of the interface: a source for promoting sensorial stimulation of individuals. This new vision of interfaces has been our guideline in terms of the methodology and the conceptualization employed in the creation and development of our learning environment. We emphasize that the interface has to be conceptualized as a communicating entity with the individual. That is, designing an interface does not simply mean to give shape and to arrange information connected to a software program, but to explore the communication mechanisms established between a virtual entity, which is revealed via the interface, and an actual entity, the individual who is interacting with the program.

In the Frame of Mind Project one way to promote such communication mechanisms is the creation of a dynamic browsing environment based on scenes. In general, the contents for Internet-based courses are simple transcriptions adapted from the written media (books, booklets, etc.) to the digital format. Our proposal of presenting the contents by means of a sequence of scenes has generated a content presentation language that is inherent to cyberspace. When the content is subdivided into scenes, there are some benefits, such as: image and text become unified information and, as in the cinema, the perceptive appeal is in the scene as a whole and not in segregated blocks of texts and images, it helps reduce the natural expectation generated before long and continuous texts on the screen, it assists in reminding the user up to where the content was covered, and it makes easier to keep track of the routes explored by the student.

As the texts are created in scenes, the student is not bombarded with an enormous amount of information packed into a single screen, instead each student can interact with the information according to his/her individual processing mechanism. Also, it is important to highlight the role of the blank spaces purposely left at the end of each text. We metaphorically call this space the student's mental space. It works as a simulation of the time-space relation that the student has to use in order to process the received information. Figure 3 illustrates an example of the sequence of the scenes.

Another important solution employed in our simulated model for a learning environment is the subdivision of content into topics and subtopics. Some of the advantages we have experienced using subdivisions are related to the promotion of different and personal ways of interaction between students and the system: a) students can access specific contents related to the course faster and more easily, b) monitoring of the procedures performed by students is made easier and c) students can easily remember how the course is structured and, consequently, better understand the relationships among the contents. In the previous figure, the left area shows us an example of a topic and subtopic structure.

Still related to the simulation and interaction created in the Frame of Mind environment, there are some topics that present exercise activities connected to the subject of the class. These exercises are generated in accordance with a procedure that suggests a step-by-step solution that enables the student to retrace the dynamics of the problem-solving

Figure 3. Example of lecture sequence

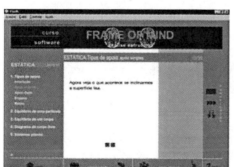 

Figure 4. Exercise being performed step-by-step

process (*Figure 4*). Inside the classroom, while the teacher solves a problem, the student copies it from the blackboard. When it is time to study, the exercise is no more than a figure and the student has lost the dynamics of how to solve that problem. Therefore, all exercises were created following a methodology that gives students a better chance of retracing the problem-solving process.

The Moon: "A Lua"

"A Lua"[9] is a site concerned with divulging scientific information about the Moon to teenagers. This site offers an online, free of charge, distance education course about the Moon and explores interactive media resources for high school students. It is a very successful experience in generating a virtual environment about science, based on the exploration of the graphic quality, along with the many possible forms of interaction either between users and models of computational simulation or between users and an online tutor. This site was the winner of the People's Choice Award for Interactive Learning at Macromedia's UCON 2001 in New York.

The site has pages with expository texts, pictures, infographs, 2-D and 3-D simulations, animations, audio narratives and QTVRs (Quick Time Virtual Reality). On this site, a group of learners (each in his/her house, workplace, or school) participates in a class conducted by a tutor in real time and all the student's progression is tracked by the system.

In the virtual (and synchronous) class, the teacher conducts the class, controlling the specific contents displayed on the student's screen and communicates with them through texts (chat). The content of the virtual class is comprised by animations, sound and a tour guided by a 3-D simulator in the system formed by the Moon, the Earth and the Sun. The site deeply explores the potential of computational simulation in a way to capture the student's attention (Figure 5).

An interesting point about this site is that it allows a diverse range of experiments that the student can reproduce at home, such as the selenoscope (Figure 6), transposing experiences learned in virtual space to the physical world. In this way, an enhancement

Figure 5. Simulation of a cosmic collision

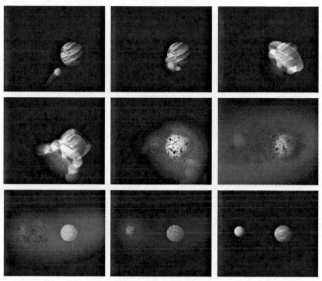

Figure 6. Guidelines to building a Selenoscope

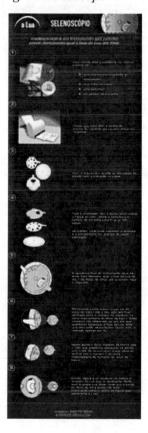

of interactive processes seems to occur, since the students interact with the system (virtual space), carry out actions in the physical space instructed by the virtual and respond to the system after observing the experiences carried out in the concrete world.

The Exploratorium: The Museum of Science, Art and Human Perception

This virtual museum[10], part of the Palace of Fine Arts in San Francisco, California, is a very good example because besides one of the earliest trials concerning the implementation of a virtual science museum it also introduces scientific issues in a creative and varied fashion (Figure 7).

It explores different interactive mechanisms, such as the Web cast, where transmissions of specific topics via the Internet can be accessed by anyone using the museum's Web site with real time presence of experts in the fields under discussion. An example of Web

Figure 7. Main page of Exploratorium

Figure 8. Web cast live transmission of the transit of Venus

cast with this kind of interaction was the transmission of the Venus transit on June 8, 2004, seen from Greece (Figure 8).

Another example of interactive proposal in this Web site is The Exploratorium Learning Studio, which receives many requests for help with science fair projects both from teachers and students. Such strategies found in the Exploratorium are examples of new references to creation based on the principles of the Web.

A significant aspect in the case of Exploratorium is that the constitution of the idea of the museum is both physical and virtual. This way of conceiving a museum in the virtual environment inserts the Exploratorium not only in the Palace of Fine Arts in San Francisco but also represents it in a way that allows interactions from visitors at any place and at any moment. What can be seen is that, although the conception of the museum results from the combination of physical and virtual spaces, there is not an actual integration between both domains because what the online visitors experience is not integrated and do not interact with visitors in the physical space. Exploratorium is still a model that splits and makes distinction between the experiences in both domains; and because of this, does not fully explore the virtual space as a possibility to enhance the cognitive experience in the broad sense, that is to say: as a possibility of transformation and reorganization of experiences outside the established borders (virtual and concrete). The environment hasn't been conceived as a living system in constant transformation and in interaction with different dimensions of space, as we propose in a previous section. Despite the implementation of many interactive strategies found in the site the interaction between subject and environment is not taken as a fundamental element to the configuration of these spaces: yet the conception of the interface is on dependence of a predefined structure by the designer. The influence exerted by the interfaces on the levels of interaction established by these spaces with their users is limited both by the technology and the concept of design used.

An example that brings light to this distinction and the determined limits between virtual space and concrete space in design is the Web site Tactile Dome, an exposition only configured at the museum's concrete space that says: "Discover the unseen world of the

Figure 9. Configuration of a Web site from concrete space alone: Tactile Dome

Tactile Dome — an interactive excursion through total darkness, where your sense of touch becomes your only guide!" (Figure 9).

This saying expresses the up to now dominant concept when designing interfaces (realized spaces): the lack of a more enlarged exploration of the virtual space merely understood as an extension of the concrete world. In the information society scenario instead, they might be thought as a single entity providing a new spatial domain where experiences can be explored.

Interfaces: Possible Spaces

What are here called possible spaces are resources that are technically and conceptually already feasible to implement. Examples of such resources are the immersive environments created by virtual reality, where the resources created building the interface are enhanced beyond the visual experience. These immersive environments thus allow the experience of our "Id" in a world of bits.

The widespread use of virtual reality environments offers the possibility of creating interfaces, where new means of experiencing open up with the use of technological devices. We define this alternative as a "possibility," not for technical reasons, because these are mostly solved, but for financial reasons, as technology still imposes high development costs. In the near future this aspect will probably be resolved and we will be able to augment our experiencing in the virtual plane.

Another example of what can be considered as a possible resource to be implemented in instructional interfaces is the insertion of agents that represent the users' interactions. The idea behind these agents is for them to work as a motivational resource for the users of environments that can now be called interactive, environments that besides exploring and facilitating interactions also incorporate a type of avatar that becomes an extension of the person interacting with cyberspace. Every action or, rather every interaction of the subject in the environment will cause an action on his/her self, represented in cyberspace. It is thus possible to explore such a resource as a mechanism that is both creative and alternative, that together with other resources allows an evaluation of both the interface's impact on the user and the use of the interface as a motivational element.

Interfaces: Imagined Spaces

At this stage of complexity — that of imagined resources to be incorporated into the creation of interfaces — we are referring to possibilities still at the stage of technological and/or conceptual development: that is, resources still under scientific research. One example would be the possibility of interfaces that adapt to the cognitive capacities of the user. We are of the opinion that in order to foresee the conceptual and technological implementation of this resource, one has to study and understand the brain's mechanisms for cognitive processing.

An aspect that will therefore probably open up future possibilities for the development of a new model for virtual interfaces is the use of neuroscience and cognitive science theory. We strongly believe that these fields of knowledge are the key to answers that will allow us to develop adaptive interfaces that take into account the user's profile. These fields of knowledge have had an ever growing presence in the techno-scientific and artistic circles, as neuroscience focuses on themes such as learning, perception, language, information processing, as well as philosophical implications, such as investigating how the visual system acquires knowledge from the outside world through the brain.

Recent technological advances allow us to see the human brain at work and consequently, reopen the discussion on many cerebral mechanisms, some of them only at an early stage of understanding, such as consciousness, semantic processing and perception itself. In this sense, both the tools, as well as the concepts originating in neuroscience are in the spotlight, moving between different fields of knowledge and serving as a basis for an attempt at capturing each person's individuality. In spite of being only recent, and still at the development stage, these concepts have helped us understand the cognitive mechanisms and strategies that are sure to be implemented in the next phase of the digital revolution of man-machine interfaces.

Final Comments

It is important to consider interfaces that are increasingly adaptable personalized, responding to the interactions of each user. These interfaces should not be considered as finished projects, but as open doors to a dialog between users and intelligent systems that transform themselves and are configured according to the actions of each specific user. Moreover, it is important to create environments that are so adaptable that the presentation of contents, the language and the type of media (text, image, animation and video) are subservient to the individual learning style of each user.

As a consequence, there is an ever-growing latent need for an inter-relation between different fields of knowledge, such as neuroscience, arts, and computer science to produce environments with interactive characteristics closer to the user's natural way of relating - a creation of environments that we call "interactionist" because they enable the user to experience and perceive him/herself in relation to his/her interactions *with* and *in* the environment.

References

Ascott, R. (1994). *The architecture of cyberception*. Retrieved March 31, 2004, from *http://www.eff.org/Net_culture/Cyborg_anthropology/cyberception.paper*

Benedikt, M. (1991). *Cyberspace: First steps*. Cambridge, MA: MIT Press.

Bergson, H. (2001). *A evolução criadora*. Lisboa: Edições 70.

Castells, M. (2000). *A sociedade em rede. A era da informação: economia, sociedade e cultura*. (Vol. 1). São Paulo: Paz e Terra, 3ª edição.

Deleuze, G. (1988). *Bergsonism*. New York: Zone Books.

Furness, T.A. (2001). Towards tightly coupled human interfaces. In R.A. Earnshaw, R. Guedj, A. van Dam & J. Vince (Eds.), *Frontiers of human-centred computing: Online communities and virtual environments*. London: Springer-Verlag.

Galofaro, L.A. (1999). *Digital Eisenman. An office of the electronic era*. Basel: Birkhäuser.

Gibson, W. (2003). *Neuromancer*. São Paulo: Aleph.

Goodbrand, A. (1997). *The art of thinking*. SENG693. Trends in software engineering. Retrieved March 31, 2004, from *http://sern.ucalgary.ca/courses/seng/693/W98/alang/minor.html*

Lévy, P. (1998). *Becoming virtual: Reality in the digital age*. New York: Plenum Press.

Merleau-Ponty, M. (1962). *Phenomenology of perception*. London: Routledge.

Mitchell, W. (1995). *City of bits: Space, place, and the Infobahn*. Cambridge, MA: MIT Press.

Prado Jr., B. (1999). *A filosofia seminal de Bergson*. Retrieved March 31, 2004, from *http://members.xoom.virgilio.it/odialetico/filosofia/Bergonseminalhtm*

Serres, M. (1998). *Atlas*. Lisboa: Ed. Instituto Piaget.

Winograd, T. (1995). Environments for software designing. *Communications of the ACM, 38*(6), 65-74.

Endnotes

[1]   Dans le premier cas, mon corps et les choses, leurs relations concretes selon le haut et le bas, la droite et le gauche, le proche et le lointain peuvent m'apparaître comme une multiplicité irréductibile, dans le second cas je découvre une capacité unique et indivisible de décrire l'espace.

[2]   La division se fait entre la durée, qui *tend* pour son compte à assumer ou porter toutes les differences de nature (puisqu'elle est dovée du pouvoir de varier qualitativement avec soi), et l'espace qui ne présente jamais que des differences de degré (puisqu'il est homogénéité quantitative).

[3]   faut comprendre le temps comme sujet et le sujet comme temps.

[4]   qu'il n'a de sens pour nous que parce que nous « les sommes.

[5]   Ce qui importe pour l'orientation du spectacle, ce n'est pás mon corps tel qu'il est en fait, comme chose dans l'espace objectif, mais non corps comme système d'actions possibles, un corps virtuel don't le *lieu* phenomenal este défini par sa tâche et par sa situation. Mon corps est là où il a quelque chose à faire.

[6]   Non plus comme objet du monde, mais comme moyen de notre communication avec lui.

Les corps chez le sujet normal n'est pas seulement mobilisable par les situatiòns réelles qui l'attirent à elles, il peut se détourner du monde, appiquer son activité aux stimuli qui s's inscrivent sur ses surfaces sensorielles, se prêter à des experiénces, et plus généralement se situer dan le virtuel.

[7]   http://www.npt.umc.br/frameofmind, retrieved March 31, 2004

[8]   http://www.mamutemidia.com.br/alua/, retrieved March 31, 2004

[9]   http://www.exploratorium.edu, retrieved September 15, 2004.

Chapter XIII

Personalization Issues for Science Museum Web Sites and E-Learning

Silvia Filippini-Fantoni, The University of Paris I Sorbonne University, France

Jonathan P. Bowen, London South Bank University, UK

Teresa Numerico, London South Bank University, UK

Abstract

E-learning has the potential to be a very personalized experience and can be tailored to the individual involved. So far, science museums have yet to tap into this potential to any great extent, partly due to the relative newness of the technology involved and partly due to the expense. This chapter covers some of the speculative efforts that may improve the situation for the future, including the SAGRES project and the Ingenious Web site, among other examples. It is hoped that this will be helpful to science museums and centers that are considering the addition of personalization features to their own Web site. Currently, Web site personalization should be used with caution, but larger organizations should be considering the potential if they have not already started to do so.

Background

In the past few years, the number of people visiting museums' Web sites has gone up rapidly. As a consequence, museums have to face the significant challenge of creating virtual environments that are progressively more adapted towards the different needs, interests and expectations of their heterogeneous users. Increasingly, museums and science centers are using their Web sites to augment their learning facilities in potentially innovative ways (Tan et al., 2003). In particular, museums need to provide for differing online requirements such as teaching, e-learning and research (Hamma, 2004). One of the solutions available to help is the introduction of personalization techniques (Dolog & Sintek, 2004) that, by providing differentiated access to information and services according to the user's profile, make facilities and applications more relevant and useful for individual users, thus improving the overall visitor's experience. Science museums, by their very technological nature, ought to be at the vanguard of applying new techniques like personalization.

Developed in the early 1990s in an attempt to try to respond to the different needs and characteristics of an ever-growing number of Internet users, personalized or adaptive Web systems have since been exploited in different sectors such as commerce, tourism, education, finance, culture and health. What distinguishes these systems from the traditional static Web is the creation of a user model that represents the characteristics of the user, utilizing them in the creation of content and presentations adapted to different individuals (Brusilovsky & Maybury, 2002). By so doing, personalization becomes a useful tool in the selection and filtering of information for the user, facilitating navigation and increasing the speed of access as well as the likelihood that the user's search is successful.

The techniques available to collect information about users, as well as the methods used to process such information to create user profiles and to provide adapted information, are varied. A brief description of the different approaches will be presented here before moving on to illustrate different application examples within the science museum world.

Personalization Techniques

A first important distinction concerning the amount of control the user has on the adaptation process can be made between customization and personalization. *Customization* or *adaptability* occurs when "the user can configure an interface and create a profile manually, adding and removing elements in the profile" (Bonnet, 2002). The control of the look and/or content of the site are explicit and user-driven; that is, the user is involved actively in the process and has direct control. In *personalization* or *adaptivity*, on the other hand, the user is seen as being passive, or at least somewhat less in control (Bonnet, 2002). Modifications concerning the content or even the structure of a Web site are performed automatically by the system based on information concerning the user stored in the so-called *user profile*. Such information about the user is provided either *explicitly*, by the user themselves, using online registration forms, questionnaires and reviewing (static profiles) or *implicitly* by recording the navigational behavior and/or preferences of each user through dynamic profiling Web technologies such as *cookies*[1] and *Web server log files*[2] (Eirinaki & Vazirgiannis, 2003).

Once the data concerning the users is collected either implicitly or explicitly, or even in both ways, as is often the case, appropriate information that matches the users' need is determined and delivered. This process usually follows one or more of the following techniques: content-based filtering, collaborative filtering, rule-based filtering and Web usage mining.

Content-based systems track user behavior and preferences, recommending items that are similar to those that users liked in the past (Eirinaki & Vazirgiannis, 2003). *Collaborative filtering* compares a user's tastes with those of others in order to develop a picture of like-minded people. The choice of material is then based on the assumption that this particular user will value information that like-minded people also enjoyed (Bonnet, 2002). The user's tastes are either inferred from their previous actions or else measured directly by asking the user to rate products. Another common technique is *rule-based filtering,* which allows Web site administrators to specify rules based on static or dynamic profiles that are then used to affect the information served to a particular user (Mobascher et al., 2000).

Last but not least, there is *Web usage mining*, which relies on the application of statistical and data-mining methods based on the Web server log data, resulting in a set of useful patterns that indicate users' navigational behaviors. The patterns discovered are then used to provide personalized information to users based on their navigational activity (Eirinaki & Vazirgiannis, 2003).

The information provided to the user through any of the above techniques can be adapted at three different levels: content, navigation and presentation (Brusilowsky & Nejdl, 2004). Adaptive content selection is based mostly on adaptive information retrieval techniques: "when the user searches for relevant information the system can adaptively select and prioritize the most relevant items" (Brusilowsky & Nejdl, 2004). By doing so, the user can obtain results that are more suitable for their knowledge capabilities. Adaptive navigation support is founded mainly on browsing-based access to information: "when the user navigates from one item to the other the system can manipulate the links to guide the user adaptively to most relevant information items" (Brusilowsky & Nejdl, 2004).

Finally, adaptive presentation is based on *adaptive explanation* and *adaptive presence,* which were largely developed in the context of intelligent systems: "when the user gets to a particular page the system can present its content adaptively" (Brusilowsky & Nejdl, 2004). The possibilities of content and presentation adaptability are a relevant element in the reuse of the same resources for different purpose, provided they have been correctly customized in advance. Considering the high cost of personalization, adaptability of resources can also offer an interesting byproduct in term of reuse of the same resources in different contexts, provided that their description is correctly defined through standard metadata applications to allow interoperability of the same service in different environments.

From the perspective of different platform services, adaptability becomes a strategic issue. It could be decided to personalize content for the relatively small screen of mobile devices, for example. Moreover, whereas personalization and adaptability on the Web is based only on the user, in the case of mobile support there is also the need for adaptation with regard to the user's environment (Brusilowsky & Nejdl, 2004).

In a museum visit, taking into account the environment where the service will be used can make a notable difference to the experience. For example, an explanation of the items kept in a single room of the exhibition can be offered while the visitor is in that room. There are some projects exploring these opportunities with special regard to mobile devices used by museum learning services (Oppermann & Specht, 1999).

Why Use Personalization in Museums?

Even if some of the techniques described in the previous section, especially the more sophisticated ones, are employed mainly on commercial Web sites, such as Amazon.com, etc., there is already some awareness of the need for their use in cultural institutions, museums, science centers, etc. Personalized access to collections, alerts, agendas, tour proposals and audio guides are just a few examples of the different applications that have recently been developed by museums all over the world (Bowen & Filippini-Fantoni, 2004). The reasons for such an affirmation are numerous, as personalization can help museums respond to various and different needs.

First of all, personalization has the advantage of improving the *usability* of a Web site by facilitating its navigation and aiding people in finding the desired information. With some knowledge about the user, the system can give specific guidance in its navigation, limiting the visitation space appropriately. The system can supply, or even just suggest, the most important links or content that could be relevant for the user, something that can help prevent them from becoming lost in a Web site's potentially intricate hyperspace.

Accessibility for the disabled (Bowen, 2004), a specific aspect of usability that concentrates in widening the number of users, can gain from personalization techniques. The ability to select the text foreground or background color, size and font, can make interfaces more easily readable for the partially sighted. A text only view of a Web site may be easier for such users and also those who are completely blind. For example, the London Science Museum has an option from the home page for a text only version of the Web site (www.sciencemuseum.org.uk). The basic content is the same, but the presentation is different. Legislation in the UK, for example, now ensures that learning materials for students in educational establishments, including those provided by university science museums, must be covered by an accessibility strategy (HMSO, 2001).

Personalized systems help to recreate the *human element* that listens to the visitor with understanding by offering an individual touch; this is another important factor that contributes to the success of Web personalization in museums. It is a particularly important element, especially for audio guides, which must offer a certain level of flexibility in order to adapt the contents to the needs and interests of the users, just like a real museum guide would do. It also helps online, making the visitors feel comfortable and oriented in the virtual space, through virtual avatars for example. Studies indicate that the "social metaphor represented through the presence of personalized animated characters (similar to real life people) can reduce anxiety associated with the use of computers" (Bertoletti et al., 2001).

Personalization could also be a useful tool in the creation and development of *online communities* for museums (Beler et al., 2004). In fact, thanks to personalized applications such as alerts, thematic newsletters, customizable calendars and recommendation systems[3] providing tailored content to people with specific interests, museums can identify homogeneous communities of users with the same concerns and needs. Once these different online communities have been identified, it is in the museum's interest to foster them by developing tools and services that aid them in their functioning, especially by stimulating communication. This is when personalization can assist once again. In particular, online forums (Bowen et al., 2003) can benefit from the introduction of personalizing features such as notification of debates or issues that might be of interest to the user, information about other users with interests on specified topics (facilitating the networking between community users), personalized news generation based on personal interests, etc. These kinds of personalized services can increase the value of the underlying museum's "e-community" beyond a social networking environment: "the website becomes an attractive permanent home base for the individual rather than a detached place to go online to socialize or network, thus strengthening the relation between the user and the institution" (Case et al., 2003).

By providing targeted information to users with different profiles and interests, personalized systems are much more likely to satisfy the visitor, who, as a consequence, is stimulated to come back and reuse the system or to encourage other people to try it as well. This is why personalization is also a fundamental *marketing* tool for the development of visitor fidelity, as well as new audiences.

Personalization and Learning

Besides helping museums to respond to their usability, marketing and accessibility needs, personalization has much potential when it comes to stimulating learning, as underlined by Brusilovsky (1994) who, early in the development of the Web, pointed out how personalization techniques could be an important form of support in education. The reasons for this are varied. First of all, visitor studies seem to confirm that learning is encouraged when the information provided is described in terms that the visitor can understand. Using different terms and concepts, that take into consideration the level of knowledge, age, education of the user, etc., can therefore improve the overall didactic experience. This is precisely what happens with personalized applications where the information delivered to the visitors often changes according to whether they are a child, an adult, a neophyte or an expert.

Research also indicates that learning is facilitated when the information provided makes reference to visitors' "previous knowledge"; that is to say, to what people already know or to concepts already encountered during navigation or exploration (Falk & Dierking, 1992). This suggests that museums should focus on how to activate visitors' prior knowledge if possible. One of the means at their disposal is personalization, which could open new and effective means for long-term learning by providing adaptive descriptions of artifacts based on objects or concepts that the visitor has already visited or explored. This is, for example, the case in projects like ILEX, Hyperaudio, HIPS and the Marble Museum's Virtual Guide — see Filippini-Fantoni (2003) for descriptions — that, through dynamically generated text, provide personalized information taking into consideration the user's history. The description of the object being viewed or selected can make use of comparisons and contrasts to previously viewed objects or concepts. By providing such coherent and contextualized information, modeled on the user interaction with the exhibition space as well as with the system itself, such applications have enormous potential from the learning point of view.

Another mechanism that can be used to justify the use of personalization to stimulate learning is "subsequent experience" (Falk & Dierking, 1992). A number of researchers have hypothesized that repetition is the major mechanism for retaining memories over a long period of time (Brown & Kulick, 1997). This is why, by allowing the visitor to bookmark objects or concepts of interest during their navigation in the virtual or real environment and to explore them more in detail subsequently (see later for further information), personalization can make it possible to further deepen and continue the learning process from home by creating continuity between the visit and post-visit experiences.

Last but not least, learning is stimulated when a person can pursue their individual interests. Researchers distinguish between "situational interest" and "individual inter-est," the first being defined as "the stimulus that occurs when one encounters tasks or environments with a certain degree of uncertainty, challenge or novelty" (Csikszentmihalyi & Hermanson, 1995). This is, for example, the case for museums where the presence of incentives like surprise, complexity and ambiguity lead to motivational states that result in curiosity and exploratory behavior (Csikszentmihalyi & Hermanson, 1995).

However this is not enough to guarantee that the visitor is actually stimulated to learn. In order for this to happen, museums have to attempt to respond to their visitors' "individual interests," that is "their preference for certain topics, subject areas or activities" (Hidi, 1990), as the pursuit of individual interests is usually associated with increased knowledge, positive emotions and the intrinsic desire to learn more. Person-alizing an educational activity in terms of themes, objects or characters of high prior interest to students should therefore enhance the overall learning experience. Take, for example, those personalized applications (see later for details) that provide tailor-made visitor plans with consideration of the individual interests of a single visitor or a group of visitors. By suggesting artifacts relating to the visitor's individual curiosity, the visit is more likely to result in fruitful learning activity.

In conclusion, by providing information at the right level of detail, stimulating subse-quent experiences and taking into consideration individual interests as well as prior knowledge, personalization represents an excellent tool for all those educators wishing to stimulate and facilitate learning. This is why personalization techniques are often exploited in the creation of *formal* e-learning applications such as long-distance courses that are able to adapt to the student's level of knowledge, cognitive preferences and interests, etc. For example, see the AHA Project on *Adaptive Hypermedia for All* [aha.win.tue.nl] at the Technical University of Eindhoven, The Netherlands, and the European IST ELENA Project on *Enhanced Learning for Evolutive Neural Architectures* [www.elena-project.org].

However, personalization can be also applied to more *informal* e-learning solutions like the ones that are often available on museums' Web sites or interactive devices, which, although not being actual lessons, represent very useful educational experiences that contribute to increasing the visitor's knowledge and understanding about a specific issue.[4]

Web Personalization for Science Museums

Until now, we have discussed more general issues concerning the use of personalization techniques in museums, focusing in particular on its potential to stimulate and facilitate the learning experience. In this section we consider some examples of how science museums in particular are applying these principles online. In fact, even if science museums are not the only cultural institutions to have experimented with personalization both online and on-site in the past few years — for a more general description of personalized applications in museums see Bowen and Filippini-Fantoni (2004) — they are among the ones that have expressed the strongest interest in these techniques. This is because science museums and science centers, whose exhibits are designed to promote playful exploration and discovery of scientific phenomena, have always been relatively aggressive adopters of information technology and innovative approaches; as a consequence, they have also been more eager to experiment with personalization.

Some museums have been focusing more on the usability and marketing aspects of personalization privileging applications such as personalized agendas, alerts and newsletters, which, although having an intrinsic pedagogical value, seem to focus more on promotion. However, science museums have been among the first to understand the real value of personalization as a learning tool, concentrating particularly on stimulating "subsequent experience," "previous knowledge," and "individual interest" in such a way as to explicitly encourage the continuity between the pre-visit, visit and post-visit experiences.

The first examples of Web personalization in a museum context were developed in the late 1990s in strict relation with the affirmation of academic research on adaptive hypermedia. Among them (Bowen & Filippini-Fantoni, 2004) was the SAGRES system (sagres.mct.purcs.br), developed in 1999 by the Museum of Sciences and Technology of PUCRS (MCT), Porto Alegre, Brazil.

The SAGRES system (Bertoletti, 1999; Moraes, 1999) is an educational environment that presents the museum's content adapted to the user's characteristics (capacities and preferences). Based on information provided directly by the user or by the teacher (for students), the system determines the group of links appropriate to the user(s) and presents them in a personalized Web page.

The principle behind the project was an attempt to overcome the limitations implicit in the one fits all approach and to take the user's individual interests as well as their level of knowledge into consideration when delivering information, with the aim of improving the overall learning experience. This is possible through an adaptation process that first generates a user model, based on information provided by the user[5]. Once these data about the user have been collected, the adaptation process can select different types of documents conforming to the visitor's model. This results in a dynamically generated HTML page with links pointing to personalized information: the page is created dynamically during the interaction of the user with the system and presents links to the documents, as well as connections to the communication mural (where users can interact with each other), to the document edition, and to the activities the user should perform (in the case of a group visit).

As well as being designed for individual users, the system is particularly meant for use in an educational setting. Through SAGRES, teachers are given the opportunity to define and register their students' profiles, to accompany them and to evaluate their performance during the visit, using reports delivered by the system. At the same time, students are allowed to interchange ideas with colleagues in their groups and to work on the activities and subjects determined by the teacher.

Figure 1. The architecture of the SAGRES system

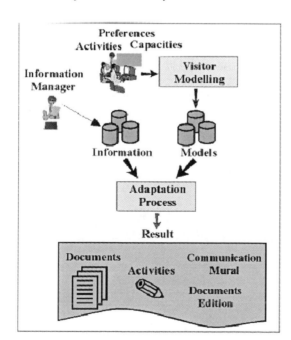

Personalized Virtual Web Spaces

The main aim of the SAGRES project was to facilitate learning through the provision of information adapted to the level of knowledge and interest of the user. Since then, other methods have been adopted to guarantee a similar outcome. Various science museums, for example, provide users with tools that allow them to save images, articles, links, search results, forum discussion topics, as well as other types of information during navigation of the Web site. By doing so, the user creates a personal environment within the museum's Web site, where they can return, find specific information of interest, and to which new items can be continuously added. This environment can be further equipped with other personalized services such as individual agendas or the ability to send personal e-cards.

Once the page has been created, visitors can log in every time they access the Web site to find all the information they need. By doing so, the user has the chance not only to find information of interest more easily, but also and especially to strengthen the learning process through reuse and repetition. The learning value of these applications for certain categories of users such as students and teachers is even greater. The personal space can offer teachers the possibility of suggesting of exhibits for their students to visit and questions that they would like the students to answer during the exploration. In response, the students can save links to the exhibits that most interest them, as well as making short notes both about questions they had at the beginning and about new questions that arise during the exploration.

One of the most interesting examples of this type of application is provided by the *Ingenious* project, undertaken by the *National Museum of Science and Industry* group in the United Kingdom and funded by the UK New Opportunities Fund (NOF) (www.nmsi.ac.uk/nmsipages/nofdigitise.asp). This project, online from mid-2004, aims at creating a learning environment for the public from the digitized collections of the Science Museum (London), the National Railway Museum (York) and the National Museum of Photography, Film and Television (Bradford) in the UK. Users of the Ingenious Web site (www.ingenious.org.uk) can explore and discover the rich collections of these museums through 50 narrative topics and over 30,000 images and other content-rich resources, such as library and object records. In addition visitors are provided with tools for entering a topical debate and personalizing their experience in the so called "CREATE" area, where registered users can save images and/or links from the

Figure 2. Ingenious home page

Figure 3. Ingenious electronic cards   *Figure 4. Ingenious selected hyperlinks*

Figure 5. Ingenious saved images

debate areas, read sections and search queries. The users can also send personalized e-cards of images by e-mail and create a personal Web gallery from their bookmarked images, including the ability to incorporate personal comments that can be e-mailed to friends and colleagues.

Figure 2 shows a general shot of the Ingenious home page. The facilities include "My E-cards" to sent electronic cards (Figure 3), selected hyperlinks (Figure 4), saved images (Figure 5) and Web galleries (Figure 6).

Even if in the wider picture for *Ingenious* users, the umbrella group is lifelong learners, the application can be particularly suitable for older age school children, teachers, and researchers who could first explore a topic in the "read" or "see" sections of the site, then use the "save image" and e-card features and gradually progress to Web gallery tools for creating a personal resource. The Web gallery outcome would be used for a project, research, shared among a group of subject enthusiasts or a class (for instance). Community building could follow from this, through the usage of the debate features available on the site.

Figure 6. Ingenious Web galleries

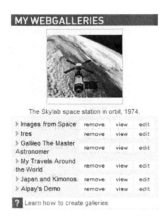

The Post-Visit Experience

In some cases, personal virtual spaces can also include information about a visitor's actual visit to museums, thus creating a direct link between the visit and the post-visit experience. Personalization is an effective tool for stimulating visitors at home to follow up on what caught their attention during the exhibition through a museum's Web site. For example, the London Science Museum's "*In touch*" project allows a record of a visitor's interaction with various exhibits in the Wellcome Wing including an eye scan, voice, face and fingerprint recognition, photo editing, etc., to be recorded using their fingerprint as an identifier, thus avoiding the need for any physical ticket (www.sciencemuseumintouch.org.uk). The results are made available as part of a personal space within the museum's Web site that can be accessed via the visitor's first name and birth date.

Figure 7. "In touch" exhibition screen shots

Figure 8. "In touch" Web pages

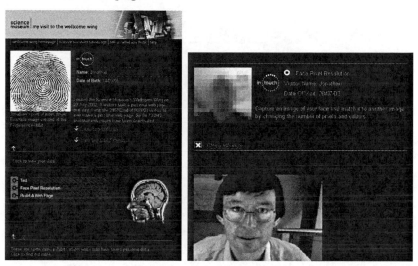

Since 2000, when the project was originally implemented, Joe Cutting of the Science Museum reports that (as of January 2004) more than 400,000 Web pages have been created, of which around 8% have been accessed at least once. In order to simplify the system, reduce the operational problems that derive from such a large database, and increase the percentage of visitors using it, the museum has decided to replace the fingerprint method (which is not completely reliable in practice) by "an email it to me" option by the end of 2004. Every time a person wants to save one of the interactions, an e-mail address will have to be provided. By doing so, there will be no more automatically generated personal pages for the visitors. However, the museum is considering the inclusion of a link in the e-mail that would allow the visitors to set up a personal page if they wish. In this way only those who are really interested will set up a page and the museum will not have to maintain a huge and largely unused database. Figure 7 shows two screenshots from the exhibition itself and Figure 8 shows example pages from the associated Web site.

In a similar manner, the *Visite Plus* service offered by the *Cité des Sciences et de l'Industrie* (www.cite-sciences.fr) in Paris, which has been used on a number of successive temporary exhibitions, "*Le Cerveau Intime,*" "*Le Canada Vraiment,*" and "*Opération Carbone,*" allows the visitor to configure a personal profile (with information on preferred language, disabilities, etc.) on an interactive kiosk placed at the beginning of the exhibition through a special bar-coded ticket or on a PDA (Personal Digital Assistant). This data can then be used to access adapted information from the different interactive devices and to play various games and quizzes in the exhibitions. The results of such interaction, as well as the path followed by the visitor, are automatically saved by the *Visite Plus* system on a personal Web page, accessible on the museum's Web site after the visit through the number of the bar-coded ticket or PDA. In this way, the visitors are able to analyze in more depth the subjects that particularly interested them during the

Figure 9. Visite Plus personalized Web site

exhibition (through the provision of additional information) and to compare results of their interactions with those of other visitors.

The fact that an important part of the content concerning the exhibition is accessible after the actual visit, at home or in another context, allows the visitor to focus more on experimentation and discovery while in the museum and to leave the more traditional didactic aspects for later. The *Visite Plus* system also offers the possibility of subscribing to a personalized periodical newsletter that focuses on a series of themes selected by the visitor at the moment of the registration. Options include selecting from a list of available subjects or receiving a complete dossier of the exhibition. See Figure 9 for an example of the view of the exhibit from the personalized Web site. Each square corresponds to a content area in the exhibition. The squares that are in full color represent the ones that have been accessed during the visit to the exhibition while the white ones correspond to the ones that have not been visited.

Similar concepts have been introduced and tested in the framework of the *Electronic Guidebook Research Project* (www.exploratorium.edu/guidebook), which began in 1998 at the San Francisco *Exploratorium* in California, in partnership with Hewlett-Packard Laboratories and the Concord Consortium. This is aimed at developing a roving resource to enhance a visitor's experience at the museum (Hsi, 2003). In particular, the purpose of the project is to investigate how a mobile computing infrastructure enables museum visitors to create their own "guide" to the Exploratorium, using a personalized interactive system. This helps in better planning of their visit, getting the most out of it while they are in the museum and enabling reference back to it once they have returned to their home or classroom. The guidebook allows users to construct a record of their visit by bookmarking exhibit content, taking digital pictures from a camera near the exhibit, and accessing this information later on a personal "MyExploratorium" Web page in the museum or after their visit (Figure 10).

The project was designed as a proof of concept study to explore potential avenues for future research and development and therefore was not envisioned to support the implementation of a fully functional system. Nevertheless, the tests that have been run so far revealed interesting conclusions. Above all, the visitors liked the idea of being able

Figure 10. MyExploratorium set-up

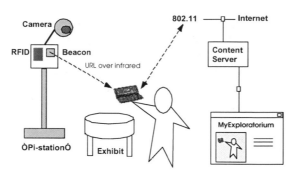

to bookmark information for later reference. Both teachers and pupils thought this feature would allow the children to play more during their museum visit, completing related homework assignments after the visit (Semper & Spasojevic, 2002).

The Pre-Visit Experience

The link between visit and post-visit experience can be also extended to the pre-visit phase through the implementation of systems that allow visitors to create personalized tours based on their interests and needs. Most museum visitors, even those who have not visited before, arrive with expectations about what will happen during the visit. Such hopes might concern specific subjects of interest that the person wants to explore, the physical characteristics of the museum, the types of activities that can be undertaken or the social context in which the exploration takes place (alone, as a family, within a larger organized group, etc.). All these factors merge to create a visitor's *personal agenda* (Falk & Dierking, 1992). The success of the museum experience is partially defined by how well it corresponds to the visitor's personal agenda.

Personalization is a useful tool to create such a correlation because it helps a visitor to find out what, within the museum, could fit better with their personal agenda or correspond more to their expectations. This can be done either from home on the museum's Web site or directly onsite through interactive devices available in the museum. Upon completing a profile, where the intending visitor must indicate different types of information such as how and when they are tentatively planning to visit, with whom, how long they plan to stay, what sort of interest(s) they have and which language they understand, the system will be able to provide a personalized plan for the visit that takes the submitted information into consideration. Personalized museum plans can be very useful, especially for large museums where visitors are likely to be overwhelmed by the number of objects or exhibits available for viewing during a single visit. In such a context, visitors are often disoriented and find it difficult to decide what they want to see or do. Answering a few very simple questions, or defining a few criteria, can help them to overcome these limitations, enjoy the visit more fully and learn more easily.

A number of museums are working on developing online and onsite applications based on these principles. The *National Museum of Ethnology* in Leiden, the Netherlands (www.rmv.nl), for example, has developed an onsite facility called *"The tour of the world in 80 questions"* that allows children aged 7 to 13 to print out a personalized tour plan of the museum based on an individual choice of subjects and continents. The tour plan, which is colorful and easy to understand for children, includes a series of maps that help locate the objects, a brief description of the artifacts and a list of questions related to the subjects chosen, which the young visitors need to answer during their museum exploration.

The *Cité des Sciences et de L'Industrie* in Paris is undertaking a project called *"Navigateur"* (Navigator), which will allow visitors to create a personalized tour based on an individual choice amid a set of criteria which include the context of the visit (alone, family, group), the language spoken, the particular interests, the time available and the type of experience desired. Once the visitor has set the criteria that are most relevant for them and has checked the offerings on the museum interactive plan, the personalized proposal can be saved on the museum bar-coded ticket, which will be used during the actual visit, when using different interactive devices throughout the museum, to obtain further assistance in finding the recommended exhibits or to reset the criteria based on new interests that might have arisen during exploration. The system will be linked directly with Visit Plus, thus creating continuity between the pre-visit, actual visit and post-visit experience, through the use of personalization.

Conclusion

The examples provided here from different science museums all over the world help to prove the potential role that personalization could play in strengthening the overall learning process before, during and after the actual visit, in advance through activities that orient visitors and afterwards through opportunities to continue reflection and explore related ideas. However, despite the obvious potential benefits that these applications can bring to the visitor's experiences, there is still very little evidence that these systems work in the terms envisaged by their promoters, especially with respect to learning. This is because, due to their relatively recent nature, most of these projects have not yet been subjected to thorough evaluations that focus on establishing, among other things, the long-lasting effects of personalization on the learning process. Until now, the very few evaluations that have been carried out have focused mainly on whether people use the systems or not, why they do so, where they encounter most difficulties and on their usability in general. Despite the fact that further studies are needed in order to shed light on the effectiveness of personalization as a pedagogical tool, the first evaluations of these early examples, as well as other similar projects, have given initial help in indicating various pros and cons related to their use.

The overall feedback concerning the introduction of personalizing applications to audio guides and virtual environments seems to be reasonably positive: visitors are spending more time in the virtual and real museum, they access information at the level of detail desired and appreciate the idea of being able to bookmark information for reference later (Semper & Spasojevic, 2002). In particular, a study by Cordoba and Lepper (1996) has evaluated the consequences of personalization with respect to stimulating intrinsic motivation and learning in a computer-based educational environment. The findings provide strong evidence that the students for whom the learning contexts had been personalized, through the incorporation of incidental individualized information about their backgrounds and interests, displayed better gains in motivation, involvement and learning than their counterparts for whom the contexts had not been personalized.

However some drawbacks have also emerged[6]. First of all, there are the issues related to the difficulty and expense of implementation and also problems in practical use by visitors[7]. So far it seems that only a limited number of visitors take advantage of the benefits available through personalization, partly because the systems are not implemented in a clear and easy manner and partly because most visitors are either not ready for technology or not willing to invest time in it. Therefore it is important to remember that personalization should not be implemented for the sake of it but when and because it brings added value to the museum for, if not all, a good percentage of visitors. Only if this occurs can the costs for investment and development be justified.

Some experts have warned against the use of personalization. Nielson (1998) has argued that personalization is over-rated, saying that good basic Web navigation is much more important. For example, it is helpful to consider different classes of use in the main home page, such as physical visitors, the disabled, children, teachers, researchers, groups, etc., and to give each of these a relevant view of the resources that are available (Bowen & Bowen, 2000). Such usability issues are certainly important, and relatively cheap to address with good design, but even Nielson admits that there are special cases were personalization is useful.

More recently, there have been further questions about the effectiveness of personalization (Festa, 2003; McGovern, 2003), despite the enthusiasm of some. For example, the costs may be up to four times that of a normal Web site, around a quarter of users may actually avoid personalized Web sites due to privacy concerns and only 8% are encouraged to revisit because of personalized facilities (Jupiter Research, 2003). This compares with 54% who considered fast-loading pages and 52% who rate better navigation as being important. However, other surveys indicate that personalization can be effective, for example in the field of downloadable music (Tam & Ho, 2003).

Another issue that needs to be stressed in personalization is related to standardization procedures and applications. This process is central both for content description and user profile definition using metadata (Conlan et al., 2002). The description process can however be very time-consuming and expensive, but if it is pursued properly it allows the resources to be reused for different purposes and a visitor profile to be created using various different sources of information following evaluation criteria. Museums are sometimes not very quick in adopting new technologies but in some cases the slow perspective allows them to make the most of other institutions' initial mistakes and thus to avoid them. Involvement with standards provides a good opportunity to share such knowledge.

Thus it is recommended for museums to use personalization on Web sites judiciously at the moment, although science museums with good funding may wish to be more adventurous. There is a place for personalization in leading-edge Web sites and for certain innovative facilities like advanced Web support for specific exhibits. It is an area that museums should certainly consider, but the costs should be weighed against the benefits. Of course, the costs are likely to decrease as commercial and open source support improves in this area. At the moment, not insignificant development effort is needed for such facilities, but in the future they could be increasingly packaged with standard database-oriented Web support software, such as content management systems, as understanding of what is useful and not useful is gained from practical experience. This is certainly an interesting and fast-moving area that should be monitored by innovative science museums, especially at a national level.

References

Beler, A., Borda, A., Bowen, J. P., & Filippini-Fantoni, S. (2004). The building of online communities: An approach for learning organizations, with a particular focus on the museum sector. In J. Hemsley, V. Cappellini, & G. Stanke (Eds.), *EVA 2004 London Conference Proceedings* (pp. 2.1-2.15). University College London, The Institute of Archaeology, UK.

Bertoletti, A.C., & Costa, A.C.R. (1999). Sagres – A virtual museum. In D. Bearman & J. Trant (Eds.), *Proceedings of Museums and the Web 1999*. Archives & Museum Informatics. Retrieved from *www.archimuse.com/mw99/papers/bertoletti/bertoletti.html*

Bertoletti, A.C. et al. (2001). Providing personal assistance in the SAGRES virtual museum. In D. Bearman & J. Trant (Eds.), *Proceedings of Museums and the Web 2001,* Seattle, Washington, March 14-16. Archives & Museum Informatics. Retrieved from *www.archimuse.com/mw2001/papers/bertoletti/bertoletti.html*

Bonnet, M. (2002, June). Personalization of Web services: Opportunities and challenges. *Ariadne,* (28). Retrieved from *www.ariadne.ac.uk/issue28/personalization*

Bowen, J.P. (2004, January). Cultural heritage online. *Ability,* 53, 12-14. Retrieved from *www.abilitymagazine.org.uk/features/2004/01/A53_Cover_story.pdf*

Bowen, J.P., & Bowen, J.S.M. (2000). The website of the UK Museum of the Year, 1999. In D. Bearman & J. Trant (Eds.), *Proceedings of Museums and the Web 2000.* Minneapolis, Minnesota, April 16-19. Pittsburgh, PA: Archives & Museum Informatics. Retrieved from *www.archimuse.com/mw2000/papers/bowen/bowen.html*

Bowen, J.P., & Filippini-Fantoni, S. (2004). Personalization and the Web from a museum perspective. In D. Bearman & J. Trant (Eds.), *Museums and the Web 2004: Selected Papers from an International Conference,* Arlington, Virginia, March 31-April 3, (pp. 63-78). Pittsburgh, PA: Archives & Museum Informatics. Retrieved from *www.archimuse.com/mw2004/papers/bowen/bowen.html*

Bowen, J.P., Houghton, M., & Bernier, R. (2003). Online museum discussion forums; What do we have? What do we need? In D. Bearman & J. Trant (Eds.), *Proceedings of MW2003: Museums and the Web 2003,* Charlotte, North Carolina, March 19-22. Pittsburgh, PA: Archives & Museum Informatics. Retrieved from *www.archimuse.com/mw2003/papers/bowen/bowen.html*

Brown, R., & Kulick, J. (1997). Flashbulb memories. *Cognition, 5,* 73-79.

Brusilovsky, P. (1994, August 17). *Adaptive hypermedia: An attempt to analyse and generalize.* Workshop held in conjunction with UM'94 4th International Conference on User Modeling, Hyannis, Cape Cod, Massachusetts. Retrieved from *wwwis.win.tue.nl/ah94/Brusilovsky.html*

Brusilovsky, P., & Maybury, M.T. (2002). From adaptive hypermedia to the adaptive Web. *Communications of the ACM, 45*(5), 30-33. Retrieved from *doi.acm.org/10.1145/506218.506239*

Brusilovsky, P., & Nejdl, W. (2004). Adaptive hypermedia and adaptive Web. In M. Singh (Ed.), *Practical handbook of Internet computing.* CRC Press. Retrieved from *www.kbs.uni-hannover.de/Arbeiten/Publikationen/2003/brusilovsky-nejdl.pdf*

Case, S., Thint, M., Othani, T., & Hare, S. (2003). Personalisation and Web communities. *BT Technology Journal, 21*(1), 91-97.

Conlan, O., Dagger, D., & Wade, V. (2002, September). Towards a standards-based approach to e-learning personalization using reusable learning objects. In *E-Learn 2002, World Conference on E-Learning in Corporate, Government, Healthcare and Higher Education.* Montreal, Canada. Retrieved from *www.cs.tcd.ie/Owen.Conlan/publications/eLearn2002_v1.24_Conlan.pdf*

Cordova, D.I., & Lepper, M.R. (1996). Intrinsic motivation and the process of learning: Beneficial effects of contextualisation, personalization and choice. *Journal of Educational Psychology, 88*(4), 715-730.

Csikszentmihalyi, M., & Hermanson, K. (1995). Intrinsic motivation in museums: What makes visitors want to learn? *Museum News, 74*(3), 34-37, 59-61.

Dolog, P., Henze, N., Nejdl, W., & Sintek, M. (2004). Personalization in distributed e-learning environments. In *Proceedings of 13th World Wide Web Conference* (pp. 170-179). New York City, IW3C2/ACM. Retrieved from *www2004.org/proceedings/docs/2p170.pdf*

Eirinaki, M., & Vazirgiannis, M. (2003). Web mining for Web personalization. *ACM Transactions on Internet Technology, 3*(1), 1-27. Retrieved from *doi.acm.org/10.1145/643477.643478*

Falk, L., & Dierking, L. (1992). *The museum experience.* Ann Arbor, MI: Whalesback Books.

Festa, P. (2003, October 14). Report slams web personalization. *CNET News.com.* Retrieved from *news.com.com/2100-1038_3-5090716.html*

Filippini-Fantoni, S. (2003). Museums with a personal touch. In J. Hemsley, V. Cappellini, & G. Stanke (Eds.), *EVA 2003 London Conference Proceedings,* University College London, July 22-26, (pp. 25.1-25.10) (Cf. Beler et al. ref).

Hamma, K. (2004, May). The role of museums in online teaching, learning, and research. *First Monday*, *9*(5). Retrieved from *firstmonday.org/issues/issue9_5/hamma*

Hidi, S. (1990). Interest and its contribution as a mental resource for learning. *Review of Educational Research, 60*, 549-571.

HMSO. (2001). *Special Educational Needs and Disability Act 2001*. UK Government, Her Majesty's Stationery Office. Retrieved from *www.hmso.gov.uk/acts/acts2001/20010010.htm*

Hsi, S. (2003). A study of user experiences mediated by nomadic web content in a museum. *Journal of Computer Assisted Learning, 19*(3), 308-319.

Jupiter Research. (2003, October 14). *Beyond the personalization myth: Cost effective alternatives to influence intent*. Jupiter Media. Retrieved from *http://www.internet.com/corporate/releases/03.10.14-newjupresearch.html*

McGovern, G. (2003, October 20). Why personalization hasn't worked. *New Thinking*. Retrieved from *www.gerrymcgovern.com/nt/2003/nt_2003_10_20_ personalization.htm*

Mobascher, B., Cooley, R., & Srivastava, J. (2000). Automatic personalization based on web usage mining. *Communications of the ACM, 43*(8), 142-151. Retrieved from *doi.acm.org/10.1145/345124.345169*

Moraes, M.C., Bertoletti, A.C., & Costa, A.C.R. (1999). The SAGRES Virtual Museum with software agents to stimulate the visiting of museums. In P. De Bra & John J. Leggett (Eds.), *Proceedings of WebNet 99: World Conference on the WWW and Internet*, Honolulu, Hawaii, USA, October 24-30, (Vol. 1, pp. 770-775). Charlottesville, VA: Association for the Advancement of Computing in Education (AACE).

Nielsen, J. (1998, October 4). Personalization is over-rated. *Alertbox*. Retrieved from *www.useit.com/alertbox/981004.html*

Oppermann, R., & Specht, M. (1999). Adaptive information for nomadic activities a process oriented approach. In *Software Ergonomie '99* (pp. 255-264). Walldorf, Germany. Stuttgart: Teubner.

Semper, R., & Spasojevic, M. (2002). The electronic guidebook: Using portable devices and a wireless Web-based network to extend the museum experience. In D. Bearman & J. Trant (Eds.), *Proceedings of Museums and the Web 2001*. Boston, April 18-20. Pittsburgh, PA: Archives & Museum Informatics. Retrieved from *www.archimuse.com/mw2002/papers/semper/semper.html*

Tam, K.Y., & Ho, S.Y. (2003). Web personalization: Is it effective? *IT Professional, 5*(5), 53-57. Retrieved from *csdl.computer.org/comp/mags/it/2003/05/f5053abs.htm*

Tan, W.L.H., Subramaniam, R., & Aggarwal, A.K. (2003). Virtual science centers: A new genre of learning in Web-based promotion of science education. In *Proceedings of the 36th Annual Hawaii International Conference on System Sciences (HICSS'03)* (Vol. 5, pp. 156-165). IEEE Computer Society.

Endnotes

¹ A "cookie" is a small piece of data sent by a website and stored on the client-side (browser) computer that can be reused later on the server-side (the Web site that sent the cookie) as unique information concerning a user.

² A Web server log is a record of each access to a Web server with information such as the name of the client computer, the date/time and the resource accessed.

³ These applications are currently available on a number of different museums' Web sites such as the Metropolitan Museum of Art, the Whitney Museum of American Art, etc. For a detailed description of these applications (see Bowen & Filippini-Fantoni, 2004).

⁴ Please note that the distinction between *formal* and *informal* education is used here in a rather loose sense. Usually, in the educational sector, classrooms are considered formal learning settings, while museums are considered informal learning settings. As an alternative, we propose here to use the term formal e-learning tools in relation to proper courses meant for students who cannot attend classes; while by informal e-learning tools we refer to online or onsite educational environments.

⁵ Note that the acquisition of knowledge about the visitor is done in an explicit way: information is directly extracted, through the filling of forms, with direct answers to questionnaires. SAGRES works with two kinds of models: individual model and group model. The group model is built by the teacher and used by students. The teacher is responsible for the definition of the students' characteristics, by the definition of the group stereotype (subject, knowledge level and language of the consultation), the activities stereotypes and the classes (name of the students presented in the group).

⁶ It is not the intention of this chapter to be negative towards the use of personalization techniques in museums, but to highlight constructively some of the questions that come to light when the social uses and design problems are considered.

⁷ For more detailed information on the problems related to the implementation and use of personalization techniques see Filippini-Fantoni (2003).

Chapter XIV

E-Learning and Virtual Science Centers:
Designing Technology Supported Curriculum

John Falco, Schenectady City School District, USA

Patricia Barbanell, Schenectady City School District, USA

Dianna Newman, State University of New York, USA

Suzanne Dewald, Schenectady City School District, USA

Abstract

A model for partnership with virtual science content providers creates technology-infused science curriculum using interactive videoconferencing technologies and supporting Web resources. The model, based on the work of Project VIEW, demonstrates the viability of videoconferencing and the integration of interactive digital technologies in K-12 classrooms as means to accessing unique science resources for the classroom to engage students in dynamic, self-constructed learning. By bringing enriched content to schools, new structures for pedagogy are emerging that motivate students to learn more, both with and without teacher assistance, effectively promoting increased cognitive development. The chapter offers research results confirming the progress of the model.

Introduction

As science centers throughout the world enter the interactive arena of digital communications, a need has emerged to design strategies of program development that seamlessly interface new technologies with existing missions and resources. To facilitate this process, Project VIEW[1] (Virtual Informal Education Web), a collaborative undertaking led by the Schenectady City School District, has bought teachers together with major centers of science learning to create templates for developing interactive point-to-point videoconferences with asynchronous Web-based resources that enhance student learning.

The integration of interactive and digital technologies into programs, delivered by science centers to K-12 schools, requires new kinds of strategies and tools to create science resources for classrooms to engage students in dynamic, self-constructed learning. The development process brought to science centers by Project VIEW enables them to design interactive, digital delivery systems of instruction that produce evidence of higher-level student learning and academic performance. In addition, Project VIEW provides unique resources to students who may not otherwise have access to them.

As centers of science learning increasingly employ innovative models to provide enriched content to schools through interactive technologies, new structures for pedagogy are emerging to motivate students to learn more, both with and without teacher assistance. Not surprisingly, to achieve this transformation, it has been necessary to change the construct of educational pedagogy and the structure of instructional content. The end result of this transformation of educational delivery systems has been to facilitate increased cognitive development among participating students.

Research Background

Underlying the work of VIEW is the premise that one of the most highly effective methods of achieving enhanced levels of learning in the sciences among students is to conduct educational pedagogy through interactive, digital technologies. The feasibility of this premise is validated by the recent International ICT Literacy Panel (2002) that proposes that, "… Because technology makes the simple tasks easier, it places a greater burden on higher-level skills." To attain this higher achievement, Project VIEW has created a model of content development that employs training and collaborative design techniques that use interactive videoconferencing and Web-based learning to bring together the needs and missions of diverse yet intersecting educational delivery systems at science centers and schools.

The theoretical ideas behind Project VIEW began in the 1980s during an era of change in American education that began to focus on the benefits of integrating technology into K-12 classrooms (including science classrooms). For example, Ragosta (1982) documented that, when compared to traditional classroom settings, students learned more quickly in a Web-based environment.

Elements of the VIEW approach integrating digital resources of science centers into K-12 education have incorporated basic knowledge gained from Forman (1982), who recognized that use of computer integration in educational contexts increased students' motivation to learn. The VIEW Model for E-Learning with science centers is reinforced by fundamental understandings of how students learn through the use of technology that were first documented in the findings of the Electronic Learning Laboratory (1982). These authors reported that computer use (compared with traditional classroom education) resulted in increased student attention span, higher levels of student questions in classroom discussion; and evidence of enhanced active participation in classroom activities.

Not surprisingly, more recent research continues to substantiate and expand these early findings. For example, Sandholtz, Ringstaff, & Dwyer (1997) have reported that instructional technology not only improves performance on standardized tests but also promotes collaborative, technological, and problem-solving skills.

As Project VIEW has sought to improve access to science centers and laboratories, it has built on research-based assumptions that have foundation in the work of Mowre-Popiel, Pollard, & Pollard (1994), who put forward the premise that use of technology supports philosophies of instruction that perceive each student as a unique learner. These assumptions have led to emergence of a student-centered pedagogy parallel to that supported in science education in which technology aids in the transformation of classrooms into student-centered learning settings.

The work of Schutte (1998) and Weiss (1998) also has been useful to the development of the VIEW model. These researchers documented that a digitally based, interactive learning environment results in "authentic" learning which includes promotion of students' cognitive growth, fostering of a deeper level of understanding of the information being discussed, and providing multiple perspectives on a given subject through the nature of the interactivity.

It also has been demonstrated that interactive, digital technologies can be used as effective tools to infuse content-rich resources (such as those offered by science centers) into curriculum (Schutte, 1998). As a result of digital and real-time videoconference access to external educational resources, students gain information from outside the classroom with the teacher acting as facilitator and science center experts serving as mentors and information sources. Further, Gernstein (2000) found that use of videoconferencing has results similar to those found for digital resources: students who participate in videoconferencing are more motivated and interested in the topic at hand, and are reported to have high levels of achievement in problem-solving and critical thinking.

At its core, the VIEW model incorporates the expertise of both primary digital and human resources of science centers into creation of curriculum based on the seamless integration of video programming and Web-based instructional resources in the classroom. To accomplish this, VIEW integrates digital technologies in content delivery by combining laboratory exploration, digital resources from Web portals, and integrated curriculum into an expansion of traditional classrooms through interaction with real-time resources.

Essential to the development of VIEW programs are pedagogical approaches that include challenges for student learning which are, in effect, "Mindtools," described by Jonassen

(1996) — "... computer [i.e., technological] applications that, when used by learners to represent what they know, necessarily engage them in critical thinking about the content they are studying." Thus, VIEW utilizes technology-based learning opportunities — interactive streaming video Web resources, videoconferencing, and interactive Web portals — that are designed to result in higher-level learning and complex critical analysis.

More recently, Pugh (2002) studied teaching elements to document their effect on the creation of "transformational" experiences in science learning. Pugh's discussion of fostering transformative experiences from "artistic crafting of content and modeling and scaffolding of perception and value" has a resonance in the VIEW model for collaborative content development with science centers. Further, the development of content from concepts learned through student-center inquiry is very reminiscent of the inquiry-based constructivist foundation of VIEW. Not surprisingly, the author's preliminary findings support the VIEW experience that such approaches result in higher student learning and performance.

VIEW's vision recognizes the critical need to prepare students with a high degree of literacy that is essential for success in the 21st century information age. To meet this need, the VIEW model provides a collaborative inquiry-based learning approach, using technology-facilitated constructivist methodology with innovative educational resources. The underpinning of this methodology is rooted in the work of Jonassen, Carr, & Yueh (1998):

> "When students work with computer technologies, instead of being controlled by them, they enhance the capabilities of the computer and the computer enhances their thinking and learning. The result ... is that the whole of learning becomes greater than the sum of its parts."

The potential application of VIEW's work is imbedded in a study by the U.S. Department of Education that has reported the potential of external educational resources in the United States — that is, 88% of museums have programs that are compatible with K-12 school curriculum (Weiss, 1998). Direct access to these external sites is often not feasible for most schools because of distance, time and lack of finances. Silverman and Silverman (1999) note, however, that through the use of technology, students and classrooms can arrange virtual visits that will yield important increases in both learning and motivation.

Determining Program Content

To reach the goal of making it possible for students to attain high achievement, Project VIEW has designed a model for program development that creates a learning community among classroom-based teachers and science center education content providers. VIEW partnerships aim to create innovative program delivery that brings non-traditional resources into the science classroom via interactive Web archives and point-to-point videoconferencing. The development process is structured to author programs that transform the way in which content is delivered and to create expanded opportunities for students to construct their own knowledge base. The overall projected outcome of all projects is increased understanding of the content, resulting in higher academic performance and greater affect toward learning.

To reach this outcome, it is necessary to engage in a series of considerations constructed to ensure that there will be synergy among the partners both in the selection of content and ultimately in its delivery. These underlying considerations for the selection of content for VIEW programs include alignment of selection with multiple factors briefly discussed below.

The Mission of the Science Institution

Program content must be aligned with the institution's core mission. Such alignment helps to ensure the basic institutional support necessary for growth or sustainability. For example, when content was selected for the Buffalo Zoo program, a major consideration was the Zoo's well-respected gorilla exhibit and the Zoo's mission to bring understanding and appreciation of the life of gorillas and the conservation of the species. By focusing on this aspect of the Zoo's collection, the program was easily integrated into the Zoo's overall priorities. (Supporting materials for the Buffalo Zoo project are found on the Project VIEW Web portal – http://www.projectview.org.)

The Responsibilities and Goals of the Schools

Alignment with educational content standards is essential for viability in the schools. If a program offers substance that assists schools in meeting mandated educational goals, and also supports and expands classroom mission and practice, administrators are likely to be open to committing valuable resources and classroom time to the program. Therefore, the selection of study of wetlands for VIEW development was a natural one, since study of wetland environments and their preservation is imbedded in the NYS and National Science Standards.

The Content Expertise of the Participating Educators

Teachers and science center professionals work most productively when they are using their own best skills, talents and abilities. VIEW projects are structured to build upon the expertise of the participants. Early in the development process, educators recognize that their best contributions to the project can come through their curriculum-based content knowledge and classroom experience. Likewise, science specialists from partner science centers utilize their in-depth knowledge of the materials and their experience interpreting that to a varied public. Together, the two points of view combine to create a foundation on which dynamic student-centered learning activities can be designed.

Establishing a Foundation

Many of the multiple components of the Project VIEW development process are not linear but rather occur in an on-demand constructivist environment, emerging both in real-time and asynchronously, as needed. To proceed with VIEW Development, it is necessary to establish a viable foundation on which to build a replicable project.

Capacity and Connectivity of the Science Center and/or Participating Schools

An underlying element in developing a program foundation is assessment of the ability among the partners to reliably deliver and/or receive programs. To ensure that the essential technological capacities are in place, initial activities require a series of collaborative planning sessions that are conducted to explore technological capacities before actual program development begins. This helps to establish an infrastructure that can support the program and allows participants to fully engage with interactive resources. A key element established at these meetings include the creation of an informal cooperative network of technology staff of the collaborating partners to assess technological infrastructure (hardware, software and connectivity) of both the science center and the schools in order to determine if the existing equipment and connectivity are sufficient for the program. In addition, a support structure protocol ("road map of service support options") among the participating partners also was developed to help users and program providers determine if they need support from their local technology specialist, VIEW's staff of highly skilled support professionals, and/or consultation with an external technical specialist.

Creation of Collaborating Teacher Teams

To maximize project effectiveness and sustainability, participating teachers are identified and organized to work in building-based, collaborative teams. The project teams consist of at least four (4) teachers from at least one (1) school. The teachers are selected so that they collectively represent at least three (3) different academic areas, ensuring an interdisciplinary approach to content development.

In reality, these building-based teams form a "Knowledge Community" — that is, an informal, self-organized network of practitioners reflecting multiple skills and competencies (Sallis & Jones, 2002). The "Community" forms the foundation of sustainability because the teachers are a working team who not only support each other in the delivery of replicable curriculum, but also create a community of enthusiastic practitioners who excite colleagues to join them.

Identification of Collaborating Science Center Teams

To facilitate institutional support and involvement, collaborating science centers also identify a team that guides the project and seamlessly integrates it into the center's content delivery structure. Usually that team (selected by the center) consists of:

- *Education staff member(s)* who work directly with the teachers to identify and develop resources which are made available through the Web portal, and content which is incorporated into the interactive videoconference;

- *Curatorial staff* who provide content validation and support for identifying and developing supporting content for the Web portal and for reference use by the development team;

- *Technology staff* or a plan for technology support to ensure the sustainable utility of both the Web portal resources and the interactive videoconference program; and

- *Administrative support* as needed to ensure that the program and Web resources are imbedded in the institutional plan and mission.

Pre-Program Orientation of Teachers and/or Science Center Staff

To assure that all participants are "on the same page" regarding goals and content focus, pre-development orientation/introduction sessions are conducted. The needs of the communities of collaborating partners are assessed and a specially tailored orientation is provided to each team. As needed, that orientation can take the form of either a short informational session, introduction to Web portal resources, or a half day or full day technological training introduction.

Designing the Program

Designing the presentation of project content and ensuring that the presentation design occurs in a useful form for enhancing education delivery are among the central activities of the development team. There are three main phases to these activities, which for the most part occur in a sequential time frame.

Phase 1. Content Immersion

A key component of the Project View development model is a constructivist blending of the knowledge and abilities of participating teams of developers (teachers and science specialists). To facilitate this process to the greatest degree, VIEW development begins with content immersion at the collaborating science center where participants receive an introduction to the resources of the institution. The primary source encounter with the authentic objects of the science center provides direct experience for the collaborating partners as they move forward in designing classroom application. Drawing on the objects and experience that they encounter in the immersion, collaborating participants in VIEW are encouraged to utilize their unique and diverse expertise as they engage in "brainstorming" about the educational application of those resources.

The development of the program is accomplished through a constructivist process, while the goal of the development is to create a program that provides a constructivist learning environment for students. VIEW has found that this goal can be more easily achieved when developers create the program through the same (or similar) method in which it will be delivered.

Phase 2. Content Refining

Within a short period of time (two to six weeks) of the content immersion, participants come together again for two (2) full-day sessions in which they participate further in a constructivist process. At these sessions, the partner teams collaboratively review the content presented during the immersion, examine materials available on the Web portal, and fine-tune possible content concepts which have been presented to them. They finalize the content selection for both the interactive videoconference and supporting Web portal, and align those selections with curriculum and standards (i.e., what students need to know and be able to do to reach high academic performance).

During this phase of development, teachers and science specialists engage in a constructivist exploration of the content, including brainstorming about the different aspects of the information and ways it could be presented to students. The first goal is for the collaborating partners to agree upon a core focus.

With this accomplished, the development team begins to review the content, to refine it and to identify central themes. Tangential areas of exploration are tagged for program enrichment and/or exploratory study. Throughout this process, it is important to remember Jeffers' (2000) words:

> "Different perspective, interpretations and criticism must be shared and creative conflicts (that lead to new discourse and new knowledge) must be engendered." (p. 29)

It should be noted that specific details of the constructivist process vary with each team of teachers and science centers, but a template for effective development methodology is emerging. Key elements of this template include facilitated discussion, by all participants, of varying points of view about content applicability and presentation; collaborative outlining of videoconference presentation and identification of experiences and resources needed to support and deliver the content; and self-evaluation of program and process and self-assignment of specific tasks for independent work.

By the end of the two days of content refinement, an over-all outline of the program emerges and the stage is set to put together the various pieces of the project.

Independent Development

As a final task at the end of VIEW's Development Phase 2 (Content Refining), participants review the work done and identify areas that need to be created to accomplish the program development. Once the project needs are collaboratively identified, the team members self-assign the various tasks attached to those needs. The self-selection aligns with the competencies and interests of each participant. In general, the tasks fall into three areas:

Authoring Supporting Lessons

Teachers design lessons that can serve as pre- or post-videoconference activities in the classroom to expand their curriculum while supporting the program that will be presented by the science center. These lessons typically access portal resources to which the provider refers during the videoconference. For example, often teachers design science activities that help students build skills in analyzing what they read and in making connections to the content of the videoconference being developed by the science specialists. Teachers working with the National Air and Space Museum designed activities (available at http://www.projectview.org) for students that access online material about Wilbur and Orville Wright as a preliminary activity to build understanding of the invention process. In designing this activity, they interwove digital, online resources from the institution Web site and other sites into the reading activities.

Identifying Resources for Student and Teacher Use

Librarians and media specialists create digital resource bibliographies and/or Web-quests to utilize material accessed through the science center Web portal. Choices by team members have included Web-centered activities that enable students to explore the science content through multiple resources. In addition, teachers have identified links that are appropriate for student use in conducting online research.

Designing a Videoconference

In general, science center staff focuses on designing the actual program presentation based on the collaboratively determined program content. During this process, they gather and share archival and collective resources with participants, and focus the videoconference presentation on key concepts that enhance understanding of essential aspect(s) of the curriculum unit.

Reflection and Completion

Following the *Content Immersion*, *Content Refining*, and the independent development sessions of the VIEW model, the teams are reassembled for a final two days of program development. During this time, the team conducts a review of their collective efforts including revisiting their goals and evaluating the results of their independent work (curriculum, resources, lesson plans, etc.). This allows the team to identify gaps in the program and to collectively identify and address unmet needs.

In addition, during this time, participating team members often conduct a pre-pilot test to try out various activities for the program with students in their classrooms. They also make use of the technology to collaborate with other teachers and with other classrooms.

When engaging in reflection, benchmarks for program excellence are discussed and the teams review assessment options and rubrics to be sure they measure achievement of the project content goals. Benchmarks for program excellence may include *seamless curriculum interface* (a clearly articulated connection to content standards required in the classroom); *effective use of interactive media* (knowledge of and skills in using the interactive technology as a tool in education); *strong use of supporting interactive technologies* (integration of interactive technologies that enhance curriculum delivery and student learning); and a*lignment of program content and activities with institutional mission* (program fits the collections and purposes of the science center).

Program Piloting

Field-testing of programs is essential to establish the sustainable viability of the content for school curriculum. It is also important in assessing the effectiveness of program structure and presentation in communicating with students in a way that has a measurable impact on their learning. There are two stages of piloting for the VIEW model.

Piloting – Stage 1 – Participating Program Developers

The first step in ensuring the educational applicability of a program occurs in the classrooms of the team of teachers who developed the program. After the development sessions, participating teachers from the team volunteer to pilot the program in their classrooms. Those pilots include a commitment among all participants to provide constructive feedback on the effectiveness of the program in the classroom with students. After responses are gathered, the science centers are charged with retooling and/or revising programs as needed.

Piloting – Stage 2. VIEW Trainees

The second stage of field-testing takes place in the context of training sessions conducted by project VIEW staff to prepare teachers to be users and consumers of the programs that are developed. As teachers learn how to design integrated classroom modules using interactive video programs and technologies, they are charged with piloting the programs developed by VIEW collaborations. Their commitment includes developing integration plans for their classrooms and providing candid responses regarding the effectiveness and utility in the classroom of the programs. The feedback is shared with the science centers and the process of improving programs continues.

Research on Outcomes

Increasing evidence gathered by VIEW's independent evaluator (The Evaluation Consortium at the University at Albany) confirms that the VIEW model of collaborative development is changing the way that schools and centers of science learning deliver content. Research results suggest that, as the discrete missions and goals of collaborating teams are merged, VIEW is achieving success by creating a learning environment that reduces constraints of schoolhouse walls to bring external expertise into the classroom through interactive technologies.

Data from the project evaluation document that substantive changes are occurring in the way that participating teachers structure science content for their students. Access to science centers through interactive digitally based programming is resulting in more student-based hands-on opportunities for learning as students are able to use data-based research and analytic synthesis with science center online resources. For example, students working with scientists at SERC are able to create science experiments based on interactive learning and collaborative gathering of information.

Lessons Learned Through Practice

VIEW is a constantly emerging model, with a commitment to on-going revision and expansion of methods and methodologies as the constructivist process moves forward. Among the more important additions to the VIEW model in the past four years are 1) the assessment of prior knowledge of technology, and 2) the necessity of a strong technical support system.

From the beginning, it has become increasingly apparent that teacher knowledge of interactive technologies is important to the early stages of program development; it allows participants to function fluidly in the knowledge environment created by technology. It also has become clear that it is essential that collaborating partners clearly understand their roles in the process to ensure that all participants use their best and most developed talents and skills.

The underpinning of technology-infused programs is the seamless functionality of the technology. Since neither institution educators nor school-based teachers are technology support specialists, it is essential that a well-structured technology support system be identified at the beginning of projects to facilitate the institutionalization of projects in the classroom.

Student and Teacher Outcomes

Assessment of classroom related outcomes is an ongoing component of the development of VIEW models. The focus of this assessment is on students' cognitive functioning, their interest and motivation in learning, and in teacher interaction with materials. As noted above, each module developed by VIEW undergoes field-testing in the classroom as the final stage of development; in addition, as part of the ongoing documentation of VIEW's impact, use by non-developer classrooms is also documented where possible. Following is an overview of preliminary findings pertaining to classroom use of science related videoconferences as they are integrated into regular, standards-based curriculum.

- *Student Affect:* A key catalyst to student learning is motivation and interest in the topic and the self-perception that learning is taking place. As part of the documentation of the impact of digital resources on student learning, Project VIEW evaluation seeks to assess the degree to which the external resources provided by the content providers affects students' interest and motivation in the topic, the resources, and future efforts to learn similar materials. Presented in *Table 1* and *2* are summary findings, reported by students who participated in science related videoconferencing as part of their regular classroom curriculum. Each of these videoconferences utilized VIEW developed curriculum and videoconference presentations.

- *Student Centered Instruction:* Students reported a high degree of affective involvement with the curriculum and the integrated videoconference. An overwhelming majority of students noted that the program was easy to understand (97%) and that the material fit in with what they were currently learning in school (81%). Self-reported perceptions of learning also were documented; 95% of the students reported that they learned a lot from the program and 93% noted that they learned more than they would have in an ordinary classroom setting. Increase in student interest also was impacted by the process; 95% of the students said the material was interesting, thereby impacting current learning, and 75% reported that they would like to learn more in the future.

Table 1. Student perceptions of the videoconference[2]

Outcome	% Positive[*]
The program was easy to understand.	97
I learned a lot from the program.	95
The program was interesting to me.	95
I learned more about the topic through the program than I would have in an ordinary class.	93
The topic of the program fit in with what I am learning in school right now.	81
I would like to learn more about what I saw or learned during the program.	75

* Percentages are based on combined "Agree" and "Sort of Agree" responses.

Table 2. Student activities during the videoconference[3]

Statement	% Engaged
Watching the program	96
Answering questions	71
Asking questions	67
Participating in an activity with the presenter(s)	56
Discussing the topic with others	53
Solving a problem with the presenter(s)	32

Each of the nine classrooms was observed by two independent raters who documented the types of interactions that occurred in the classroom. In addition, each student was asked to report the types of activities in which he or she was engaged during the process. As noted in Table 2, students reported being engaged in a diversity of tasks that ranged from direct instruction to advanced problem solving. At least two thirds of the students reported being directly involved in guided inquiry during the videoconference process. This included either asking questions of the instructor or answering questions posed by the provider or by other students. In addition, over half of the students expanded this direct questioning methodology to include discussion of why information presented was true.

Findings by the external observers complimented students' reports. Observers noted the majority of students to be actively engaged in watching the videoconference, taking notes, sharing information with peers during problem solving sessions, and assisting each other in formulating questions for peer or presenter review. The majority of students also collaborated with each other on hands-on activities and advanced students were noted to be assisting students who had difficulties. Observers also reported that those students who had been noted as at-risk appeared to be more attentive to the presenter and the material and were involved in the learning process to a deeper degree than when observed during prior non-videoconference visits.

Concluding Comments

Using interactive videoconference programs and supporting Web-based resources, VIEW offers a model for making new and innovative integration of science expertise and collections into curriculum delivery. Science centers that have participated in VIEW are realigning their content presentation to interface and enhance curriculum in the schools. The result is a new, emerging reality in classroom pedagogy and a new, exciting extension of interpretive activities of science centers.

In general, participants in VIEW perceive that interactive educational technologies (such as point to point videoconferencing and digital archives) serve to benefit schools by increasing their access to authentic and exceptional resources from scientifically unique locales.

The VIEW model makes new and innovative use of institutional expertise and collections while structuring content delivery that can be seamlessly integrated into K-12 curriculum. The multifaceted nature of VIEW seeks to both increase access to videoconferencing as an educational tool and improve the use of that tool, and has served as a foundation for initiating comprehensive change in educational technology integration. As content providers, such as centers of science education, realign their content presentation to utilize interactive digital resources and communication tools, a new reality is emerging in education through new and innovative models for use of instructional time.

References

Digital transformation: A framework for literacy. (n.d.). Princeton, NJ: Educational Testing Service.

Electronic Learning Laboratory. (1982). *On task behavior of students during computer instruction vs. classroom instruction.* New York: Teachers College, Columbia University.

Forman, D. (1982, January). Search of the literature. *The Computing Teacher,* 37-51.

Jeffers, C.S. (2003). Gallery as nexus. Art Education, *56*, 19-24.

Jonassen, D.H. (1996). *Mindtools for engaging critical thinking in the classroom* (2ⁿᵈ ed.). Columbus: Prentice Hall.

Jonassen, D.H., Carr, C., & Yueh, H. (1998). Computers as mindtools for engaging learners in critical thinking. *Tech Trends, 43*, 24-32.

Mowrer-Popiel, E., Pollard, C., & Pollard, R. (1994). An analysis of the perceptions of preservice teachers toward technology and its use in the classrooms. *Journal of Instructional Psychology, 21*, 131-138.

Newman, D. (2004). *Research on how technology enhances standards-based teaching and learning: Points of VIEW, a Project VIEW collaboration with C-Span.* Albany, NY: University of Albany, State University of New York, Evaluation Consortium.

Pugh, K.J. (2002). Teaching for transformative experiences in science: An investigation of the effectiveness of two instructional elements. *Teachers College Record, 104*, 1101-1137.

Ragosta, M. (1982, Spring). *Educational Testing Service Bulletin.*

Sallis, E., & Jones, G. (2002). *Knowledge management in education: Enhancing learning & education.* London: Kogan Page.

Sandholtz, J.H., Ringstaff, C., & Dwyer, D.C. (1997). *Teaching with technology: Creating student-centered classrooms.* New York: Teachers College Press.

Schutte, C. (1998). Videoconferencing: Expanding learning horizons. *Media & Methods, 34*, 37.

Silverman, S., & Silverman, G. (1999). The educational enterprise zone: Where knowledge comes from. *T. H. E. Journal, 26.*

U.S. Department of Education. (1998). *True needs, true partners, museums serving schools: 1998 survey highlights.*

Endnotes

[1]   U.S. Department of Education Technology Innovation Challenge Grant, Project No: R303A000002

[2]   Data are based on surveys from 148 students in six classrooms at grade seven, 26 students in one classroom at grade six and 41 students in two classrooms at grade four; videoconferences were from Buffalo Zoo, SERC, and Cincinnati Zoo sponsored programs.

[3]   Data are based on surveys from 148 seventh grade students and 41 fourth grade students.

Section III

Case Studies

Chapter XV

A Virtual Museum Where Students Can Learn

Nicoletta Di Blas, Politecnico di Milano, Italy

Paolo Paolini, Politecnico di Milano, Italy

Caterina Poggi, Politecnico di Milano, Italy

Abstract

SEE, Shrine Educational Experience, represents an example of how Internet and multimedia technologies can effectively be exploited to deliver complex scientific and cultural concepts to middle and high school students. SEE (a project by Politecnico di Milano and the Israel Museum, Jerusalem) is based on a shared online 3-D environment, where students from four possibly different countries meet together to learn, discuss and play, visiting the virtual Israel Museum with a guide. The educational experience combines online engagement and cooperation to "traditional" off-line learning activities, spread across six weeks. Data from an extensive two-year-long evaluation of the project, involving over 1,400 participants from Europe and Israel, prove the educational effectiveness of this innovative edutainment format.

Introduction

SEE — Shrine Educational Experience — is an e-learning project based on a shared online 3-D environment, where students from different countries meet to learn, play, and engage in a high-level scientific debate about the Dead Sea Scrolls, one of the major archaeological discoveries of the 20th Century. The Dead Sea Scrolls were written by a Hebrew community who lived in the archaeological site of Khirbet Qumran between 170 BC and 68 AD (Roitman 1997). They represent the earliest known version of books from the Bible, and a precious source to understand the roots of Western civilization.

SEE is the result of cooperation between the Politecnico di Milano and the Israel Museum, Jerusalem. As part of its educational mission, the Museum wished to make its large body of knowledge and artefacts upon the Dead Sea Scrolls accessible to the public at large, and to open issues of scientific research to a broader public, with respect to the small group of scholars to whom the discussion is usually restricted.

Thanks to Internet technologies, providing simultaneous access to users independently from their geographical location, (middle and high school) students from all over the world can visit the virtual Shrine of the Book (Figures 1-2), and take part in discussions, games, and debates with international experts, discussing state-of-the-art research about the Dead Sea Scrolls.

Each SEE experience involves four classes of students between 12 and 19 years of age, located in different geographical areas: they meet, in the online virtual world, four times (over a period of six to seven weeks). Through the online meetings students get acquainted with each other, discuss, play, answer quizzes, present their social and cultural environment, etc. Students, in addition, cooperate off-line, under their teacher's supervision, studying background material (based upon interviews with leading international experts) and carrying on their own homework.

This innovative learning experience aims at four major educational goals:

1.   Providing rich, in-depth knowledge about the Dead Sea Scrolls and related issues, including the scientific methods of philological/archaeological research.

2.   Favouring a truly international, cross-cultural approach, where students of differ-

Figure 1. A screenshot of the virtual Shrine of the Book, reproducing the wing of the Israel Museum where the Scrolls are preserved

Figure 2. The "real" buildings of the Shrine of the Book at the Israel Museum, Jerusalem

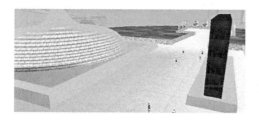

ent countries can understand/compare their tradition, their background, their views and beliefs, etc. Developing a better understanding of differences and respect for the "other" is the underlying goal.

3.   Fostering the use of information and communication technologies for educational purposes, with innovative teaching-learning paradigms.

4.   Offering interaction, fun and engagement (i.e., powerful motivators, encouraging students' active participation, even in the context of a demanding learning activity).

A massive field-test evaluation, involving over 1,400 students and teachers between November 2002 and May 2004, has proved the effectiveness of SEE: the experience achieves all the main goals (above mentioned) and, in addition, it produces a wide range of beneficial side-effects and generates an overall strong educational impact.

How the SEE Experience Works

Activities

A SEE experience consists of 4 cooperative sessions (i.e., online meetings in a 3-D virtual world) spread across 6 weeks, and of several learning activities, that participating classes (four at a time) perform in the intervals between a session and the next one (Figure 3).

Cooperative sessions are not expressly meant to be a learning moment; they are devoted to social activities, such as students introducing themselves, discussions upon the themes of the experience, and games testing the students' knowledge of content. In order to be prepared for the sessions, students must study detailed material in advance. For this "traditional" learning activity, "old fashioned" methods remain the most effective:

Figure 3. Schema of SEE learning activities

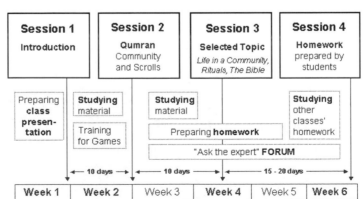

students download documents in printable format from SEE Web site, and study them at home or in the classroom.

The four sessions are organized in order to bring pupils from a basic knowledge of the Dead Sea Scrolls to a more in-depth analysis of some particular topic of interest. At the moment of registration, teachers can select from a set of available topics the one they wish to study in depth. Sample topics are: "Life in a Community," "Rituals in Qumran," and "Canons and the Holy Bible." Students are then required to do further research on this topic, investigating its links with aspects of their own environment and searching through local sources. In the final session they will present in turn to each other the results of their research.

All the considerations that follow are the result of direct onsite observation of over 70 online cooperative sessions performed during SEE evaluation, between November 2002 and May 2004 (see the third section of this chapter for details). More than 1,400 students and teachers from over 30 schools in Europe and Israel were involved: in addition to making the project real, they have been an invaluable source of insights, anecdotes and information about the actual educational impact of the SEE experience.

What Happens Online

Users in SEE online environment are represented by avatars (Figure 4). Two students per class connect to the 3-D world, and therefore have an avatar to control: they move around, see other users, and communicate with them in real-time via chat. The rest of the class supports them in various ways, either grouped around the two computers or following the session with the help of a projector, connected to one of the monitors.

During the **first session** students introduce themselves, their school and their country to each other (Figures 4-5). Then they are briefly introduced by a "museum guide," that

Figure 4. Avatars introducing themselves in the Shrine virtual environment. The boards in the background, once clicked, activate a pop-up window showing classes' presentations

Figure 5. An example of class presentation. Before the first session, classes are asked to send a HTML page with a picture and a short self-presentation

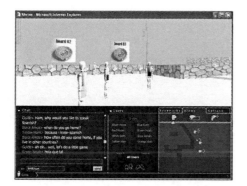

is, a member of the staff (also represented by an avatar), to the historical — geographical context of the experience. The goal of the first session is to stir the students' interest and motivate them to study the first set of documents. The evaluation tests in schools showed that the engagement of exploring a new virtual world and the excitement of meeting peers from faraway countries are extremely powerful motivators, able to capture the students' interest.

During the **second session** students enter the museum (Figure 6), where they are shown pictures of archaeological findings and other historical evidences (Figure 7). Discussions start concerning both the background material and the most interesting issues surfaced in the session itself. The "museum guide" moderates the discussion, asks questions (also to test the students' knowledge) and encourages participants to think about their own experience related to the issues being raised.

The guide coordinates every cooperative session, directing the activities (Chang, 2002), stimulating the discussion, assisting students with technical problems, assigning scores for the games and even assigning penalties for improper behaviour. In order to avoid waste of time, disorientation, and ineffective interactions, each slot of time in a session is dedicated to a specific activity: the guide makes sure that everyone knows what to do and does it. This is crucial for keeping the experience fast-paced and educationally effective; the guide is also the ultimate referee for the games.

Figure 6. Avatars in the corridor of the virtual Shrine of the Book; boards (once clicked) show significant images of objects preserved in the museum, or of the place where these were found

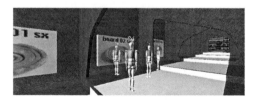

Figure 7. A board showing rests of a pool for ritual baths at Qumran

Figure 8. An avatar performs an ability game in the Quiz. If he reaches the top of the stairs before his opponent, he earns the right to answer the Quiz question first.

Figure 9. Avatars reflect on the answer to Quiz Question 1. When their team member finishes the ability game, they must counsel him about the answer he should choose.

Figure 10. Matching Pairs Game. Avatars try to reconstruct meaningful pairs among the objects found in the labyrinth

Figure 11. The vault space inside the Shrine of the Book white dome. Discussions of the third and fourth sessions take place here

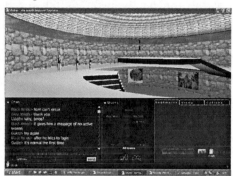

The guide awards scores to students depending on their contributions to the discussion. We could observe that assigning scores highly increases their participation and commitment to answer as correctly as possible. Classes are paired to form two teams: competition between them is strong from the very beginning. However, it touches its highest point in the games (Figures 8-9), where students demonstrate both their "physical" skill in controlling their avatar and their knowledge of the study material. Games in fact combine "movement" and "thinking:" despite their ability to move in the 3-D world, students earn no point if they cannot solve the riddles based on the content they should have studied.

The second session includes one game: a Treasure Hunt, where students are given clues to identify four particular museum objects among a set of 20 hidden in a labyrinth.

In the **third session**, the discussion about the topic of interest is followed by two games: a quiz, based on multiple-choice questions (Figures 8-9), and a matching pairs game, where students have to identify meaningful associations between couples of objects, again scattered in a labyrinth (Figure 10).

Finally, the **fourth session** is entirely devoted to the presentation of the students' homework: this is definitely, from a cultural point of view, the most intense moment of the experience. Students have the opportunity to explain their work, confront their views with the others', and answer to peers' critiques with passionate argumentations (Figure 11). The possibility of being confronted with different points of view is always interesting and valuable: when the discussion is about socio-cultural phenomena (such as aspects of the students' everyday life somehow related to the Scrolls' world), a cross-cultural approach bringing together people from very different backgrounds becomes intrinsically informative and enriching.

At the end of the fourth session, the guide announces the final scores — taking into consideration also the quality of the teams' homework — and proclaims the winner.

What Happens Off-Line

The effectiveness of a SEE experience from a learning point of view is determined, in large part, by the involvement of the teachers, especially in supervising the activities performed by each class in the intervals between a session and the next one.

Teachers are able to transmit their enthusiasm to the students, motivating them to study and do their homework accurately. When the teacher is not motivated (no matter how perfectly the technology works, how smart, responsible, hardworking the students are) the experience will probably be a failure. Well-motivated teachers, on the contrary, are able to make the best of a SEE experience, even with their most disaffected students.

Class work can be organized in different ways. While 4th year high school students are able to organize themselves quite autonomously, middle school kids need more directions and support from their teachers. In both cases, classes participate as a group, not as individuals: students split in sub-groups, divide the material among the sub-groups, and everyone takes charge of one particular task. They know that everyone's contribution is important for the team, and therefore they do their part with strong commitment for the team's success. Students, moreover, are implicitly taught how to collaborate in a group (Vygotsky, 1978): very important skills nowadays, when working environments frequently imply teamwork.

Teachers may decide to act only as supervisors, encouraging the students' initiative and personal responsibility; or they may exploit the interest stirred up by the project, either treating more in-depth the parts of the curricular program more related to the core theme, or basing lessons, class activities and exercises on the project's material, or taking advantage of its multidisciplinary character to involve as many colleagues as possible, and offering a multi-perspective approach to the subject matter.

Students, on their part, enjoy the game-like approach of the project and the use of technologies — which in some cases they know even better than their teachers. When they see the teacher at a loss with technical problems, they take the initiative and try to find a solution: their responsibility and resourcefulness are stimulated.

The project's offline activities include: downloading, printing and studying the contents, keeping in touch with remote team members, and collaborating with them to prepare the research homework.

The peculiar format of contents and the means for asynchronous communication with other participants are described in detail in the next sections.

What It's All About

The Interviews: A Dialogic Approach

Detailed content is offered to students in the format of interviews to leading international experts on subjects like Dead Sea Scrolls, Holy Scriptures, Ancient Literature, Hebrew,

Christian and Middle-East Culture in general. Unlike school textbooks, interviews provide a faceted, thought-provoking overview of the current state of research at academic levels, in a readable, straightforward style. Since debate over some issues is still open, students are startled to find (in the interviews), sometimes, totally conflicting assertions by different experts. They realize that historical and archaeological researches are not as linear and problem-free as they appear in history schoolbooks. They become curious of finding out which is the most convincing hypothesis. Teachers are thrilled to see how eagerly they engage in further research, investigating the criteria on which each hypothesis is based.

Interviews are integrated by a rich set of auxiliary material, including summaries, maps, anthological excerpts from the Scrolls, the Bible, or other sources quoted by the experts in the interviews, and editorial insets on relevant historical characters, peoples, or events. These also help integrating the backgrounds of students from different countries and of different ages.

The Experts' Forum: Debating with Researchers

If the dialogic format of the interviews gives the flavour of — as a teacher defined it — "a debate at academic level" reflecting "the state of the art of research," an even more exciting opportunity is offered to the students: one or more of the experts interviewed is available during the experience to answer the participants' questions about the issues presented in the interviews, or that emerged during online discussions. Experts communicate with the classes via a shared online message board provided on SEE Web site, allowing every user to see public messages and keep track of message threads.

This is a wonderful opportunity for middle and high school students, who would hardly ever have a chance to reach high-level scholars directly and ask them questions, not to speak of engaging in a serious discussion with them.

The message board is also the place where discussions started during cooperative sessions (and cut off at a certain point for lack of time) can be resumed and continued in a less hectic style. Although extremely stimulating, the chat is often frenetic and confusing: a sort of forum on the online message board allows users to post their contribution, somewhat lengthier and deeper than a chat message, possibly after having thought about it for a while. Even the experts might be involved in the discussions started online.

The Virtual Museum

Every cooperative session starts either outside or inside the virtual Shrine of the Book. A SEE experience is a totally new museum experience. Many factors influence a museum visit, the social aspect being not the least important (Falk & Dierking, 1992). Moreover, museum visits are far more significant from an educational point of view when preceded and followed by activities enhancing the comprehension of the objects exposed (Falk & Dierking, 2000). A SEE experience is a highly social activity, requiring participants to

Figure 12. The Shrine buildings

Figure 13. The virtual Shrine – external environment

Figure 14. Inside the white dome

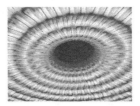

discuss together, play together, work and learn together. Additionally, all activities are aimed at enhancing the comprehension of the museum's content — which would appear rather enigmatic, even when seen for real, if not accompanied by explanations.

The virtual museum therefore reproduces only those aspects that, in the real Shrine, are meant to help visitors entering in the "Scrolls world", recreating the atmosphere of Qumran. For example, the members of the community who wrote the Scrolls referred to themselves as "the sons of light," and to their enemies as "the sons of darkness:" this opposition is symbolized by the contrast between the white dome and the black basalt wall, forming the architecture of the Shrine of the Book (Figures 12-14). The white dome emerges from a pool of water, representing the bathing pools (Figure 7) used by the inhabitants of Qumran for purification rituals. Furthermore, it is shaped like a lid of the jars inside which the Scrolls were stored. The corridor (Figure 6) reminds of the caves where the jars were found, and so on...

Rather than striving to create a perfect virtual reproduction of the "real" museum, SEE aims at helping students to "get into the atmosphere:" they absorb a great deal of information almost effortlessly, by simply looking around and talking about what they see. The engagement of interaction, discussion, and competition stirs their interest in the subject matter: at the end of the experience, many students — who did not even know about the Scrolls' existence before — were fascinated by them, and wished to visit the real museum.

How We Know It Works

The educational effectiveness of the experience has been assessed through massive on-field experimentation (*Table 1*) started in November 2002 and continued through three different phases.

Figure 15. Orientation of the schools involved

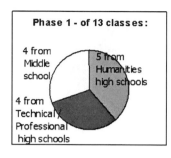 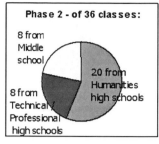 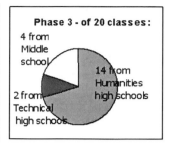

Table 1. Schools involved in SEE Experimentation. Students' ages ranged from 12 to 18. Computer expertise varied with the kind of schools; however, skilled computer-users were present in every class.

Phase 1: Nov. -Dec. 2002	Phase 2: Mar.-May 2003	Phase 3: Mar.-May 2004
7 schools in Italy	22 schools in Italy and Israel	9 schools in Europe and Israel
13 classes	36 classes	20 classes
15 teachers	44 teachers	21 teachers
Over 200 students	Over 700 students	Nearly 500 students
15 sessions performed	36 sessions performed (each one involving 4 classes)	20 sessions performed (each one involving 4 classes)

Phase 1 aimed at testing the effectiveness of the **technological platform and interaction dynamics**: a restricted number of schools simulated the sessions (with mock-up contents), first within a single class, and then with students of different schools logging in simultaneously, playing the games and chatting together. Observers in schools witnessed every session, detected problems, and collected comments and suggestions from students and teachers, on the basis of which several minor adjustments were introduced. On the whole, however, Phase 1 showed that the experience is definitely engaging: students enjoyed the interactive and competitive aspects very much, and wished that sessions were longer.

Phase 2 was an extensive, full-scope testing of the whole **educational experience**, assessing contents, subject-related discussions, off-line activities, games, homework, timing, organizational aspects and coordination with teachers. Observers monitored each cooperative session in every single school, registering notable aspects on written reports, and offering assistance to teachers if needed. Questionnaires were sent to students and teachers after the end of the experience, to evaluate its educational effectiveness and the users' satisfaction.

Phase 3 focused especially on the international, **cross-cultural value** of the experience: a larger proportion of participants outside Italy were involved, bringing fascinating multi-perspective contributions to online discussions. Scheduling, instructions for teachers and the workflow of activities were refined, enabling schools to manage all the complex organizational issues of the project without the assistance of a person from the staff physically present in the school. Again, feedback was collected through questionnaires before, during and after the experience, and an accurate analysis of chat flows disclosed many interesting insights about the educational value of online cross-cultural cooperation.

Direct Observation

Monitoring cooperative sessions in presence proved a rich and reliable source of information, necessary to understand and complete questionnaires results. Observers could easily assess the content's level of complexity, the effectiveness of interaction dynamics, and understand the nature of those problems, which caused a generic sense of frustration in students and teachers, without them being able to explain why.

After witnessing a session, observers had to write an overall evaluation on a report, with precise feedback concerning interaction, content, and technology.

Many refinements and improvements were made to the experience, basing on the experience thus accumulated and on the observers' reports. Their analysis revealed several interesting aspects; for example, even significant technical problems (such as users repeatedly disconnecting) did not hinder the students' enthusiasm if they were able to take part in the interaction and play at least part of the games. On the other hand, students get really upset when they are involved in some engaging activity (either game or discussion) and suddenly they have to stop. They are also disappointed when they do not get as high scores as they thought they deserve. Observing the users, we realized that what they really appreciate is the engagement of interaction, the excitement of competition, and the gratification for the guide's rewards (for their correct answers or contributions). Despite technical and organizational problems, when these elements are present the session is successful.

Students' Questionnaires

After the end of the second experimental phase, students were asked to fill in a questionnaire, sent to the teachers via e-mail and then distributed in the classes. 226 questionnaires were filled in and sent back. Some of the questions had a 4-scale predefined set of answers (such as "a lot, enough, not so much, not at all"); others were just open questions, to collect opinions and suggestions. We shall focus now on a few of them:

Q 6:   *Would you be happy to repeat the SEE experience?* 50% of the students declared themselves very keen on repeating the experience; another 40% said they would be glad to do it again. Only 1% answered that they wouldn't. The main reason to repeat the experience was the intercultural aspect, that is, the "possibility of meeting other students from different countries and cultures;" the second reason was the interest in the topic and the third one was the fun of the experience.

Figure 16. Data from students' questionnaires

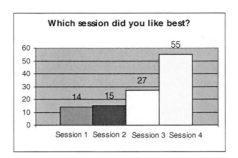

Actually, as also the observers had reported, the most appreciated session turned out to be the fourth one, during which every group presented to the others a research and discussed it, comparing different cultural points of view.

Q 11 and 12: Which game did you enjoy best? Do you have any suggestions or critiques about any game? The most appreciated game turned out to be the quiz, that is, the one in which the rules were the clearest and the "physical" and cultural part most clearly distinct (while the avatar was performing an ability game, the rest of the class tried to find the correct answer to the quiz). Students suggested making questions more difficult: they had studied hard and felt underestimated when questions were too easy.

Q 17: Which of the 4 sessions did you like best? The outcomes (Figure 16), apart from confirming the fact that students liked discussing their homework in Session 4, also show an ever-increasing interest and involvement in the activity, with a climax in the last meeting that evidently left them, so to speak, with a "good flavour."

Data concerning the outcomes of the third experimental phase have been compared with the expectations collected before its beginning.

Expectations on the overall impact of the experience were actually a bit higher than the final outcomes, although the difference is very slight. This clearly proves that both students and teachers are eager and ready to exploit innovative educational tools; this outcome clearly shows that "traditional" e-learning is probably already outdated and encourages us to keep in this path.

Teachers' Questionnaires and Focus Groups

Teachers were involved in the design and assessment processes from the very first steps of the project. Five focus groups were held to choose the cultural topics, structure online activities, define the format of the background material and of the homework; three additional focus groups provided precious insights about the project's educational impact. Their contribution to the tuning of the project was invaluable.

At the end of the second experimental phase, questionnaires were distributed to all the teachers who had taken part in the project. 19 questionnaires were filled in and given back to us. On the whole, scores ranged from "good" to " very good," never scoring "very bad." About students' interaction (during the cultural discussions, the games and the preparation of the homework), teachers would have appreciated more communication outside online sessions among the schools involved; they suggested that schools should be helped to keep in touch after the project's end, possibly meeting in the "real" world. In the third experimental phase, more attention was devoted to encourage cooperation among schools.

Teachers found the interaction very engaging and a powerful stimulus for studying; moreover, the use of new technologies was a good opportunity to couple "diligent" students with those more apt at interacting in a virtual environment, thus emphasizing their different skills (Gardner, 1983). Particularly rewarding was the outcome of Question 3 (*How do you evaluate the educational impact of the experience?*): two teachers judged the educational impact of the experience "excellent," eight scored it as "very good," another eight as "good;" only one scored it as bad, and none as "very bad."

The outcomes of the focus groups will be discussed in detail in the next section.

In Which Sense It Works

Teachers were constantly consulted before, during and after every experimental phase. Their contributions, collected through interviews, personal communications and focus groups, were illuminating — especially as far as the educational benefits and unexpected effects of the experience are concerned. A second valuable source of information were the chat flows, registering students' conversations, and showing how each one's remark stimulated the others' thoughts and provoked reactions, in a progressing discovery process.

SEE educational impact was observed on three fronts: 1) Content; 2) Students' motivation and attitude; 3) Learning methodologies.

Learning the Content

Teachers particularly appreciated the interview format, enhancing **critical thinking** and stimulating students to evaluate the experts' different contributions, possibly assuming an opinion of their own. They said: *"The interview approach is extremely interesting: it shows the state-of-the-art of the research, a debate at academic level."*

Quotes from teachers' statements collected during focus groups are reported in italics between quotation marks.

The strong **interdisciplinary** character of the experience was also appreciated. Students could see how many different disciplines, with their diverse criteria and methods, may converge on one single issue, each one bringing its special contribution. *"Among teachers, great emphasis was put on the interdisciplinary quality of the project. We discussed on how to involve as many subjects as possible: Italian Literature, English,* *and Religion. It was important for us to involve not only our students, but also our colleagues. If the class coordinator feels involved in the project, and works in collaboration with teachers of other subjects, this becomes a real strength. Students realize that the Experience is multidisciplinary."* In some humanistic high schools, teachers exploited the interest stirred by SEE introducing new topics related to the subject: *"I slightly modified the curricular program of History, Latin and Greek Literature, studying in depth the authors which had more to do with the project. We read Greek excerpts from the Genesis and Josephus Flavius [...]; students were very curious about him, because they knew he had to do with the project."*

Finally, the **cross-cultural** exchange is one of the most fascinating aspects of the experience. Students discuss in real-time with peers located in distant countries, discovering how different their perspectives can be. While Italian students, for instance, tend to regard the Dead Sea and Jerusalem as remote, almost fabulous places, Israeli pupils are much more concrete about them, because many have been there: they described the archaeological site of Qumran, the heat, the rocks, the bathing pools; they talked about Qumran religious feasts that are still celebrated today; when someone suggested that the inhabitants of Qumran used to eat fish, they immediately pointed out that "there is no fish in the Dead Sea." After studying the *Rituals in Qumran*, each class presented its research on a particular rite or feast celebrated in its local area: it was wonderful to observe the variety of uses and traditions, and to discover how all of them were originated by the same need for celebrating important events, that is shared by human communities of all times and places. *"The 'otherness' element, the meeting with other countries and cultures, is always stimulating. During the first session, it was exciting for the students to see themselves and the boys and girls of the other class. They are right there, and so faraway at the same time..."*

Enhancing Students' Motivation

SEE's unusual approach, dealing with complex and serious matters in a playful, engaging way, had strong effects on motivation and attitude: we observed a general increase in students' care and attentiveness, the occasion for "bringing out" problematic students and the improvement of discipline.

Competition, the desire of winning the games, and the engagement of communicating with peers of different countries, are powerful motivators for studying both the material and the language in which it is discussed during the sessions: *"They realized the importance of learning English." "They read the interviews at home, then we discussed them in class; we inquired on the historical perspective, doing further research on the different opinions of the experts. I never threatened examinations; nevertheless they studied with great care to be prepared for the Experience. They probably wouldn't have been so committed, without the games' spur."* It was amazing to see how even very young students became deeply knowledgeable about "difficult" subjects, such as the history of Middle East, religious issues, etc. In SEE, interaction drives the content: the thrill of meeting peers from far away, the impression — given by the 3-D virtual world — of "being there" with them in a remote country, the desire to win the games, are crucial for arousing the students' interest.

Figure 17. Scenes from the experimental phases

Teachers typically chose for experimenting SEE their most motivated, hardworking students; some teachers instead made a different choice: *"We selected our least motivated students: we thought that, if something could "rescue" them, it would be a project like this. And we were right."* The innovative teaching-learning style proposed by SEE offered disaffected students an opportunity to show their **commitment**: *"All of them participated with enthusiasm. Even two kids with comprehension problems had studied well and knew everything."*

Of the many different skills pupils possess, only few are evaluated in "traditional" school activities. A different learning approach gives these '**hidden talents**' a chance to emerge, and such abilities are extremely appreciated by the class. Teachers, on their part, are glad to reward the kids' keen involvement and commitment: *"One of my students had never been outstanding in Greek; however, being good at using computers (and being the only guy in a class of girls), he was chosen to play all the games, and to supervise any activity involving the use of technologies. He worked very seriously and accurately. I gave him a good mark to reward his active participation in the project."*

Some **discipline** rules had to be followed during online meetings: no offence was tolerated, and scores were taken from teams for misbehaviour. Sometimes students, while cheering for their team, tended to address the opponents via chat in rather unfriendly terms; when necessary, the guide admonished and punished them, yet, most of the times, it was the students themselves who urged their classmates to be disciplined, and restrained each other from reacting to provocations. In classes with discipline problems, teachers regarded this fact as a huge improvement.

Learning How to Study

There are many ways to organize class activities related to the experience. Once agreed on few essential guidelines, necessary for coordinating work among different schools, much is left to the creativity and initiative of the teachers. Without their passionate and professional work, the SEE experience would never work, no matter how carefully prepared by the staff. On the whole, the project had two basic effects concerning learning methodology.

1.   All the students of a class felt as a "team," understanding that they had to cooperate in order to successfully participate in the experience.

2.   Students learnt how to work in groups, autonomously and responsibly (although with the fundamental supervision and support of the teacher).

The project has a sort of **"team spirit building"** effect: all the students felt "as one," knowing that *"each one's skills were resources for the class. They understood that, by playing their role well, the whole team would benefit." "I saw none of the usual jealousy for those who controlled mouse and keyboard: they stood together, united to win."*

In order to better organize their effort, they usually split in groups, each one taking charge of a specific part of the study material, sharing their knowledge with the others and answering the cultural questions during the online experience. While sometimes it was the teacher who formed the groups and assigned materials, in most cases the kids organized **autonomously** and **responsibly**. *"Students worked a lot by themselves. They really had the idea that this was* their *project. "*

They had to work in group for preparing the homework: they met together after school hours, doing research, interviewing local experts and even working at the teacher's house, who put his/her library at their disposal: *"They live in the same neighbourhood and met in the afternoons to study together, whereas usually they work alone."* Again, teachers' supervision played a crucial role: *"They worked hard, preparing schemas and conceptual maps. We worked on two fronts: knowledge and method; we wanted them to learn not only the contents, but also how to distinguish important information from secondary aspects."*

On the whole, they learnt how to **effectively collaborate** in view of a common goal – a very precious skill in today's society — and the experience left its mark: *"One of my pupils of last year, who now attends a senior high school with two other ex-students of mine, came to see me and told me about her new schoolmates. She said: 'we had to do some work in groups, and the others are* so *clumsy! You know, they didn't do the Scrolls'... "*

Conclusion

We are convinced today that the "format" of SEE is effective and applicable to a variety of subjects (including scientific and technological ones), but also we have derived a number of "lessons" about what to do and about what not to do. We do not claim that we have completely understood what happened during the experimentation, but we can provide the readers with a few (possibly) useful hints:

A.   Trying to "reconstruct" a museum (and this was our starting point in 1999, when the project begun — see the works of Barbieri et al., Di Blas et al.) in 3-D is of little relevance. A 3-D virtual museum never conveys the magic of "being there" in the real place, and therefore it can't emotionally influence the students. 3-D can be used, however, for recreating the general "atmosphere" (e.g., Jerusalem, the main architectural features of the Israel Museum, etc.), but success or failure do not depend upon the quality or faithfulness of the reproduction.

B.   Showing, in a virtual world, all the objects on display in the real museum is useless: a virtual museum cannot be "used" as a real museum (e.g., for a group visit). In order to emphasize the "objects" (always an important goal for a museum) a different approach is needed.

C.   The "virtual visit" must be compelling, engaging, and fast-paced; we decided that its main purpose, in our case, was to meet other people "there," in the virtual museum.

D.   Motivations and interaction dynamics typical of the real world do not always work properly in the virtual world. The social sense and warmth of being "in a group" in a real museum is not easily conveyed in a virtual one; the natural engagement of a guided tour (with a good guide) in a museum is hard to replicate in a virtual one. On this ground, we have used games and competition as a key factor for creating engagement, stimulating social interaction, and motivating the students. Moreover, the 3-D environment is not the place where substantial learning happens; yet, the activities in the virtual environment are tremendous "motivators" for the learning process, which mostly takes place off-line, in the classrooms.

E.   Competition, involving the whole class, in cooperation with another class, acts as a global strong motivator both for teachers and students: they want to "win." Competition builds up a strong feeling of team-ship within the class and produces long lasting beneficial effects.

F.   Games, within the sessions, involve two students per class only, and we were afraid that the rest of the class would not be involved. As a matter of fact, instead, classes followed their "champions" playing for them, cheering and trying to help.

G.   Interviews have played an important role: their format, their natural way of exposing difficult problems, their "state-of-the-art" quality, and their mutual inconsistency (they often contradict each other) have captured the interest of the students and teachers and stimulated their critical thinking.

H.   The discussions, in the 3-D environments, have been alive and vibrant: more than we expected. The fast pace and the action of the "guide" played a crucial role in this.

I.   Homework was felt as very important: students showed their own specific traditions and cultural background, illustrating their discoveries — with great satisfaction — to the other classes during the homework's comparison.

We can ask now, to ourselves and to the readers, two fundamental questions:

A.   Does an experience like SEE provide benefits of any kind for a museum?

B.   Is SEE applicable to any kind of museum, and to scientific-technical ones in particular?

As far as the first question is concerned, the answer depends, of course, on what a museum perceives as its mission: if a museum conceives itself as "objects-holder" and "objects-displayer" for the public, then an activity such as the one described in this paper has little to do with it. Many museums, however, are meant to be "culture facilitators or mediators," that is, a means through which culture is popularised. The main difference with respect to universities or research centres is that museums are places where anybody can go, and interaction with the museum content is at the centre of the experience: "abstract" knowledge, unrelated to a sort of "physical experience," is not appropriate for a museum. Scientific museums in particular consider themselves as a place where visitors do learn something (rather than simply looking at objects). If the goal of a museum is to facilitate learning, then we can consider our "format" as a novel way to achieve its goal.

The second question requires a little bit of thought: gaming, competition, quizzes, etc., would certainly work with scientific and/or technological topics; nonetheless, two crucial factors would be missing: the display of local traditions and culture (we assume that science and technology are "the same" everywhere) and the consequent discussion, that create an atmosphere of cross-cultural environment.

Interviews again would work as content's format, but they should be carefully crafted: they should present state-of-the-art research, but with terminology and concepts acceptable to students and teachers. The attitude of teachers could be also a source of problems: our long experience (since 1996) in introducing technologies in public schools has shown that teachers of scientific-technological subjects are often afraid of state-of-the-art content, since they do not understand it, in most cases.

The above said, we are not afraid to take up a new challenge and try to adapt the SEE format for a science museum: candidates are welcome.

Acknowledgments

We wish to acknowledge the work of all the people who contributed to make SEE a successful experience (details can be found in www.seequmran.it).

References

Barbieri, T. (2000). Networked virtual environments for the Web: The WebTalk-I and WebTalk-II Architectures. In *Proceedings of IEEE for Computer Multimedia & Expo 2000* (ICME). New York.

Barbieri, T., & Paolini, P. (2000). Cooperative visits to WWW museum sites a year later: Evaluating the effect. In D. Bearman & J. Trant (Eds.), *Museums and the Web 2000: Selected Papers*. Pittsburgh, PA: Archives and Museum Informatics.

Barbieri, T., & Paolini, P. (2001). Cooperation metaphors for virtual museums. In D.

Bearman & J. Trant (Eds.), *Museums and the Web 2001: Selected Papers*. Pittsburgh, PA: Archives and Museum Informatics.

Barbieri, T. et al. (1999). Visiting a museum together: How to share a visit to a virtual world. In D. Bearman & J. Trant (Eds.), *Museums and the Web 99: Selected Papers* (pp. 27-32). Pittsburgh, PA: Archives and Museum Informatics.

Barbieri, T. et al. (2001). From dust to StardDust: A collaborative virtual computer science museum. In *International Cultural Heritage Informatics Meeting: Proceedings from ICHIM 2001*. Milano, Italy.

Bowman, D., Hodges, L., & Bolter, J. (1998). The virtual venue: User-computer interaction in an information-rich virtual environment. *Presence: Teleoperators and Virtual Environments, 7*(5), 478493.

Bowman, D., Kruijff, E., LaViola, J., & Poupyrev, I. (2001). An introduction to 3D user interface design. *Presence: Teleoperators and Virtual Environments, 10*(1), 96-108.

Bowman, D., Wineman, J., Hodges, L., & Allison, D. (1999). The educational value of an information-rich virtual environment. *Presence: Teleoperators and virtual environments, 8*(3), 317-331.

Chang, N. (2002). The roles of the instructor in an electronic classroom. In *Proceedings of ED-Media 2002*. Denver, CO: AACE.

Di Blas, N., Paolini, P., & Hazan, S. (2003a) Edutainment in 3D virtual worlds. The SEE experience. In D. Bearman & J. Trant (Eds.), *Museums and the Web 2003: Selected Papers*. Pittsburgh, PA: Archives and Museum Informatics.

Di Blas, N., Paolini, P., & Poggi, C. (2003b). SEE (Shrine Educational Experience): An online cooperative 3D environment supporting innovative educational activities. In D. Lassner & C. McNaught (Eds.), *Proceedings of ED-Media 2003*. Norfolk, VA: AACE.

Di Blas, N., Paolini, P., & Poggi, C. (2003c). Shared 3D Internet environments for education: usability, educational, psychological and cognitive issues. In J. Jacko & C. Stephanidis (Eds.), *Human-computer interaction: Theory and practice. Proceedings of HCI International 2003* (Vol. 1). Mahwah, NJ: Lawrence Erlbaum Associates.

Di Blas, N., Paolini, P., & Poggi, C. (2004). Learning by playing: An edutainment 3D environment for schools. In *Proceedings of ED-Media 2004*. Lugano, Switzerland.

Falk, J.H., & Dierking, L.D. (1992). *The museum experience*. Washington, DC: Whalesback Books.

Falk, J.H., & Dierking, L.D. (2000). *Learning from museums: Visitor experiences and the making of meaning*. Walnut Creek, CA: Altamira Press.

Gardner, H. (1983). *Frames of mind: The theory of multiple intelligences*. New York: Basic Books.

Roitman, A. (1997). *A day at Qumran: The Dead Sea sect and its scrolls*. Jerusalem: The Israel Museum

Vygotsky, L.S. (1978). *Mind in society*. Cambridge, MA: Harvard University Press.

Chapter XVI

Open Learning Environments:
Combining Web-Based Virtual and Hands-On Science Centre Learning

Hannu Salmi, Heureka, The Finnish Science Centre, Finland and
University of Dalarna, Sweden

Abstract

This chapter describes the changes in the role of informal learning education in science centres. It shows by several cases how the rapid development of modern information and communication technologies after the mid-1990s has influenced the traditional hands-on exhibitions to move towards open learning environments. The reported experiences of the different types of Web-based solutions in science canters provide evidence and practical hints for further development of traditional exhibitions towards open learning environments. Results underscore the role of intrinsic motivation as the key element for learning. The prices and other thresholds for using existing ICT-based learning solutions have decreased considerably, and now the main consideration is whether there are enough social innovations, that is, are there meaningful content and use for the innovative technology? To create an open learning environment from the elements of the exhibition and the Internet is clearly one of the main challenges of science centres.

Introduction

Using the Internet has become a common practice in education, both formal and informal, during the past few years. According to a recent Europe-wide survey (Ilomäki et al., 2004), searching for knowledge and information as well as surfing freely are the most common activities. The use of e-mail in education has also become a more common feature according to a survey among European teachers in 2003. However, other possible advantages, such as discussions, building knowledge structures using net applications and tools, making contact with experts or using the net as a publishing forum for pupils, are very rarely used in formal education. (Ilmomäki et al., 2004).

The strength of science centres' role in e-learning is the willingness of teachers to exploit opportunities to get involved easily with free-ware or moderate-price new digital learning materials which then can create new pedagogical practices. This matters also to individual learners in informal education.

New educational software can be divided into two main types (see Collins & Strijker, 2003): 1) Large, complicated systems functioning mainly as distributing databases, and 2) Small digital learning materials, which can be chosen from several sources, and which can be combined with regard to the situational needs of the learner. The most common term related to this kind of educational software is "Learning Object" (LO). It began to be used widely in Europe, especially after several projects partly funded by the EU (see www.eun.org). The definition of the term is very broad, meaning practically any material, in digital or non-digital form, to be used for learning, education, teaching, or training (Duncan 2003). Characteristic features for successful and effective LOs are: 1) easy access, 2) usability, 3) readiness for use by teachers without extra work, and 4) pedagogical flexibility (Collins & Strijker 2003).

Although ICT-based applications can provide many opportunities for science centres, these institutes have not been early adaptors of cutting-edge technologies (Severson et al., 2003). One reason is that until recently there have been problems with the usability of these technologies — they have not been user-friendly and are not particularly suited for mass use such as in exhibitions. However, there are clear signs that the blended experience of both exhibitions and ICT-based solutions can fulfil a variety of learning-style needs and would appeal to both formal and informal learning.

The combination of an exhibition and the Internet forms an excellent example of an *open learning environment*, a term that has been used in the e-Learning literature since the mid 1990s, especially in the context of life-long learning and distance education (see, for example, EURYDICE — the information network on education in Europe: *www.eurydice.org*). The open learning environment consists typically of a combination of real physical environments and ICT-based learning. The model by Falk and Dierking (1992) describes well the physical, social and personal context inherent in this kind of situation.

When we consider the role of the Internet in science centres, we have to first define the place of science centres in informal education. Then, we have to define the role of the Internet, both in the science centre and in the wider field of education.

Informal Learning and the Internet

Informal education has often been regarded as the opposite of formal education. Even the names of classic books, *Deschooling Society* by Ivan Illich (1971) and *The Unschooled Mind* by Howard Gardner (1991), have been provocative. These books also contain harsh criticism of the failures of schooling which have alienated students from meaningful learning. Moreover, they argue that learning from informal sources is effective and motivating. These books have had a significant bearing on education and research.

In addition to his sharp criticism towards schooling, Illich also proposed new solutions for learning. The positive ideas for learning are the "Learning Webs", which consist of four kinds: 1) Reference Services to Educational Objects; 2) Skill Exchanges; 3) Peer-Matching; and 4) Reference Services to Educator-at-Large (Illich, 1971). As a matter of fact, some of Illich's basic concepts have become reality, for example in the World Wide Web and e-mail. However, it must be stressed that Illich (1971, p. 115) does not use the term "network" (preferring "opportunity Web" instead), because "network is often used, unfortunately, to designate the channels reserved to material selected by others for indoctrination and entertainment".

To advance public understanding of science, new forms of education are actively being sought. A large amount of information, especially about modern phenomena, can be obtained in a personal way from family, friends, and peer groups. The roles of television, libraries, magazines and newspapers are also important. More significantly, museums and science centres have seen increased attendances during the last decade. Most of these forms of education can be classified as informal learning: either focused on young people via informal, out-of-school education programmes or as clearly informal learning occurring totally outside of any educational institutions for young people or adults.

The model in Figure 1 (Salmi, 1993, 2003) forms the basis for the role of Internet learning related to informal learning. Since the 1990s, informal education has become a widely accepted and integrated part of school systems (Crane et al., 1994). In recent times,

Figure 1. Informal and formal learning (Salmi, 1993, 2003)

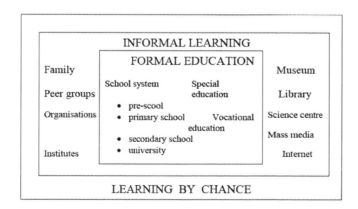

informal learning has become a more accepted part of educational science, and there are nowadays considerably more literature and research reports about topics such as learning via the Internet (Clark & Slotta, 2000). The role of the Internet is a clear example of a learning source that was originally created for other purposes. The Internet is now an effective informal learning source, and is commonly used by teachers, students and other formal learning institutions. In other words, the Internet can be described by the term 'out-of school education', meaning schools using informal learning settings and sources as a part of their curriculum.

The educational role of science centres, as well as computers and the Internet, has been regarded as being more or less self evident. However, some classical educational theories can be detected in the principles underlying science centre education, although few educators in these institutions have been explicit in their approach.

Computers in Science Centres

The proposed model is extended by the theoretical approach of Hawkey (2002, p. 7), in which the continuity of "real vs. virtual" is added to the model of "formal vs. informal" learning.

The model shows the characteristic of the Internet; being both informal and virtual, it has certain strengths and limits. The main difference between the "new" science centres set up in Europe in the late 1980s and the "Oppenheimer" model based on physical "hands-on " centres in northern America (Hein, 1990), is the wide and open use of audiovisual and computer technologies. The key element for the new science centres set up at the beginning of the 1990s was on interactivity. The new computer and audiovisual laser-disc technologies allowed the provision of large amounts of data to visitors in a totally new form. The new technology was seen as an attractive tool, and was also itself fascinating for many visitors — and exhibition planners. Unfortunately, this has led to some overrated solutions with colourful designs but with practically no content. Nonetheless, the beginning of the personal computer era was important for science centres and their audiences. For example, in Finland, in Heureka's 1985 test exhibition "Pulssi", 21 percent of the visitors indicated that it was the first time ever that they had used a personal-computer by themselves. This percentage stayed high until the beginning of the 1990s, especially among the elderly generation.

However, the Hawkey model does have some weaknesses, especially as the placement of television in the upper right (formal/virtual) quadrant of Figure 2 raises some questions: What makes television a "formal" medium as compared to the Internet? At least the ordinary TV programmes such as documentaries are clearly informal learning sources according to the nomenclature of learning sources (see Bitgood, 1988; Salmi, 1993, 2003; Crane et al., 1994; Hofstein & Rosenfeld, 1996). Even educational TV programmes, or related VHS or DVD videos on the TV-screen, fulfil more the definitions of informal than formal learning. Indeed, television fulfills the definition of "free-choice learning" (Falk & Dierking, 2002). Furthermore, recent research related to closed societies indicates clearly that informal learning via Western style television during the Soviet

Figure 2. Persistent dichotomies or blurring boundaries? (Hawkey, 2002, p. 7)

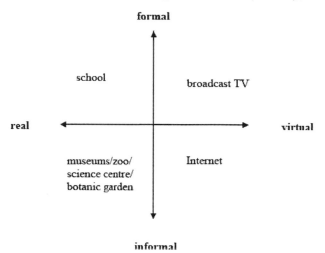

period was the most difficult problem for the KGB (Salmi, 1993; Graf & Roiko-Jokela, 2004) as well since the Internet is out of reach in otherwise strictly closed and controlled societies.

The dilemma of virtual and real science centres can be stated very clearly in the following interview reply by the children's programme television producer, Mr. Orvo Kontio from Finland: "When we found an excellent exhibit unit at the science centre exhibition and started to film it for our television programme, we soon found out that it lost its effectiveness when seen on television. In addition, when we decided to make rather passive and least popular exhibit units as a topic for the programme, the reactions of the audiences (and ourselves as professional television producers) were eager and positive!" (Salmi, 1991). The weakness of television is that it cannot move the tactile information that the visitor receives immediately while using a hands-on exhibit. On the other hand, the visual sense dominates the human brain, and this is capitalised by the power of television to good effect. Even exhibits rated "below average" might be impressive on television if the visual component is strong enough and provides impact for situation motivation, or mediates some new information (Salmi, 1991).

The same dilemma is also present in other screen-based media: video, CD-ROM, DVD, and the Internet. In some aspects, the Internet as a screen-based learning source can avoid part of the lack of tactile information because it is at least semi-interactive. If the Website of a science exhibition contains several options and alternative approaches instead of the usual way of presentation, it might even compete with the exhibition itself.

Bevan and Wanner (2003) have, as a result of their surveys, presented the main elements of the relationship between hands-on science exhibitions and the Web: 1) The best Web solutions have been created in a very similar way to the best exhibits; i.e. as original ideas by exhibition planners that allow the learner to direct the process, accrete the experiences, and further his/her own free-choice learning; 2) The main difference is the lack of physicality of the Web; 3) The exhibits that can be converted most easily and directly to the Web are those exhibits relating to perception and vision; and 4) The growing multimedia capacity of the Web has led science centres (like the Exploratorium) to create more Web casting experiments related to exhibitions.

However, to produce suitable Web casting material as an alternative to established broadcasting programmes, professional skills and funding for hardware and software are needed. The expertise for producing meaningful content remains the main consideration.

Live Discussion and Interaction

A pioneering trans-Atlantic Internet interaction experiment between science centres was developed and administered by the Minnesota Science Center and Heureka in 1998. This project allowed students visiting science centres in Europe and the US to communicate on-line with each other. This was possible during the late afternoon (European time), when the visitors at Minnesota had arrived at their centre early in the morning. The senior high school students from Finland were able to communicate in English. The content matter of the project was fairly easy to finalize because the centres had the same kind of interactive exhibitions and this provided a context for the discussions. The main topic was on ancient cultures: students were comparing and reporting facts related to Egyptian mummies, graves and the everyday lives of ordinary people. Physically, the interaction and communication via the Internet took place from the science centre exhibitions. However, the main interest was in the brand-new technology, which did not cause major problems. This project greatly resembled the experiments with video conferencing a decade earlier; the situation motivation of people attending the virtual meetings was at a high level but after the project, it did not create new permanent solutions for virtual science centres.

Sleep Laboratory:
A Combination of Exhibition, Media, Television, and the Internet

How can added value for an exhibition be provided via the Internet? It is frequently the case that the exhibition receives more visibility in the media than in the science centre itself. An example of this was *Illusions* — the brain exhibition in 1996 at Heureka which presented the latest brain research. One of the main attractions of the exhibition was the Sleep Laboratory. For a period of 200 days, volunteers were sleeping in shifts during the day time in a quiet, darkened room located in the middle of the exhibition. While the visitors were able to see these 'sleepers' through semi-transparent glass, on-line data was being displayed by EEG equipment, which registered the different phases between dozing and REM (rapid eye movement) sleep.

This data, combined with interpretation, provided a great deal of knowledge about the basic functioning of the brain. Students had an opportunity to do real science research projects, both at the exhibition and by following the sleep process on the Internet.

Although the main image on the Website was certainly passive — only part of the face on a pillow, it somehow fascinated Web users. In any case, it was also showing the slow development and change in the EEG signals on the screen. One of the reasons for the novelty of the programme was that people very rarely see other people sleeping, either because of darkness, or because of intimacy and privacy considerations.

The project was administered as part of European Science Week in 1996, and it aroused a great deal of interest in the media. Over a period of three months, several newspapers and magazines had written articles about the project, and after three television programmes on European satellite channels were screened, the number of visits to the Heureka Sleep Website was overwhelming the server!

Evaluation data from schools, students and teachers on the effectiveness and popularity of this programme indicated that the key element was content matter: the topic was meaningful for the students. People sleep for one third of their lives, and sleep and dreams are an essential part of our culture. The history of scientific sleep research is rather short, although there is a great deal of recorded anecdotal experiences dating back to ancient history. The evolution of sleep research is directly related to the development of technology and computers. There are several phenomena that were first recognised centuries ago but have only become measurable in recent years.

The content of the project supported the curriculum of school education. As an informal learning source, it gave much-needed information about the various phenomena that was not available in ordinary text books. According to the evaluation, the most active users of the Website were the 5th grade students in primary school and the 9th grade students in high school — the levels where human biology is one of the subjects taught in the science curricula. In addition, senior high school biology and psychology teachers were actively contacting the sleep laboratory. According to the survey, senior high school students (n =180) attending the sleep laboratory during their visit to the centre with preparatory Web-based material stayed for a noticeably longer time at the exhibition area. The difference was clear when compared to the control group which used textbook based materials. As can be seen from Table 1, the intrinsic motivation level of the Web user group is at a higher level, and the difference is statistically significant ($p<.05$). However, there was practically no difference in the situation motivation level. It seems as if that the sleep laboratory had an effective motivation effect on anyone seeing it 'live' at the exhibition.

Table 1. Intrinsic motivation: Pre- and post-test

Group	Score	
	pre	post
Web preparatory	25.7	26.8*
Textbook preparatory	25.2	25.3

However, the core focus of the project was the real sleep laboratory supported by additional information from newspapers, television, and other media. This made the project well-known and people started to access it via the Internet. Although the evidence from this project comes from the early years of the Internet, it is still relevant today. When creating a meaningful Web site, it can only become popular outside the circle of normal active users only by informing people via other media. This is very important since, although search engines are widely used, the users are usually people who already have an interest in the topic and have at least some information about the possible existence of the site.

Bionet

The topic that visitors seems to be very interested in all science centres and cultures is that relating to the visitor him/herself, i.e., human biology. The rapid developments in life sciences has made it a necessity to deliver information about more complicated topics too, and not just focus on the basics of human biology and body, so popular in most of the science centre exhibitions. In this context, science centre planners face a number of issues that need to be addressed for the benefit of the public: should we clone human embryos to cure diseases? Would you choose your child's genes? Would you eat genetically modified food? Will new drugs keep you healthy and make you live forever?

The above considerations led eight European science centres and museums to create a Website called *Bionet* (*http://www.bionetonline.org*), which is presented in nine languages. The site lets visitors explore and debate on the latest discoveries in the life sciences. Visitors can explore the science in question, look at the ethical issues, compare the regulatory laws in different countries, play games and express their own opinions.

The concept and content of the Bionet Website was the result of brainstorming sessions with experts from both research centers and science centres. As a result, the "E-Exhibition" was structured as a series of "scenarios" (Bandelli, 2003), e.g. challenging situations which provide access to the following interactive zones: social implications, ethics, overview of risks, research, toolkit, legal issues, European vs. National viewpoints, interviews, glossary, and debates.

The content available on Bionet is comprehensive compared to many other sites. It provides information for different types of audiences, and this has been documented exceptionally well in the final report of the project (Bionet, 2004): The site had 119,000 visitors in the first six months starting from November 2002. The average time spent on the visit was 11 minutes. The most common language of the users was English (24%); 11 percent of the users chose Portuguese and 12 percent Swedish. One of the strengths of the site was that it caters to the user's native language. This is extremely important when the content is both *new* and *complicated*. Especially for youngsters, the language barrier might be the reason for skipping the topic.

Analysis of the amount of visits to the various content hosted on the site of Bionet does provide evidence of rather content-orientated users: 63 percent of the viewed pages were content-based; 15 percent was on legal issues, 15 percent was on tools, and 13 % was on ethics (Bionet, 2004). Video-based content was chosen very rarely (1%). Reasons for this might be technical: watching video type of file imposes software and bandwidth constraints on a user's PC.

When the visits are analysed by scenarios, stem cells was the most frequently (30%) chosen topic. Another supplementing factor for the content and scenario was the age of the users. In the voting session, 28 percent of the users were 16-19 years old, and 12 percent of the users were 10-15 years old. The majority of the users were young people. This might explain why the theme on ageing was not so popular! It seems that the users are mostly high school and senior high school students. Sixty-seven percent of the visitors came to the Website directly, 30 percent via search engines, and 3 percent via links (mostly from science centre sites) (Bionet, 2004).

The feedback for this site has been encouraging for the planners. An evaluation of the Web pages also indicates that the users of these pages come more often from places other than science centre Web sites. Many students trying to find the latest information related to biotechnology have found the Bionet site via search engines. The opportunity to use it in different languages as well as the platform to compare regulatory issues and views in different countries have made Bionet a useful tool and resource in European debate. The project was co-funded by the European Commission.

Though it was conceived of as a temporary project, it became very popular among audiences and was such a useful tool for science centres that it became a permanent platform for information and discussion. It is a good example of how the Internet can provide added value for science centres and their visitors.

The Web element has to be meaningful, and the meaningfulness has two components. First, the content should be meaningful to the learner. Second, the learning process should be structured pedagogically in a meaningful way (according to age, prior knowledge and skills of the learner, and structure of the topic to be taught).

If science canters cannot respond to social change and update themselves, they may very easily lose their ideological credibility and sources of financial support. Controversial issues (such as gene technology) are all too easily omitted from the list of subjects covered by science centres, but the Bionet has shown that visitors do rely upon and respect institutions trying to address these kinds of topics.

Via the Internet, it is possible to react to the rapid developments of science and technology in a more cost-effective way than by creating huge and static exhibitions. Unfortunately, Websites too can get out-of-date very quickly by content and design. This is because the audience has become used to demanding the latest on-line information from the Internet. The dilemma is that this non-adequate (or even erroneous) information can circulate for years linked to many sites which are out of the control of the original producer of the information. The reliability and validity of Internet information are among the most difficult challenges that face informal learning. The Bionet project succeeded because it combined exhibitions, live happenings, discussions, and the use of the Internet. This combination was novel and attracted current interest. The sequel to Bionet is DECIDE - DEliberative CItizens Debates, in European science centres and museums. Earmarked to run from 2004 to 2007, it will be based on the earlier experiences and will concentrate more on citizen participation.

Improving Science Communication

ISCOM (Improving Science Communication) was a three year project from 2000 to 2002, and was led by ECSITE, the association for European science centres. It sought to sensitize leading science centres to the potential of new information and communication technologies for promoting the public understanding of science. The common misconception was that these institutions, according to their mission and practice, would have been the very first to apply ICT as an effective tool in their daily functions; however, regular use of electronic networks was not then widespread in European science centres and museums. It also appeared evident that the new technologies had been utilised to maintain the traditional functions and practices, rather than changing them, or creating new innovative forms (ISCOM Memo, 2001). This critical analysis clearly shows that science centres have not been among the leading developers either of the technology or the content of the Internet.

Although the analysis above is harsh, there were some early experiments. One goal was to create a science communicators' forum to exchange and diffuse best practices from science centres/museums. This initiative occurred under the aegis of the Laboratory for Science Communication Research at the University of Poitiers. Although the main focus was on the public's understanding of science, it soon became evident that the use of Internet and other ICT-based methods are some of the key elements in the discussion among experts, decision makers and their audiences. The ISCOM project also revealed a lack of use of the Internet as a means of contact between science centres and their audiences.

There were several occasions when the Internet was utilised at science centre exhibitions. At Heureka, there was free use of the Internet in the Children's Gallery in 1995. The main objective of the exhibition's Internet station was to give an opportunity to ordinary visitors to get acquainted with the Internet. The exhibit contained only a short introduction of the technology involved, and was more focused on the software and content which characterizes the Internet. The exhibit area became rather popular in the science centre. It was not only the opportunity to use it free, but for more than 20 percent of the visitors it was the first time that they had ever stepped inside the world of the World Wide Web, not to mention the fact that for 1 percent of the visitors it was still — in 1995, the first time that they had used a personal computer!

Context E-Learning with Broadband Technologies

CELEBRATE (Context e-Learning with Broadband Technologies) is a project with the ambitious goal of creating the first Europe-wide demonstrator portal that will provide up to 500 pilot schools   access to high quality Learning Objects and a virtual learning environment. During the period 2002-2004, it was being funded by the "Cross Programme

Action" of the European Commission's IST (Information Society Technologies) programme related to European youth in a digital age.

The principal objective of the initiative was to establish a small number of strategic, large-scale experiments in learning based on pedagogically sound principles while addressing operational and scalable technologies, applications and services. The project document describes it as follows:

> CELEBRATE will outline a pedagogy for e-learning in European schools based on a vision of what 'content' (resources + services + communication spaces) may look like in the future and how this will be created and delivered in online environments. In particular the project will provide schools with access to a large-scale demonstrator of an online content repository that includes the ability to share a critical mass of Learning Objects (LOs) and 'components' (understood as combinations of multimedia assets) that can be used to create LOs. It is intended that CELEBRATE will act as a catalyst for the European e-learning content industry (the entire value-chain including content owners, publishers, broadcasters, national school networks and ICT platform vendors).
>
> The project will explore how a LO methodology can be applied to educational activities and services as well as learning materials. A key aim of the project is also to investigate how Learning Objects can be handled by a new generation of Learning Content Management Systems (LCMS) from a variety of different vendors and to test the interoperability of these systems in a real-life demonstrator.
>
> CELEBRATE will particularly analyse the extent to which new, more flexible forms of content development and distribution (based on reusable LOs) impact upon the learning process and support a new pedagogy for e-learning in schools based on constructivist learning models. In particular, the project will assess how the use of LOs encourages the development of key digital age skills such as collaborative working, creativity, multidisciplinarity, adaptiveness, intercultural communication and problem-solving.

In addition to the different languages used in the European community, different school systems as well as various national and local curricula constitute formidable challenges for the development of pan-European learning materials. The funds invested during the last decade in ICT education has recently led national educational administrators and also the EU to focus on the results obtained in this field. More than on hardware and infrastructure, resources have also been focused on teacher training, computer courses, and further education of publishers.

E-Learning materials produced until recently have been based on practical experiences from early PC-based and CD-ROM learning materials. There are some excellent examples of cost-effective and well planned educational materials, but the overall concept in creating these materials is rather weak. Most of the materials repeat and replicate conventional didactic models. "ICT has been considered as an "add-on" to traditional forms of teaching and learning. It also assumes that the concept of the school and the role of the teacher will remain essentially unchanged and that pupils will remain basically consumers of pre-digested resources that have been tightly designed to meet specific curricular objectives" (de Figuero, 2001).

The CELEBRATE consortium consists of 27 organisations, and is co-ordinated by the European Schoolnet, a permanent organisation for all European ministries of Education that has strong experience in e-learning projects in Europe. The evaluation project was done by five universities in Europe. The main producers were formal education institutions and some leading publishing houses of Europe. In addition, two science centres, Heureka from Finland, and La Cité des Sciences et de l'Industrie from France supported CELEBRATE by producing Web-based learning objects (LOs) for use. These science centres were representing informal learning sources which were not so closely bound to national curricula.

Ilomäki et al. (2003) have published (in Finnish) a "Handbook for teachers, trainers and producers", which defines the main principles of creating Digital LOs. The book also underlines the principles of constructivist learning theory. It describes the flexible role of Digital LOs on the Web, and one of the key suggestions was to activate the prior knowledge of the learners through the supporting of conceptual changes in the learning process. The components of meaningful learning through digital LOs contain elements of problem-based learning, which benefits learners.

Research in Learning and Motivation in Open Learning Environments

Common sense and experience tells us that students are eager to learn in informal settings. Field trips, school camps, visits to workplaces, museums and science centres, or even having an art lesson in the school yard, evoke positive memories in students' minds. The Internet is a new challenge and possibility in this regard. Can the motivating effects of freedom and physical context be taken advantage of? After all, this is an important aim of science centre education. Evidence for this was sought in a study related to CELEBRATE and which was administered at Heureka in 2003.. It was expected that the study would provide useful indicators about learning in informal settings and open learning environments. In fact, most studies about motivation orientation have been conducted in the classroom-learning context.

The following theoretical model with definitions (see Salmi, 1993, 2003) describes different types of motivation.

A: Extrinsic Motivation

Situation motivation: motivation grows from a new situation. Temporary, external factors are important. Social relations are often an affecting factor. Entertainment is always a significant factor.

Typical features:

* short-lasting motivation
* learning is easily disturbed
* learning is orientated to irrelevant subjects

Instrumental motivation: the basis of this motivation is to get a reward and/or to avoid punishment. The main stimulus is 'to get things done' rather than being interested in the deeper meaning of the subject.

Typical features:

* the goal is often to pass an examination
* the learning of isolated facts, but not common principles
* connections or the theoretical background are less important for the learner
* facts are very quickly forgotten after an examination

B: Intrinsic Motivation

Intrinsic motivation: The basis of this motivation is a real interest in the topic studied. No other person persuades the learner to learn. The student sees the value of the studies and forms plans to use the knowledge or skill in the future. Curiosity, exploring and problem solving are key elements of this motivation.

Typical features:

* a critical and open-minded attitude towards learning
* seeing the connection between isolated facts and the topic area as a whole
* connection between theory and practice
* curiosity, interest, problem-based learning

In particular, this study aims to ascertain 5th graders' (aged 11 years) experiences of studying in an informal learning environment at the Heureka Science Centre. The study also sought to discover what kind of impact an Internet-based pre-lesson before the actual visit to the science exhibition might have on this experience. Further, it looked at the different types of motivation the pupils had. As the experiment took place in an open learning environment, it provided an excellent opportunity to describe and measure the qualitative differences in each pupil's situation motivation.

The subjects (n=135) consisted of six comprehensive school classes of 5th graders in the Greater-Helsinki area. Four of the classes (n=101) participated in the preparatory lesson and two classes (n=34) did not. The latter two classes formed the the control group for the motivation tests. The data was collected using written questionnaires before and after the visit to the science centre, and also by interviewing 16 pupils immediately after the visit.

The students had an opportunity to use the Web-sites in the "Open Questions" exhibition during a school lesson one week before the visit to the science centre. They were allowed to make their own plans concerning how to visit the exhibition. With the support of the Website they collected information, created their route through the exhibition, and structured their own timetable according to their interests and motivation.

The main results confirmed the fact that studying informally in a science centre is very positive and motivating for 5th grade pupils. The results of the situation motivation test also confirmed that the students attending the preparatory lesson had much stronger situation motivation in the science centre visit than the control classes.

The results showed that there were statistically significant (.05) differences between boys and girls in their motivation in school. Girls' motivation in school was more intrinsic whereas the boys' motivation was more extrinsic. The results also showed that boys are more dependent on rewards and that they are not ready to put in as much effort into working at school as the girls.

One very simple but crucial factor showing their interest and motivation was the time spent in the exhibition: for most of the students who had experienced the preparatory Internet lesson, one hour was all too short a time, and they would have liked to stay there for a longer time. On the other hand, the control group did not enjoy the exhibition for more than 30 minutes on average.

Another aspect that showed the free-choice learning element of the pre-lesson via the Internet was the structure of the visit. In earlier experiments (see Salmi 1993, 1996, 2003) the pre-lesson by teacher, textbooks and written materials gave the students a feeling that they have to explore the exhibition according to the plan given to them by the authorities. Now they were able to choose the topics themselves based on their own activity via the Internet. Furthermore, results of the interviews showed that more than 15 percent of the students had visited the pages in their leisure time too. Another important factor was that the students became interested in the other Web-pages too — pages that were not specified by either the teacher or the science centre.

Open Learning via Information Technology

Over the years, science centres have learnt a lot from each other while creating exhibitions. This open tradition was started by Frank Oppenheimer in the Exploratorium in the late 1960s, and sharing ideas has been one of the strengths of the science canter movement (Hein, 1990, Bewan & Wanner, 2003). There have also been some cooperative projects for creating new solutions for ICT-based learning in science centres: four European science centres (Heureka, Finland; Deutsches Museum, Germany; at-Bristol, UK; Ciência Viva, Portugal) have developed interactive exhibitions towards open learning environments for use also in distance learning via information and communication technologies. This project called OPEN SCIENCE INFO was supported by the EU-Minerva programme over the years 2002-2004.

The target audience was school groups, teachers and individual learners. The development process was supported by an action research of school groups and observations of visitors. The research was based on theories in informal learning, out-of-school education, motivation and usability. The project involved close cooperation with local schools, and the results are now disseminated both by ECSITE (European science centre organisation) and ESHA (European Secondary Heads' Association).

The topics of the exhibitions and Websites were different in each science centre according to their specialities: Chemistry, human physiology and pharmacy (Deutsches Museum), Wild Walk – biology and environmental education (atBristol); Hands-on physics (Ciência Viva), and Open questions – future of science (Heureka). The project was ambitious and it was known right from the start that, while it might solve some practical problems related to ICT/ODL education, it would certainly create new questions, too.

In informal education, and especially in Web-based learning, it is nearly impossible to create standardised and fully-controlled learning situations. The whole idea of informal e-learning is that learners can create their own approaches.

The purpose of the survey related to the Open Science Info was to investigate learning in science centre exhibitions in order to develop the use of the Web. The motivation model that was used as the theoretical background of the study separates intrinsic, instrumental and situation motivation. The design of the study was quasi-experimental: it comprised two different pre-treatment groups with different types of pre-learning, both with and without the Internet.

Design of the Study

The subjects (n=628) come from four countries (UK, Germany, Portugal, and Finland). In each of the countries, four classes (n=496) participated in the Internet-based advance education, and two classes (n=132) did not. The subjects were from secondary school classes (students aged 13-15).

The classes were tested by 1) the standard intrinsic/instrumental motivation test, 2) a situation motivation test which measures and compares the exhibition experience, 3) the ICT habits questionnaire related to the use of computers and the Internet, and 4) the standardised Raven-test defining the cognitive thinking level of the students.

Tests were administered in autumn 2003 and winter 2004 according to the following timetable: the pre-test a month before (T1), the post-test a week after (T3), interviews during the exhibition visit day (T4), and the delayed post-test three months after (T5) the exhibition visit (Figure 3) The data related to ICT experience and usage habits were collected during the pre-test. The situation motivation test was administered only after

Figure 3. Design of the study

	time
Motivation test	T_1
• intrinsic / instrumental	
Pre-lesson at school	T_2
• computer aided	
• "exhibition map"	
EXHIBITION VISIT	T_3
	Interviews of randomly selected pupils
Post-test	T_4
• motivation (intrinsic / instrumental / situation)	
• knowledge test	
• computer questionnaire	
• exhibition questionnaire	
Delayed post-test	T_5 motivation
• cognitive thinking & reasoning (Raven)	

the exhibition visit, and the best timing for the cognitive test was during the delayed post-test.

The Internet pre-lesson treatment group was the most positively motivated towards the exhibition. The difference from the other groups was statistically significant ($p<0.05$), and in addition some long-lasting effects could be observed. Most notably, the comments of this group in the recorded interviews gave added information about the role of pre-learning. The free-choice element was found to be important: the pupils felt that the exhibition was more meaningful since they could choose their route in the exhibition independently.

According to earlier results in the literature, situation motivation is activated when the outer attractions are strong enough: new settings, strong stimuli, interesting people, social relations, humour, etc. All this and more are readily available in any science centre. So this study clearly indicates that situation motivation is effective. Quite apart from being at the science centre, the pupils were out of school, and it was pleasant to leave the four walls of the classroom for some enriching settings.

The exhibition itself has such a strong *situation motivation* effect on pupils that it easily overshadowed other motivational effects, for example pre-visit lessons. It may be surmised that the most important benefit to be gained from a visit to a science centre is the change in the visitor's easily-aroused situation motivation into intrinsic motivation and deep-learning strategy. Results indicate that the intrinsic motivation of the Internet pre-learning groups grew during the project. This is in contrast to the control group, whose intrinsic motivation actually decreased or stayed at the original level.

Clear differences in the level of overall motivation were evident from the pre-tests conducted in the different countries. The level of intrinsic motivation was highest in Portugal, second in the UK, third in Germany, and lowest in Finland. The difference was statistically significant (.05) between Portugal and Germany, and Finland. However, the limitations in the number of subjects used in this study do not permit an overall comparison across the countries, although the school motivation in Finland has been lower than average in other studies, too.

The meta-results of several earlier studies (see Salmi 2003) show that the level of intrinsic motivation is highest during the primary school years, second highest in sixth form students, and lowest in secondary school. In light of this, the present results are encouraging, although it is not possible to induce intrinsic motivation simply through external means. Intrinsic motivation is acquired over a long period when the personality develops, and meaningful learning conditions in both content and context are essential.

However, the fact that the study took place in an open learning environment such as that which prevails in a science exhibition caused several background variables related to school and home to be beyond our control. Findings suggest that the students' situation motivation can be changed into intrinsic motivation through well-organised programmes linking schools to the informal learning settings of science centres.

The higher the students scored in the cognitive reasoning test, the stronger was their intrinsic motivation. However, there was no link between the cognitive level of the students and the situation motivation. The attitude towards the exhibition was not especially gender-dominated either, instead, girls in all the countries had a slightly more positive attitude towards the exhibition, but the difference was not statistically significant (.05) except in Portugal.

The hypothesis according to which earlier experience and activities with computers, and easy access to the Internet should have correlated with motivation and learning proved unsupported by the empirical data. Only in Finland did the Internet/computer-orientated students feel that the exhibition visit supported by the Internet pre-lesson was a deep-learning experience as compared to non-computer orientated students. In other countries, it seemed that most of the students who had less experience in e-learning or using the Internet seemed to feel that they had learnt more compared to the others. However, there was one clear exception: those students who had visited the science centre Website during their spare time at home, in the library or in some other place between the school lesson and the exhibition visit, felt that the process was a deep-learning experience.

The survey of Open Science Info contained several uncontrolled factors, such as different countries and cultures, different types and sizes of exhibitions, and different content of the Websites. The common characteristic of the Websites was the "bird's eye map", which allowed the students to have an overall vision of the exhibition, and gave them the opportunity to plan their own route for the visit.

The results strongly indicate the ability of high-school students in the four countries to independently apply their own learning strategies by using a combination of Web-based learning and an exhibition. The survey also gave comparative results for the differences and similarities of using computers and the Internet in different countries.

Conclusion

The main question of the survey done in this study was whether it is possible to find any common features and similarities in motivation and Web use. This is significant since using programmes and approaches such as the linking of schools and science centres together in meaningful learning initiatives, teenagers' decreasing motivation for learning can be minimised. The right cognitive level is important, but the same exhibit units and exhibitions apply for different students with different motivational and knowledge backgrounds if the introduction is carried out in the proper pedagogical context. The meaningful content of the exhibition is the principal factor. The visit to the science centre supported by Internet-based learning evidently created a worthwhile and valuable learning experience for the students.

In the CONNECT project 2004-2006, four European science centres (atBristol, UK; Växjö, Sweden; Heureka, Finland; and Euginides, Greece) are cooperating with universities, ICT experts, researchers and companies to create a new type of a collaboration involving schools, science centers and the Internet.. As part of this initiative, the classroom of tomorrow will be designed using advanced technologies (Williams, D. et al. 2004) so as to connect the formal and informal environments. An interesting development stage is the step from virtual reality (VR) to augmented reality (AR), where the latter not only conjures up virtual reality through special spectacles or a helmet, but also combines the real environment and the virtual, for example by giving the user of the AR some add-ons to view the scenery around (see *http://hci.rsc.rockwell.com/AugmentedReality*). These technologies already exist, mainly in research laboratories but are likely to enter the market in due course.

Some existing solutions, such as the "Foot driven navigation interface for a virtual landscape walking input device" by Barrera, Takahashi &Nakajima (2004), "Drawing, painting, and sculpting in the air" by Mäkelä & Ilmonen (2004), are undergoing serious development work by universities and researchers (see *http://www.tml.hut.fi/Research/ HELMA/*; and *http://www.vr2004.org/program/Workshop_BF_Proceeding.pdf*). The applications of the latter project in a science center setting are currently under intense planning.

The prices and other thresholds for using existing Internet and other ICT-based learning solutions have dropped considerably in recent years, and now the main consideration is whether there are enough social innovations, that is, are there meaningful content and use for the innovative technology? (Collins & Strijker, 2003). This development work has already started to create new principles for the pedagogical use of educational software. To create an open learning environment from the elements of the exhibition and the Internet is clearly one of the main challenges for science centers.

References

Bandelli, A. (2003). *ECSITE Newsletter, 54*, 7-8. Retrieved from *www.bionetonline.org*

Barrera, S., Takahashi, H., & Nakajima, M. (2004). Foot driven navigation interface for a virtual landscape walking input device. In *Beyond Wand and Glove. Proceedings of IEEE VR Workshop* (pp. 35-39). Chicago, IL.

Bevan, B., & Wanner, N. (2003). Science on the screen. *International Journal of Technology Management, 25* (5), 427-440.

Bionet. (2004). *BIONET final report*. An EU-project. Contract HPRP-CT-2001-00008.

Bitgood, S. (1988). *A comparison of formal and informal learning*. Technical Report No 88-10. Jacksonville, AL: Center for Social Design.

Clark, D., & Slotta, J. (2000). Evaluating media-enhancement and source authority on the Internet: the knowledge integration environment. *International Journal of Science Education, 22*(8), 859-871.

Collins, B., & Strijker, A. (2003). Re-usable learning objects in context. *International Journal on E-Learning, 2*(4), 2-16.

Crane, Y., Nicholson, H., Chen, M., & Bitgood, S. (1994). *Informal science learning. What the research says about television, science museums, and community-based projects*. Dedham, MA: Research Communications Ltd.

Duncan, C. (2003). Conceptions of learning objects: Social and economical issues. In A. Littlejohn (Ed.), *Re-using online resources: A sustainable approach to e-learning*. London: Kogan Page.

Falk, J., & Dierking, L. (1992). *The museum experience*. Washington, DC: Whalesback Books.

Falk, J., & Dierking, L. (2002). *Lessons without limit*. Walnut Creek: AltaMiraPress.

Gardner, H. (1991) *The unschooled mind: How children think and how schools should teach*. USA: BasicBooks.

Graf, M., & Roiko-Jokela, H. (2004). *Vaarallinen Suomi*. Jyväskylä, Finland: Minerva.

Hawkey, R. (2003). The lifelong learning game: Season ticket or free transfer? In *Computers and Education* (pp. 5-20).

Hein, H. (1990). *The Exploratorium: The museum as laboratory*. Washington, DC: The Smithsonian Institute.

Hofstein, A., & Rosenfeld, S. (1996). Bridging the gap between formal and informal science learning. *Studies in Science Education, 28*, 87-112.

Illich, I. (1971). *Deschooling society*. New York: Harper and Row.

Ilomäki, L. et al. (2004). *Opi ja onnistu verkossa – aihiot avuksi*. Opetushallitus - The Finnish Board of Education. Helsinki, Finland: Hakapaino.

Mäkelä, W., & Ilmonen, T. (2004). Drawing, painting, and sculpting in the air. Development studies about an immersive free-hand interface for artists. In *Beyond Wand and Glove. Proceedings of the IEEE VR Workshop* (pp. 88-89). Chicago, IL.

Salmi, H. (1991, April). *Television as informal learning source*. Paper presented in Colston Symposium, University of Bristol, UK.

Salmi, H. (1993) *Science centre education: Motivation and learning in informal education*. Research report 119. University of Helsinki, Department of Teacher Education.

Salmi, H. (1999, January 6-9). *Opportunities for meaningful contents in digital*-TV. Paper presented at the 2nd International Congress on Electronic Media & Citizenship in Information Society, Helsinki, Finland.

Salmi, H. (2001). Public understanding of science: Universities and science centres. In *Managing University Museums* (pp. 151-162). Paris: OECD Publications.

Salmi, H. (2003). Science centres as learning laboratories: Experiences of Heureka, the Finnish science centre. *International Journal of Technology Management*, *25*(5), 460-475.

Severson, J., Cremer, J., Lee, K., Allison, D., Gelo, S., Edwards, J., Vanderleest, R., Heston, S., Kearney, J., & Thomas, G. (2003). Exploring virtual history at the National Museum of American History. In *VSMM – Proceedings of the 9th International Conference on Virtual Systems and Multimedia* (p. 10). Quebec, Canada, Oct. 15-17, 2003.

Williams, D. et al. (2004). *A reinspection of HCI evaluation methods*. CONNECT Project. Department of Electronic, Electrical and Computer Engineering, School of Engineering, The University of Birmingham, UK.

Chapter XVII

Use of Log Analysis and Text Mining for Simple Knowledge Extraction:
Case Study of a Science Center on the Web

Leo Tan Wee Hin, Nanyang Technological University, Singapore

R. Subramaniam, Nanyang Technological University, Singapore

Daniel Tan Teck Meng, Singapore Science Centre, Singapore

Abstract

Log analysis of server data has been used to study the Web site of the Singapore Science Center, which is the largest Web site among all science centers in the world. This has yielded a wealth of data, which has been useful in assessing the effectiveness of the content hosted on the site. Additionally, the use of text-mining to structure an effective query interface for the Science Net database, which is an online repository of over 6,000 questions and answers on science and technology, is assessed. A commentary on the use of log analysis for virtual science centers is also presented.

Introduction

Data archived in the Web logs of servers represent a potentially rich source of information (Jones et al., 2000; Ren et al., 2002; Zhang, 1999). However, the raw data recorded on server logs are not of much use. It has to be first cleaned to remove redundant information such as, for example, sound, graphics, video and image files embedded on a Web page for which a hit has already been registered, and then stored in a data warehouse before using appropriate techniques to extract useful information (Joshi et al., 1999). Log analysis software is needed to unravel the data, and the information mined can help to evaluate the efficacy of a Web site through the identification of patterns and trends embedded in the data.

Published studies on the effectiveness of commercially available log analysis software are rather sparse in the primary reference literature, not surprisingly since the use of such software may be construed as endorsement of the product. Since such software represents, to a good extent, the state-of-the-art development in the field, its use can permit an evaluation of its effectiveness in real life settings. This can help to further bridge the chasm between theory and practice as well as provide a basis for further research. It was with this objective that the present study was undertaken.

More specifically, this study aims to use log analysis software to study the access patterns and trends from data archived in the server logs of the Web site of the Singapore Science Centre, which is the largest Web site among all science centers in the world. It is recognized that administrators and stakeholders would need reassurance that investments in a virtual science center are reaping dividends in the form of increased Web traffic. Log analysis software aims to contribute towards this. Sophisticated analysis of data archived in server logs is not the object of this study, and is therefore not addressed here. Only simple statistical analysis is used, as this is adequate for the needs of most science centers. An additional objective of this study is to use a commercially available text-mining software to construct a query interface for information retrieval by users of the Science Net database on this Web site.

Related Work and Rationale for Study

Science centers are institutions for the popularizing of science and technology among the masses (Danilov, 1982). The need for people to be cognizant of science and technology is a given in today's age of globalization and rapid scientific advances, and science centers do this in a way that makes people realize the impact of science and technology in their everyday lives. Commonly, this entails the use of science exhibitions and various promotional activities. Raising popular science literacy levels of the people through such initiatives empowers them to make informed decisions as well as become meaningful participants in science and technology driven nation-building efforts.

The Exploratorium in San Francisco pioneered the science center movement in 1968 (Oppenheimer, 1972; Delacote, 1998) and made a success of it for others to emulate the concept. In due course, science centers started mushrooming in many parts of the world. There are more science centers in the Western world, especially in the USA and in Western Europe, than in other parts of the world. In South-East Asia, the Singapore Science Center has been the pioneer (Tan & Subramaniam, 1998, 2003a).

With the arrival of the Internet in the 1990s and its phenomenal growth thereafter (Tan & Subramaniam, 2001), science centers have been compelled to turn to this new media as a way to enhance their mission objectives further and reach new audiences. In the early years, the Web content hosted by science centers was more of a static nature and limited to providing information to the public about their opening hours and educational programs as well as the hosting of some science-based text resources. Even then, in 1997, when the virtual science center movement started to pick up momentum, a total of 195.3 million hits were registered by 77 science centers and science museums (Association of Science-Technology Centers, 1998). With advances in Web-related software technologies, the offerings of virtual science centers have grown in sophistication, so much so that this now constitutes a new genre of learning in informal science education (Tan & Subramaniam, 2003b).

Virtual science centers now host a range of content, including exhibitlets and science resources (Honeyman, 1998; Jackson, 1996; Orfinger, 1998; Tan & Subramaniam, 2003b). In fact the Web site of the Association of Science-Technology Centers (http://www.astc.org) has links to over 200 science centers, science museums and other institutions of informal science learning.

The challenges faced by science centers in competing for the online attention of surfers, who have a gamut of options to turn to in cyberspace, means that they need to not only strategize their range of Web offerings but also configure these in a way such that a decent measure of customer satisfaction is promoted. Web audiences are different from visitors entering the gates of these institutions. Whilst in the latter case, visitors will dwell for a significant length of time to savor the various attractions after having purchased a ticket, in the case of the former, there is a choice of just clicking away to another site if the initial experience is not up to expectations. Tools that aid in the study of online visitor characteristics are thus imperative.

Though log analysis has attracted significant attention (Ren, 2002; Zhang, 1999), it has not been applied to a number of settings. Applying such tools to virtual science centers would thus make an interesting case study. By presenting a context for the application of such tools, a platform is presented to help bridge the divide between theory and practice. No previous systematic study of a science center Web site using log analysis has been reported in the literature. However, some mention about the use of log analysis for studying the Web sites of museums have been reported (Cunliffe et al., 2001; Heinecke, 1995; Jackson, 1996; Streten, 2000).

It is of interest to note that even with museum Web sites, analyses tend to rely more on qualitative forms such as user surveys and observational analysis (Peacock, 2002). Log analysis is generally not used because of the (misplaced) pronounced emphasis on its deficiencies when used in such settings; this can generally be traced to the discrediting of "hits" as a means to gauge the popularity and effectiveness of a site (Peacock, 2002).

The Web site of the Singapore Science Center is hosted at http://www.science.edu.sg. It is the largest among all science centers in the world in terms of number of pages. Rich knowledge fields abound in this site on many aspects of science and technology. More specifically, this study aims to look at the following:

1. Analysis of server logs through log analysis software to study user access patterns and efficacy of the offerings.

2. Analysis of information retrieval using text-mining techniques from the Science Net database, which is a repository of nearly 6,000 questions and answers on various aspects of science and technology.

3. A commentary on the importance of log analysis for science centers

The present study is the first systematic research on the use of log analysis software to study the Web site of a science center.

Mining of Server Logs

Server logs contain a mine of information about activity levels in various parts of the Web site. Each time a server processes a request from a computer user connected to it, a transaction record is made in its log. Typically, the log entries would include number of hits, number of page views, user's IP address, timestamp, byte size of data transferred from page, type of browser used in accessing the Web site, volume of activity on each page, URL of page accessed, referrer status, errors encountered by users, and traversal patterns (Joshi, 1999). At the Singapore Science Center, a dedicated server is used for its Web site and this makes it easier to study the Web logs using appropriate software.

At the Singapore Science Center's Web site, a program called WebTrends (http://www.netiq.com/webtrends) is used for mining the Web log entries. As the field of log analysis is still evolving, this commercially available program represents to a good extent the state-of-the-art in Web log analysis. WebTrends produces detailed profiles of activities occurring on the server in a format that is easy to comprehend. Filters are provided to weed out certain entries, which skew the distribution of data or cause extraneous predispositions.

Figures 1-9 and Tables 1-2 show a range of descriptive statistics from records produced by WebTrends after mining raw entries from the server log for the year 2001/2002. Though these statistics are simplistic reports, they do provide a good operational perspective on the performance of the website.

The data in Figure 1 shows that the site attracts a good number of visitors, a key consideration in justifying the investment and recurrent expenses on the site. November is the month with the most number of virtual visits. This could be due to the onset of the school holidays in Singapore, which frees up more time for surfing by students. The data also shows that February/March are the months with the least number of visits. With the academic term well in progress and students pre-occupied with their studies, there is probably little time for surfing the site — hence, one of the possible causes for the

Figure 1. Number of virtual visits for the year 2001/2002

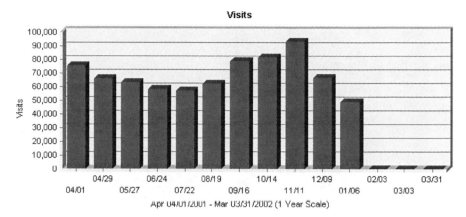

diminution in Web visits. It needs to be noted that in Singapore, home PC penetration rate is about 60%, while Internet penetration rate is over 90%.

The data in Figure 2 shows that the home page is the most frequently accessed document. This is not surprising as the home page is well structured, contains up-to-date information on the programs at the Singapore Science Center, and permits easy navigation to other parts of the site. In updating the Web site, particular attention is often paid to the home page to ensure that visitors can access any page on the site with minimal difficulty by starting on the home page.

According to the data in Figure 3, the top entry page has consistently been the home page. We believe one reason for this is that the URL address is indexed by all search engines. The top entry page is the first page that a visitor accesses when he enters the Web site. Usually this is the home page but it can be other pages as well. The former is an indication that the user has probably bookmarked this page and accessed it directly through a URL link from his browser. Presumably, the user has endowed the page with sufficient value for a re-visit.

In Figure 4, the temporal activity as a function of data transfer is depicted. The volume of data downloaded peaked in mid-December before tailing off in February/March. Besides the reasons addressed earlier, this also provides an indication that the server architecture needs to be able to cope with more than the recorded maximum amount of traffic at all times.

The data in Figure 5 shows that the distribution of activity remained fairly constant throughout the week, with a peak during mid-week.

The virtual science center appears to be well utilized throughout the day, as there is not much variation in activity levels over the hours (Figure 6). The peak occurring at 4:00 a.m. can most likely be attributed to hits from overseas visitors (see Table 2). Data on the peak activity level is important for Web site administrators as they need to ensure that the capacity of the server is adequate to meet these demands and, if necessary, to install upgrades.

Figure 2. Top documents accessed for the year 2001/2002

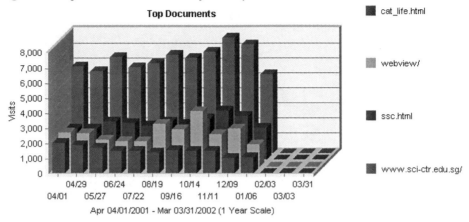

Figure 3. Top entry pages accessed for the year 2001/2002

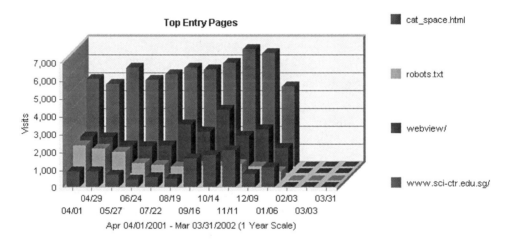

Figure 4. Activity level by time increment for the year 2001/2002

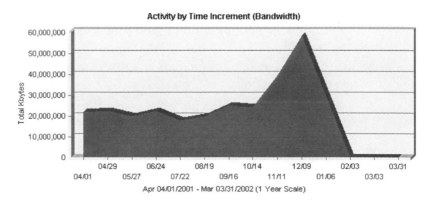

Figure 5. Activity level by day of week for the year 2001/2002

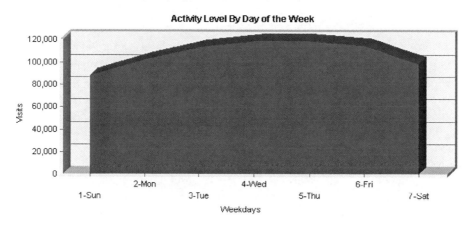

Figure 6. Activity level by hour of day

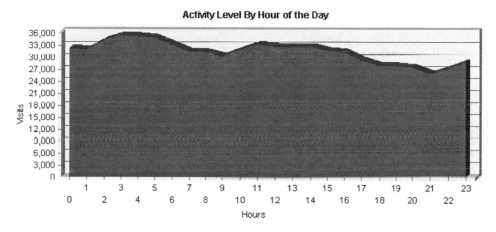

Figure 7. Activity level by visit length for the year 2001/2002

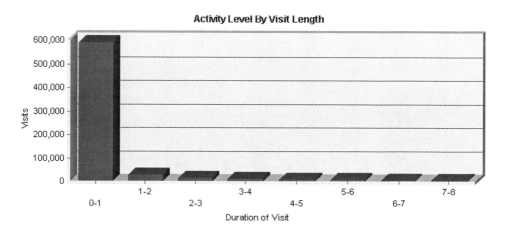

As evidenced in Figure 7, most Web users spend about an hour or less at the site. Longer duration sessions of up to eight hours are also evident. The latter is especially an indication that the content on the site is of a compelling nature.

Figure 8 shows the number of page views per visit – this is a numerical measure of the number of html documents retrieved during a Web session. Most views per visit are for two pages or less. The maximum number of pages viewed per session is 12.

Analysis of error data can provide insights into the reliability of the site as well as discover the occurrence of possible systemic problems. Pages not found, as shown in Figure 9, can translate into user frustration, which may affect re-visits. Disparate links on the site can also become apparent through a study of such data. Also, error analysis helps to enhance content delivery and monitor system performance. All this can help to fine-tune the topology of the Web site. It is thus a good diagnostic tool for use by Web masters. In an earlier study of museum Web sites, Peacock (2002) has cautioned on the risk of disenfranchising potential users by pitching the site design beyond the capabilities of the software, network connection or file size. His comments are equally valid for virtual science centers.

Figure 8. Number of views per visit for the year 2001/2002

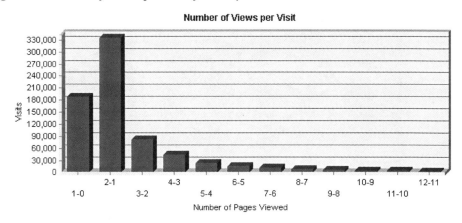

Figure 9. Error data for the year 2001/2002

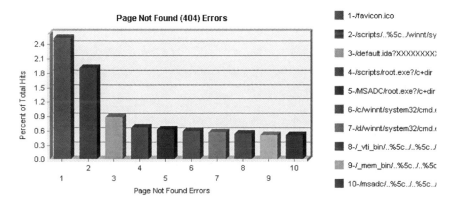

Table 1 encapsulates in a nutshell the salient statistics of the site and also presents a general overview of the effectiveness of the site. With over 60,000 hits and over 12,000 page views recorded per day, the site is certainly attracting Web traffic, a testimony to the effectiveness of its range of offerings.

The audience segmentation profile in *Table 2* shows that of the top 20 countries accessing the Web site, the USA ranks first. Ironically, Singapore is quite a distance away even at number 2 position! This is a good indication that the site is a compelling attraction, an important factor in reaching out to new audiences as well as a diversity of users.

All the above metrics confer an evaluative dimension as well as useful insights on the effectiveness of the Web site. The discovery of patterns and relationships through log analysis allows for improvement of system functionality of the Web site and also helps to initiate measures for enhancing the quality of experience of surfers. Demonstrable advantages have been achieved by the Singapore Science Center through an examination of these profiles, and all these have helped to better position its virtual science center for attracting online traffic. The macro trends encapsulated in Figures 1-9 and Tables 1-2 are generally sufficient for science centers to have a good overview of site effectiveness.

Table 1. General Statistics for the Virtual Science Center for the year 2001/2002

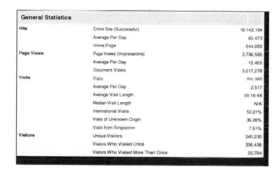

General Statistics		
Hits	Entire Site (Successful)	18,142,164
	Average Per Day	60,473
	Home Page	244,806
Page Views	Page Views (Impressions)	3,736,565
	Average Per Day	12,455
	Document Views	3,217,278
Visits	Visits	766,980
	Average Per Day	2,517
	Average Visit Length	00:16:48
	Median Visit Length	N/A
	International Visits	53.21%
	Visits of Unknown Origin	39.26%
	Visits from Singapore	7.51%
Visitors	Unique Visitors	240,230
	Visitors Who Visited Once	206,436
	Visitors Who Visited More Than Once	33,794

Table 2. Visitations by country for the year 2001/2002

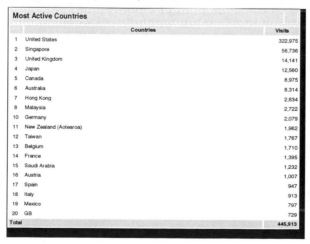

Most Active Countries		
	Countries	Visits
1	United States	322,975
2	Singapore	56,736
3	United Kingdom	14,141
4	Japan	12,560
5	Canada	8,975
6	Australia	8,314
7	Hong Kong	2,834
8	Malaysia	2,722
10	Germany	2,079
11	New Zealand (Aotearoa)	1,962
12	Taiwan	1,767
13	Belgium	1,710
14	France	1,395
15	Saudi Arabia	1,232
16	Austria	1,007
17	Spain	947
18	Italy	913
19	Mexico	797
20	GB	729
Total		445,913

We now discuss a few issues in relation to the hit counts extracted through log analysis.

(a)   IP addresses – this is used to identify a computer linked to the Internet. Each computer linked to the Internet has a unique IP address assigned to it. Thus, the number of IP addresses captured by the server represents the number of unique visitors accessing the site. Cognizance must, however be borne of some of the limitations in treating the number of IP addresses as unique visitations. This is due to the nature of the topological configuration of many Web repositories. For example, when a user connects to the Internet through dial-up access, he is assigned a dynamic address – when the same user logs on to the Internet at a different time, the IP address can be different. This has the effect of increasing the number of IP addresses captured by the server, even though the number of distinct users has not increased. Note that in access by cable modem, the IP address is fixed since there is a dedicated connection to the Internet. Also, when corporate users link to the Internet through a proxy server, which has a fixed address, the server records cause an undercounting of the number of IP addresses captured since all the computers linked have the IP address of the proxy server.

(b)   Caching – this is a unique mechanism to speed up the retrieval of commonly accessed documents by storing them on the proxy server. In this way, network latency and server load are minimized, all of which help to promote speedier downloads and effective utilization of spectrum bandwidths. For example, if there are 100 distinct users accessing a Web site from a corporate network, caching by the proxy server means that it fetches the page(s) only once and stores it in memory — this is recorded as just a single hit in the server logs. Moreover, the user's browser can also cache recently visited pages — if he revisits a page during the same session by clicking on the front/back buttons on the monitor screen, the server log does not record this as an extra hit. Caching thus contributes towards diminution of the hit counts.

Every user to a Web site leaves an electronic "fingerprint" in the Web log, and an analysis of user access profiles can thus shed useful information about his motivation for visiting the site and provides some indication about the quality of his experience. This constitutes a basis for conducting visitor studies in the cyberspace of science centers which, in turn, can help to fine-tune service delivery levels of the institution.

Text-Mining of the Science Net Section

On the Web site of the Singapore Science Centre, the Science Net is a massive database containing nearly 6,000 questions and answers on various topics in science and technology (Tan & Subramaniam, 2003c). The genesis of this science knowledge repository can be traced to the need for the public to have a service whereby their scientific queries can be addressed. No proper institutional mechanism currently exists to service such learning needs of the public on a regular basis. Recognizing this as a learning opportunity for the public, the Singapore Science Center launched this service as a Web initiative. Addressing the scientific queries of the public constitutes an important aspect of furthering the public understanding of science, which is an important mission objective of science centers.

The global public can post any scientific question on the site and, provided it fulfils certain criteria — for example, the question has not been answered before, it is not a text book problem, it is not a homework assignment of a student, and it is not a question which can be answered by consulting standard text books in the library — the question would be answered. Constraining the posting of questions within the bounds of these parameters has been found to be essential in order to ensure that the service is not abused and that frivolous requests are not received. In this way, the service can cater to the larger interests of the general public. A battery of over 100 scientists from the National University of Singapore, the Nanyang Technological University and the Singapore Science Center, representing a formidable concentration of expertise and resources, helps to answer the scientific queries. Figure 10 shows a sample page from the Science Net section.

Since its introduction in 1998, the database of Science Net has expanded tremendously. Content is organized according to multi-dimensional hierarchies based on subject interests — six sections and over 70 sub-divisions in science and technology (Table 3). To ensure that the global public does not post questions which have already been answered and that are available in the database, text-mining software is available to help them with their search queries as well as find answers.

Figure 10. Sample page from Science Net

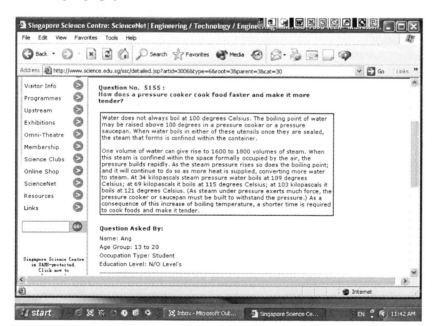

Table 3. Categorizing of questions and answers in Science Net

Section	Sub-sections	Sub-sections
Computer Science/Information Technology/Mathematics	Computer systems Computer vision & machine Intelligence Database Internet	Mathematics/Algorithms Network & Communications Programming languages/Computer software Robotics/Automation Security & Encryption
Earth Science	Agriculture/Farming Auroras/Northern lights Geology & Geophysics	Meteorology Natural resources Oceanography/Hydrology
Engineering/Technology/Engineering Materials	Acoustics Aviation Biotechnology/Bioengineering Civil/Structural Engineering Electrical Engineering Electronic Engineering	Food Technology Industrial/Production Engineering Materials Science/Polymers Mechanical Engineering Optical Engineering/Photography
Life Sciences	Animal behavior/Zoology Biochemistry/Biophysics Botany Ecology/Environment General Biology Genetics/Reproduction Genomics/Bioinformatics Human Anatomy Marine Biology	Microbiology Molecular & Cell Biology Neuroscience/Vision Pharmacology/Medicine/Disease Physiology Safety/Health Systematics/Taxonomy Human Behaviour/Psychology
Physical Sciences	Analytical/Clinical Chemistry Fluid Dynamics General Chemistry General Physics High Energy/Particle/Plasma Physics Lasers/Optics/Photonics	Magnetism/Electricity Mechanics/Waves/Vibrations Organic/Theoretical Chemistry Relativity Theoretical/Quantum Physics
Astronomy & Space Science	Astrophysics/Cosmology Comets/Asteroids/Meteors General Astronomy Observatories/Telescopes Planetaria/Constellations	Milky Way/Galaxies Radio Astronomy Search for Extra-Terrestrial Intelligence Space Exploration Satellites Solar System
Others	Science policies Tips on passing science examinations Etcetera	

That the service is well utilized can be seen from data presented in Table 4.

Accessing a question-and-answer of interest in this massive database is a formidable task to an online visitor. With over 6,000 questions and answers, there needs to be a convenient mechanism to facilitate easy retrieval of the question-and-answer of interest. Whilst the stratification of the database according to topics (Table 3) is a useful mechanism to guide this exploration, a simple query interface will be of great value and convenience. In this context, text-mining software called *Webinator* is used. Text-mining seeks to extract textual information of interest from a repository of documents and present them in a coherent format. *Webinator* is a Web crawling and indexing software that also

Table 4. Distribution of questions posted to Science Net by year

Year	Number of questions posted
1998	2,064
1999	3,686
2000	5,664
2001	4,382
2002	3,208
TOTAL	**19,004**

provides a retrieval interface for Web documents stored in a database (http://www.thunderstone.com/site/webinator4man/node3.html). Written in the Web script language called Vortex, Webinator comprises the Texis binary program and three Vortex scripts which are run by the Texis CGI program on the Web server, and are accessed from the Web browser. Each of the Vortex scripts has specific functions: one provides the administrative interface, another provides the site walker and indexer, and a third provides the search function.

The Texis search engine uses a SQL (Structural Query Language) relational database to store and search documents. Figure 11 shows the Web query interface of the program. As can be seen, the query interface allows for judicious strategizing of a search followed by browsing of the documents returned.

The fill-in field in the query interface requires the input of key words for searching the Science Net database. This can be done by simply inputting the necessary key words or, for more complex searches or accurate returns, by activating the available options in tandem. Pull down menus can be accessed on the next line to identify the disciplinary database in which the search is to be conducted. The query can also be entered as a question, for example, "What is the temperature of the Sun?" The Texis program will then search for "temperature" and "Sun" while ignoring the noise words embedded in the question. If the proximity option is activated, there is provision to search the database either by line, sentence, paragraph or page for the occurrence of the keyword(s). By enabling the concept expansion option, the search can be enlarged to include ontological equivalents of the keyword – this is done by prefixing a tilde character (~) to the word or phrase. This is an especially important feature since the lexicographic base for a

Figure 11. Web query interface of Science Net

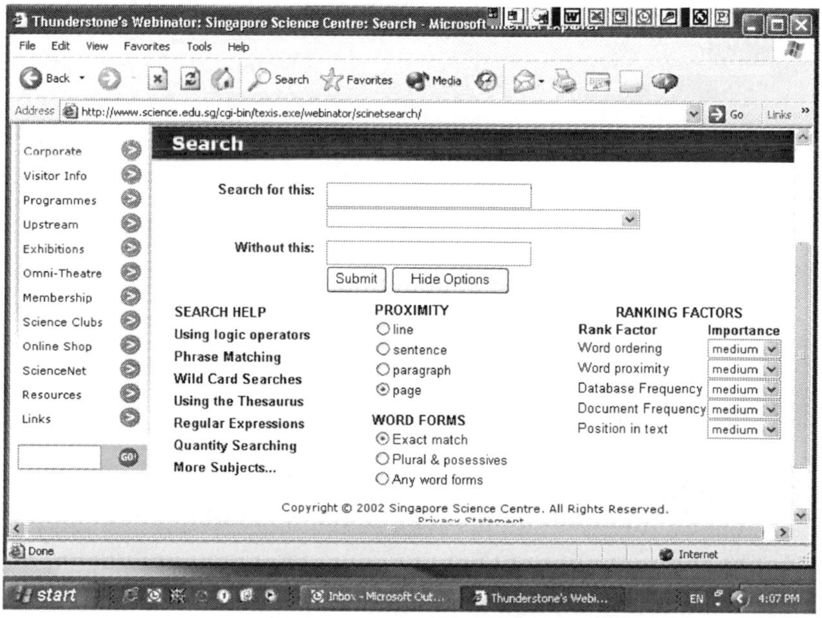

Figure 12. Extract of a search on the Science Net

number of scientific concepts admits of a plurality of expressions — variants of the same word can be expressed differently by different persons. The semantic connections that are established for the keywords then drive the query search.

The design of this schema provides the user with the best search results for a query, which is necessarily constrained by the retrieval and extraction dynamics of the algorithm driving the search on the database. Whilst the retrieval step selects the relevant documents, the extraction step analyses the structure of the data embedded in the selected documents.

When a search returns a selection of documents with the URL and an extract of the text, it also indicates the overall quality of the match through a percentage rating (Figure 12). More precise searching can be done by turning on the "search for similar records" option. Commonly known as "query by example," this is a powerful feature of the mining program to narrow or refine the search field further. Typically, this will take cognizance of the text within quotation marks as a phrase and the * sign is used as a wild card.

Another powerful feature is the Word Forms availability. This helps to determine the number of variations of a term in the search query — it could be an exact match, include plural and possessive forms, or as many forms as can be derived from the root term in the query.

The effectiveness of the search engine used for mining the Science Net can be traced to the fact that Texis employs set logic to the search query rather than Boolean logic. This has the advantage of offering a greater number of functionalities compared to the use of Boolean operators. When querying using set notation, the elements are bracketed and separated by commas. For example:

'-' (without) – use of this logic symbol in the search query will return entries sans the item in question

'+' (mandatory) – prefixing this symbol to the search word or phrase will retrieve entries with the item in reference

"@N' (permute) – prefixing @ to a number will fetch entries including this number of confluences of the term in reference.

All these help to find answers to questions with a greater degree of precision.

To illustrate the use of the text-mining algorithm, we attempted some simple experiments that require the finding of answers to a few questions available in the database. Different versions of the question all produced correct results, thus attesting to the efficacy of the search algorithm.

Discussion

With the proliferation of online content on the Web, science centers face the challenge of embracing new paradigms in evaluating the content of their offerings and building an online community of users. Log analysis contributes toward this as it furnishes a rigorous approach to test the relevance and effectiveness of their Web attractions. A pro-active stance by science centers in fine-tuning content on the basis of such studies permits enhancement of their outreach effectiveness on the Web. Though the use of log analysis software has yet to catch up with science centers on the Web, the metrics enunciated earlier do provide a simple basis for assessing site effectiveness and thus the usability, navigability and visibility of the site.

The use of log analysis techniques to assess the efficacy of the Singapore Science Centre's virtual science center has contributed significantly towards strategizing its range of offerings. Important Web access patterns and trends have been uncovered, and the analyses have shown that it is a well-utilized site. Data obtained have also been used to improve the design and functionality of the site as well as improve its information structure. The practicality of using log analysis has thus been demonstrated. We cite five instances where the use of log analysis has led to the instituting of pro-active measures to enhance user needs, thus leading to better decision making for the Singapore Science Center.

1. Traversal patterns extracted from server logs indicate that about 25% of the hits to the virtual science center are for the Science Net database (Table 5). These hits come from across the world. It therefore made sense to devote extra effort for this section as a means to draw further Web traffic to the site. A direct result of this finding has been that the section is updated almost everyday, and newer versions of the Webinator software are regularly installed to facilitate the effectiveness of search queries. Also, hyperlinks have been appended to many of the answers in order to add further depth to the answers.

2.   The traditional offerings in the galleries of science centers are interactive exhibits. When visitors interact with these exhibits, it contributes towards expanding their conceptual horizon of a science topic. It is the multi-sensory experience that is evoked when visitors interact with exhibits that contributes towards learning in science centers. A direct consequence of this has been the addition of interactive virtual exhibits (also called exhibitlets) on the Web site. Special attention is devoted to building a user-friendly interface for learning. Regular surveys of Web logs indicate that these are popular, as evidenced by their page views and length of activity sessions, and this has been taken into consideration to populate the site with more exhibitlets. One reason for the popularity of exhibitlets is the presence of game elements in their operation — this stems more from the science center philosophy, which stresses that learning has got to be fun. It needs to be borne in mind that exhibitlets are expensive to fabricate because it involves inputs from designers, educators, curators and software specialists. The number of virtual exhibits on the Web of any science center is thus limited.

3.   The increase in number of hits attest to the effectiveness of the resources offered on the site. This has made it somewhat easier to obtain funding from government and corporations for some of the Web programs of the Singapore Science Centre since quantitative data to attest to site efficacy is available, and this is free of observer or survey bias. Increased number of hits can be correlated with visibility, which is especially important for corporate sponsors.

4.   In view of the dynamic nature of the virtual science center and its effectiveness in reaching out to the online public, the need for a fulltime Web master and support staff was recognized. This has greatly contributed towards enhancing the dynamism and vibrancy of the site.

5.   Without the information provided by log analysis, the need to resort to regular upgrades of server hardware and software in order to cater to the growing online traffic would not have been recognized.

Table 5. Web site statistics of Singapore Science Center for the period from January 1, 2001 to December 31, 2002

Section	Total hits	Average hits per day	Total number of page views	Average number of page views per day
Virtual Science Center	52,377,282	71,749	10,208,973	13,984
Science Net	12,799,165	17,533	2,753,308	3,771

Visitor studies on people coming to science centers are well established in the literature. The use of structured and validated survey instruments as well as the employment of site staff to administer these is a common practice. However, this is time-consuming and subject to questioner bias. In contrast, mining of Web log data residing quiescently on the server provides a spectrum of information in relation to virtual user access. Discounting the cost of the server and log analysis program, log analysis posts cost advantages and facilitates the retrieval of a medley of patterns since visitors leave an electronic trail of their activity, and this is captured by the server. Peacock (2002) has observed that log data is free of observer and questioner bias, and is thus a cheaper and effective form of evaluation compared to observation and survey research.

A useful measure of online visitor satisfaction can be gauged from an analysis of pathing. On the WebTrends program, the entry occurs as "paths through site" attribute. Tracing the genesis of a visitor's entry into the site and monitoring his web of explorations before exiting the site reveals useful data about customer preferences and satisfaction. In ensuring that the site remains vibrant and dynamic, such evaluations are not only a prerequisite but also a practical necessity. It needs to be borne in mind that whilst Web log transactions are an accurate record of user sessions, they cannot provide indication of user's actions, especially those relating to intangible attributes. These are worthy of further study.

The fixed reporting format adopted by WebTrends is adequate for the needs of most science centers. Having a Web presence, though necessary for science centers in today's networked world, is not sufficient: the site need to draw traffic and contribute towards realizing the mission objectives of the science center. If the site is not attracting enough traffic, as benchmarked with pre-determined norms, then it indicates a problem that needs to be addressed. It could mean that the content on the site is not compelling enough or that there could be access problems due to the server.

A major contribution of this study has been the demonstration that a number of parameters can be extracted using data mining techniques to gauge the effectiveness of the range of offerings hosted in the virtual science center. Previous studies of the virtual science center by other authors have focused more on hit counts as a measure of site effectiveness (Honeyman, 1998; Orfinger, 1998). Judging from the scarcity of published literature related to mining of log entries on the Web sites of science centers, it appears to be an under-utilized tool to evaluate the effectiveness of their Web sites. Cognizance of the trends amassed from mining Web log data can help science centers to better position themselves for attracting cyber traffic and ensuring the effectiveness of this new initiative for the public understanding of science. Updating of the site with new content can thus be aligned with the directions uncovered by log analysis.

In summary, the results reported in this paper have significant implications for site management by Web administrators and the management of science centers

Conclusion

Log analysis is a powerful utility tool for tracking the effectiveness of a virtual science center. It is expected that the use of such tools will become more prevalent since the information they extract can provide strategic value to stakeholders and site administrators. Benchmarks for site performance can be crafted using these indicators, and this can serve as a basis for setting performance indicators. Text-mining of the Science Net database has also been facilitated through a good query interface.

In the case of the virtual science center in Singapore, mining of log records has been helpful in strategizing its range of offerings on the Web site, understanding the impact of the Web site vis-à-vis other competing attractions, and in pro-active positioning of the site, all of which have been helpful in drawing more online traffic to the Web site. Being an under-utilized tool among Web-based science centers, it is suggested that this study has implications for other science centers that wish to track the performance of their Web sites.

Acknowledgments

We thank Mr. Edwin Teng, Technical Manager at the Singapore Science Centre, for his assistance in generating statistical profiles from server logs for this study.

References

Association of Science-Technology Centers. (1998). *Yearbook of science centers statistics.* Washington, DC: ASTC.

Cunliffe, D., Kriton, E., & Tudhope, D. (2001). Usability evaluation for museum web sites. *Museum Management and Curatorship, 19*(3), 227-25.

Danilov, V.J. (1982). *Science and technology centers*. MA: MIT Press.

Delacote, G. (1998). Putting science in the hands of the public. *Science, 280*, 252-253.

Heinecke, A. (1995). Evaluation of hypermedia systems in museums. In D. Bearman (Ed.), *Multimedia computing and museums* (pp. 67-68). Pittsburgh: Archives and Museum Informatics.

Honeyman, B. (1998, April). Real vs virtual visits: Issues for science centers. *Australasian Science & Technology Exhibitors Network.*

Jackson, R. (1996). The virtual visit: Towards a new concept for the electronic science center. In *Conference on Here and How: Improving the Presentation of Contemporary Science and Technology in Museums and Science Centers*. London.

Jones, S., Cunningham, S.J., McNab, R., & Boddie, S. (2000). A transaction log analysis of a digital library. *International Journal on Digital Libraries, 3,* 152-169.

Joshi, K., Joshi, A., Yesha, Y., & . Krishnapuram, R. (1998). Warehousing and mining weblogs. In *Proceedings of the Workshop in Web Information and Data Management* (pp. 68-72).

Oppenheimer, F. (1972). The Exploratorium: A playful museum combines perception and art in science education. *American Journal of Physics, 40,* 978-984.

Orfinger, B. (1998). Virtual science museums as learning environments: Interactions for education. *The Informal Learning Review,* 1-10.

Peacock, D. (2002). Statistics, structures and satisfied customers: Using web log data to improve site performance. In *Proceedings of the Museum and the Web Conference.*

Ren, K.H., Kwakkelaar, R., Min, T.Y., & Chun, C.L. (2002). Exploring behavior of e-journal users in science and technology: Transaction log analysis of Elsevier's ScienceDirect OnSite in Taiwan. *Library & Information Science Research, 24*(3), 265-291.

Streten, K. (2000). Honored guests: Towards a visitor-centered web experience. In *Proceedings of Museum and the Web Conference,* Minneapolis.

Tan, W.H.L., & Subramaniam, R. (1998). Developing nations need to popularize science. *New Scientist, 2139,* 52.

Tan, W.H.L., & Subramaniam, R. (2001). Wiring up the island state. *Science, 288,* 621-623.

Tan, W.H.L., & Subramaniam, R. (2003a). Science and technology centers as agents for promoting science culture in developing nations. *International Journal of Technology Management, 25,* 413-426.

Tan, W.H.L., & Subramaniam, R. (2003b). Virtual science centers: Web-based environments for promotion of non-formal science education. In A.K. Aggarwal (Ed.), *Web-based education: Learning from experience* (pp. 308-329). Hershey, PA: Idea Group Publishing.

Tan, W.H.L., & Subramaniam, R. (2003c). Science Net: A virtual school for the extension (science) education of the public in Singapore. In C. Cavanaugh (Ed.), *Development and management of virtual schools: Issues and trends* (pp. 244-261). Hershey, PA: Idea Group Publishing.

Zhang, Z. (1999). Evaluating electronic journals and monitoring their usage by means of WWW server log analysis. *Vine, 111,* 37-42.

Chapter XVIII

The Development of Science Museum Web Sites:
Case Studies

Jonathan P. Bowen, London South Bank University, UK

Jim Angus, National Institutes of Health, USA

Jim Bennett, University of Oxford, UK

Ann Borda, The Science Museum, UK

Andrew Hodges, University of Oxford, UK

Silvia Filippini-Fantoni, The University of Paris I Sorbonne University, France

Alpay Beler, The Science Museum, UK

Abstract

Science museums have embraced the technology of the Web to present their resources online. The nature of the technology naturally fits with the ethos of science. This chapter surveys the history, development and features of a number of contrasting pioneering museum Web sites in the field of science that have been early adopters of the technology. This includes case studies of Web sites associated with the Natural History Museum of Los Angeles, the Museum of the History of Science in Oxford, the Science Museum in London and the completely virtual Alan Turing Home Page. The purpose is to demonstrate a diverse set of successful scientifically-oriented Web sites related to science museums and the history of science, giving an insight into Web developments in this area over the past decade.

Introduction

The idea of using computers for education is not new. Seymour Papert (www.papert.com), an Artificial Intelligence pioneer with Marvin Minski at the Massachusetts Institute of Technology (MIT), first mooted the idea of using computers for learning in the 1960s. For example, he held a symposium at MIT in 1970 entitled "Teaching Children Thinking," proposing that children could learn by teaching computers, developing the Logo programming language to help in this quest (Papert, 1993, 1999). Of course it then took some years for computers to become widespread and cheap enough to make such ideas a reality in everyday life. Papert has continued his interest in learning in the context of the Web. In 1996 he conducted a tour including locations such as the Smithsonian in Washington, D.C. and the Boston Computer Museum to promote a book (Papert, 1996). He has also helped in the development of *MaMaMedia*, a Web site with over 4.5 million registered users that provides free activities for children to learn through the playful use of technology (www.mamamedia.com). This is the type of well-funded e-learning resource with which museums are now in competition for their offerings on the Web.

Science museums should, by their nature, be aware of technological developments and use these appropriately in a timely fashion. As an example, the Science Museum in London held two exhibitions on the *Challenge of the Chip* (Maynall, 1980) on microprocessors and *This is IT* on information technology in the early 1980s. In these exhibitions, the technology being presented was used to present itself. For example, the *Challenge of the Chip* included a PET microcomputer that illustrated the manufacture and workings of Field Effect Transistors (FET). This was among the earliest examples of using computers in museum displays, employing the animated computer screen in an educational context to augment the more traditional static displays. The use of a computer display meant that the operation of an FET could be illustrated in a dynamic and apt manner. The PET computer was itself also an appropriate exhibit in the context of the subject matter being presented. Nowadays, of course, such use of computers in museum galleries is commonplace, even expected, and is certainly far more sophisticated. In addition to in-gallery use, computers can now readily be connected via networks; access through the Internet, normally using the World Wide Web, is an expected mechanism for disseminating information resources available from museums in general, and perhaps especially so for science museums (Díaz & del Egido, 1999).

In this chapter we survey science museum Web sites in the context of e-learning. In particular, we give several personal accounts of specific Web sites and their development by people who have been involved directly. Finally, some general conclusions are drawn.

Survey of Web Sites

The annual *Best of the Web* competition at the *Museums and the Web* conference, established since 1997, includes an explicit section on the *"Best Museum Web Site Supporting Educational Use,"* providing a snapshot of the state-of-the-art each year.

For example, in 2004, the Smithsonian National Zoological Park's online educational program, Conservation Central (nationalzoo.si.edu/Education/ConservationCentral), was a finalist. This includes a number of multimedia interactive resources, including, for example, one that allows Web visitors to design a panda habitat. Nowadays, relatively sophisticated animation is expected, within the confines of generally available Internet access speed, which is still limited by the quality of telephone lines for many.

The *Resources for Learning* section of the American Museum of Natural History Web site was among the *Best of the Web* finalists in 2003 [www.amnh.org/resources]. This is a collection of activities, curriculum materials, articles, exhibition materials, reference lists, etc., for educators, families, students and anyone with an interest in teaching or learning about science. It is possible to search for resources, browse by topic or explore special collections based around a particular theme. The material is explicitly aimed at different age levels that are selectable when searching (primary, upper elementary, middle school, high school and up). Different completions times, varying from less than a typical lesson period up to more than a week, can also be selected. In the previous year, 2002, the same museum was a finalist with its "Ology" Web site [www.ology.amnh.org]. This includes subject areas such as archeology, astronomy, biodiversity, Albert Einstein, genetics, marine biology and paleontology.

A search for "science museum education" under Google at the original time of writing produces a rather eclectic collection of links. For example, at the head of the list, the Franklin Institute in Philadelphia provides some "Educational Hotlists" in the form of an organized set of links to online resources under over thirty topic areas [www.fi.edu/tfi/hotlists]. *TryScience* provides a gateway to current science and technology using both online and off-line interactive resources in conjunction with a large number of science centers around the world [www.tryscience.org]. It includes information for parents and teachers. The Science Museum of Minnesota has a Computer Education Center, established in 1983, that has its own Web site [comped.smm.org].

The Science Learning Network (SLN) links a number of science museums around the world (www.sln.org/museums), but the information available on the Web site appears to date from 1996. An important part of any online initiative is that it remains up-to-date, and this is especially true for children's resources that can very quickly appear dated if not maintained regularly. The San Francisco Exploratorium provides some excellent "tools for teaching" (www.exploratorium.edu/educate). The Web site includes a list of "Ten Cool Sites" for science education, started in 1995, that seems to continue to be updated regularly, making the site dynamic and encouraging repeat visits (www.exploratorium.edu/learning_studio/sciencesites.html).

Accessibility issues are increasingly important for the museum sector, for both moral and legal reasons (Bowen, 2003; Bowen, 2004), to ensure resources are available to all, including the disabled. In the 2003 Jodi Mattes Access Award, the first year in the UK in which it was awarded for accessible museum Web sites, there were nine sites that were nominated. Only one site, the Australian Museum spiders Web site (www.amonline.net.au/spiders), was science-related. The site includes facilities to change the size of the text to be larger for partially sighted users at the click of a hyperlink, located at the bottom of each page.

The Association of Science-Technology Centers (ASTC) includes a "TEXT ONLY MENU" link at the top of its main Web page (www.astc.org). This is hidden as small white text against a white background for sighted people, but will be read immediately for blind people using text to audio software. The blind normally scan Web pages sequentially and will find such a link quickly and easily. ASTC also provide online advice on accessibility for science centers and museums (www.astc.org/resource/access).

The Museum of Science & Industry in Manchester, UK have a "Text Only" link as part of their standard set of navigation links at the top of each page (www.msim.org.uk). Thus this can be found easily even by users entering the site from pages other than the main home page (e.g., via a search engine). In 2004, the Science Museum in London had a similar a "Text Only (Beta)" link in operation (the "Beta" indicating that this facility was under test), available under an "options >" link (www.sciencemuseum.org.uk). On this site, the text-only versions of pages also include a link back to the graphics versions at any time, thus giving good flexibility of navigation.

Later in this chapter we consider some personal views of individual science-oriented Web sites, both associated with real science museums and of a completely virtual nature, written by people directly involved with their development. In each case, a brief history is given, together with some of the more interesting features of the site, especially with respect to e-learning aspects. Here we briefly look at a couple of examples, one associated with the leading science center in France and the other providing a well-established completely virtual resource on the history of computing.

La Cité des Science et de l'Industrie

The first version of the *La Cité des Science et de l'Industrie*'s Web site (www.cite-sciences.fr) was developed in 1994 and consisted of a very few pages presenting general information on the museum. It was nothing more than an online brochure (sometimes dubbed "brochureware"), updated by the Communication Department as needed. In 1998 it had around 300,000 visits per year. The second version, which is still online today, was introduced in 1999; it was inspired by a "mediation approach" rather than simple communication/transmission of information and content.

The progressive growth of the renewed site in terms of content (temporary exhibitions, learning material) and services (*Visite+*, educational tools, etc.) in the past four years has contributed to creating a rich and varied offering, which attracts a significant number of visitors. In 2003, for example, the museum counted 3,161,000 visits, that is to say a 122% increase in comparison to the previous year. The next reorganization of the site is scheduled for 2005. The idea is to continue the mediation approach. This implies taking visitors into consideration as much as possible and creating online services for them, especially when it comes to the more pedagogical aspects of the site (Cité des Savoirs). For example, the different sections of the existing Web site are not conceived from the visitor's point of view; that will change in favor of a much clearer segmentation approach and solutions that include more semantic navigation.

From an educational viewpoint, the site already offers many different types of services and tools for the general public, as well as for teachers and students/children. An explicitly educational Web site is available (education.cite-sciences.fr). Educational material such as reading matter, interactive facilities, quizzes and animations are available, relating to both temporary and permanent exhibitions. An entire section of the Web site, the scientific lab, is dedicated to manipulation and interactivity with games and quizzes for all ages. For professionals and more expert visitors, text, video and audio versions of conferences and presentations that take place in the museum auditorium are available online, as well as a section dedicated to the latest news and developments in science and research, with articles, dossiers and even a specialized online journal on astronomy and space.

Special sections of the site are also dedicated to teachers (Cité des Enseignants) and children or students (Site pour les elèves). In the "Cité des Enseignants," information is available on general and specific activities for groups and schools as well as tools and material to prepare for the visit, while in the student section, direct access to manipulation games and quizzes targeted for specific age groups is provided. There is also access to Web pages that have been created by students as well as children's technical workshops ("*ateliers*"), normally held at La Cité. In spite of the already conspicuous material and tools available on the site, the idea is to move further and create a real educational platform ("platforme educative") that will be used by teachers and students to download material for the visit or use in the classroom. This will enable the museum to create a community of users that will use the site as a proper working tool and that might be able to exchange opinions and comments through a series of forum activities (Beler et al., 2004).

Virtual Museum of Computing

Virtual museums, without a physical counterpart, are by their very nature a relatively new phenomenon whose form is still developing (Schweibenz et al., 2004). As a fairly early example, the Virtual Museum of Computing (VMoC) (vmoc.museophile.org) was originally set up on a whim one Monday morning in 1994 when the founder and subsequent maintainer was a computer science researcher at the Oxford University Computing Laboratory. This grew out of an early online museum directory, the Virtual Library museums pages (VLmp) that also started shortly before this in 1994 (Bowen, 2002) and was subsequently adopted by the International Council of Museums (ICOM) (icom.museum/vlmp). In those days (and even now) it was relatively easy to set up a Web site in an academic environment, with few bureaucratic or technical barriers to those determined to do so. The initial Web site provided a small number of links to history of computing resources then available on the Web. By the end of the first week the site was already receiving around a hundred visitors each day. It quickly gained international prominence online (Bowen, 1996) and has continued to form a nucleus of online computing history information ever since. An important aspect of success is stability and continued maintenance of educational resources online, as in this case.

The VMoC website consists of an eclectic collection of links to online history of computing resources, categorized in a number of broad types, together with a selection of local virtual galleries. For example, there is a resource presenting *A Brief History of*

Algebra and Computing, based on an article (Bowen, 1995), together with additional external hyperlinks and visual material from the MacTutor History of Mathematics archive (www.groups.dcs.st-and.ac.uk/~history). This is specifically linked from and recommended by a number of educational Web sites. Despite its simplicity, quality of content is of prime importance in e-learning resources.

VMoC also includes links to important computer pioneers. One of the major "virtual galleries" is a resource on the computing pioneer, Alan Turing, maintained by Andrew Hodges, Turing's definitive biographer (Hodges, 1983). More detailed information on this resource can be found later in this chapter.

Case Studies

In the rest of this chapter we present a number of examples of Web site case studies associated with both real museums and in one case a virtual museum. The first, the Natural History Museum of Los Angeles County, was one of the first museums to start a major Web site, as is demonstrated by its enviable Web address [www.nhm.org] that is no doubt coveted by other natural history museums around the world. Secondly, the Museum of the History of Science in Oxford is located in one of the earliest purpose-built museum buildings in the world (originally for the Ashmolean Museum). It was also able to initiate a Web site relatively early because of the advantageous networking facilities and expertise available in a university environment. Thirdly, the Science Museum in London is one of the major science museums in the world. Again it was able to establish an early Web presence partly due to the proximity of Imperial College, but also spurred on by the fact that the Natural History Museum next door actually established the first dedicated museum Web server in the United Kingdom just before them. Finally, a completely virtual Web site is presented, again established relatively early due to the support of a university environment and the enthusiasm of an individual for the project. Since this Web site is a personal project, in contrast to the others that are associated with actual museums, the section describing it is written by the originator in the first person.

All the Web sites described in these case studies are early pioneers in the field in different ways. Thus it is apt that their various histories should now be reflected upon in this chapter. It is hoped that other later adopters can learn from some of the lessons illustrated here.

Natural History Museum
of Los Angeles County

The story of the Natural History Museum of Los Angeles County's Web site (www.nhm.org) is in many ways the story of the Web itself (Angus, 2000). It is a journey of discovery that starts before the advent of the Web at a time when only universities and governments were using the Internet. It is a story of the particular needs of an organization and how the Internet and later the Web provided answers. It is a journey that continues to this day as the medium itself evolves and adapts to the community's needs and aspirations.

Super Science, Kiosks, Gophers and the Web: The Genesis of One of the First Museum Web Sites

In 1991, the museum sought to establish a comparative genetics laboratory that would enable museum researchers to more easily determine evolutionary relationships between different species of plants and animals. By 1991, the Internet was highly utilized by the academic community enabling a degree of collaboration that would lead to a rapid increase in the pace of scientific discovery. Genetic sequence repositories, established on the Internet, were key resources to help researchers determine evolutionary relationships.

The need to access these repositories drove the museum to seek funding from the National Science Foundation (NSF). The museum's application was successful and with the NSF's help they established a direct connection to what was then known as NSFnet. This always "on" connection operated at speeds of 56 kb/sec, the speed of a standard PC modem today. This connection was the first building block in a foundation that would place the museum in a position to rapidly deploy a Web site at a time when the Web was new and audiences were demanding substantial content.

The second block was provided by the museum's educational outreach program. The museum held an annual open house for members, who were interested in the comparative genetics laboratory, but the lab was understaffed. The museum turned to multimedia, another relatively new innovation, to address the issue. Several animated presentations were developed and installed on laboratory computers. These presentations substituted for docent led explanations and established the necessary expertise to later develop Web-based presentations.

The third block was set down when the museum developed software for informational kiosks that were to be placed in local businesses, most notably airports. The programming included all the basic facts about the museum and its programs and mirrored what would be needed in a basic Web site.

In 1993, the museum began to investigate the possibility of developing a "Gopher" site to provide a presence on the Internet. Gopher, a precursor to the Web, offered a text-based interface with an innovation: hyperlinks. But already, this technology was old. A newer application called the World Wide Web was gaining popularity. It offered the same text-based interface and hyperlinks, but included an exciting option: the ability to link to pictures. There was only one experimental browser available and that browser could not display the images directly. Text could only be crudely formatted and position control of various page elements was minimal. But by the end of 1993, the museum had launched one of the first museum Web sites, using material that had been developed for the informational kiosks. The first "online" exhibit consisted of educational presentations derived from the multimedia developed by the genetics lab for the annual members' open house. The University of Southern California hosted this site and it is still operational, though transformed (www.usc.edu/lacmnh).

Building Audiences: From Tricks to Substance

By 1995, the enormous potential of the Web was becoming clear to a number of people. The University of California Berkeley Museum of Paleontology had launched a major Web site and a content starved community came to the site and kept coming. Because there is no "home page" for the World Wide Web, no "site map," the location of a Web site was passed by word of mouth, or more accurately, by e-mail. So two graduate students from a San Francisco Bay Area university (Stanford) came up with an idea, why not build a directory for Internet Web sites? And the brainchild that would eventually become "*Yahoo!*" was born. People quickly discovered the directory and came back again and again. Any site that was listed was certain to get visitors.

Museums understand the need for audiences. Exhibits and educational programs both build and reach out to audiences. The trick was for museums to do the same on the Web. Clearly the key to building an audience was to have content and to be listed in a directory. The question was how to open that door. The museum decided to host a directory of cultural sites called "*The Guide to Museums and Cultural Resources.*" It was structured along the lines of the Stanford site but focused on the cultural sector. The museum actively collected links to cultural resources through a reciprocal exchange with the managers of other directories and by setting up a system where community members could add listings to *The Guide* using a simple Web-based application. In a very short time, *The Guide* had links to thousands of new sites worldwide and more importantly, had tens of thousands of links back to *The Guide*. And each of those visitors to *The Guide* was able to easily click over to the museum's Web site.

The museum also recognized the importance of a *name*. In 1996, the museum moved to secure "naturalhistorymuseum.org," "nhm.org," and the ".com" variations. So in addition to referrals from *The Guide* and other Internet directories, any time a person entered "natural history museum" into one of the new search engines, the museum's site was likely to come up, if not first, then within the first half dozen listings.

Within a short time, the museum was getting tens, then hundreds, of thousands of hits per month to their Web site. However, aside from a few simple online exhibits, substance was seriously lacking. This changed in 1996 when the museum received funding from the National Science Foundation to produce an exhibit on the natural history of cats. Because of the success of the museum's Web site, at least in terms of hits, it was easy to convince museum administrators to invest the resources to develop an online version of the exhibition. The *Cats! Wild to Mild* Web site (www.nhm.org/cats) was born and the museum committed itself to producing a series of online exhibits (www.nhm.org/exhibitions/online.html) that included fully developed lesson plans, classroom activities and other curricula for schools. A balance between content and directory referrals was achieved.

The success of the *Cats!* exhibit was not assured. The museum made an effort to actively involve the wider cat research community in the production of the Web site. Draft pages were offered up for review to a variety of Internet-based e-mail distribution lists and ideas were solicited. This resulted in "buy-in" from these communities even before the launch of the Web site and with the launch, immediately resulted in links that referred more visitors to the Web site (Angus, 1998).

In 1997, the museum's Web site was voted "Best Educational Use" at the first international *Museums and the Web* conference held in Los Angeles. This firmly established the success of the museum's Web site in the minds of the museum's administrators and the wider community.

Building Audiences: Make Their Web Sites Accessible

The museum fell upon hard times. Several scandals rocked the Los Angeles cultural community and helped to dry up sources of funding for the museum. The budget for the Web site, never generous, took a beating. With available resources, the museum would not be able to compete for new visitors against other, better-funded organizations. Nor would the museum be able to invest in any of the promising new technologies that could help leverage and re-purpose existing Web-based content. How could the museum maintain its lead with limited funds? The key was discovered while attending an *American Association of Museums* annual meeting in 1998. A visually impaired woman suggested to a panelist (the author of this case study) that museums could reach out to another audience if they would make their Web sites accessible to persons with disabilities. *Make their Web sites accessible.* This comment resonated and inspired the museum to move forward in a direction that was both on the leading edge of Web development and was ethically sound. Another benefit was that it would keep the museum's Web site in the spotlight for several years to come. The museum led the way in accessible Web design and sought to promote the use of standards that would provide equal access to all audiences, including those with disabilities (Angus, 2001; Bowen, 2003; Bowen, 2004).

A Single Content Repository: Multiple Audiences, Multiple Devices and Multiple Uses

During the course of seven years of growth, the Natural History Museum's Web site grew until it consisted of over 10,000 files. How could all that content be managed? How could it be updated as technology changed? How could it be used to reach new audiences via new devices? Simply put, how could it be used over and over in new ways?

The Web grew explosively because it was technically easy to write Web pages in HTML (HyperText Markup Language), the "language" of the Web. Everyone knows that your friend's 16-year-old son or daughter can have a Web page up and running in an hour or two. However, HTML has a hidden and fatal flaw. It blends *content* and *presentation*. How can material be presented in a new context if the format, the style and the association with other pieces of content cannot be separated? A new language of the Web, a new standard is required, and that standard is XML (eXtensible Markup Language), XSL (eXtensible Stylesheet Language) and XSLT (XSL Transformations) (www.w3c.org).

Information or *content* can be placed in a single location, a *content repository*. That information can be marked with XML so that a computer "knows" what that piece of content is and how it relates to other content. For example, information about a painting can be marked so that a computer knows what parts are the name of the painting, its description and the artist's name, and it will know much about the object's relationship to other paintings. Other standards allow Web developers to access specific pieces of content and present them with a particular "look and feel" or a particular *context*. The information itself remains untouched within the content repository and is available to be reused in different ways. This allows visitors to request specific kinds of information and have it returned in the context of the request. Many organizations are using these standards to leverage their content, to repurpose the collective efforts of hundreds of staff, to reach the public with educational materials that can provide a teacher with new options and in many instances change a student's life.

The separation of presentation and content can also allow the museum to reach new audiences. The museum's content can be presented using more than one *presentation template*. A template can be designed for the "typical" Internet user, a template that easily meets the marketing goals of the museum. A second template can be created for individuals who are blind. The information can be presented in a format that allows a visually impaired person to use assistive technology that will read the page out loud. A third template can be created and used to serve information to visitors using hand-held portable devices. *The same content that is on the Web site can be used to guide visitors within the museum.*

The volume of electronic content continues to grow and museums must find ways to manage the content. Software that allows museums to manage Web content is available. Many of these systems use the new standards that allow the separation of presentation and content. Although the software is expensive, in the long term the museum will find that it is able to better serve the public with accurate and up-to-date information. For example, the name and telephone number of the museum's outreach coordinator may appear in as many as 20 places on a Web site or within a variety of electronic documents. A single edit within the content repository will update the information wherever it appears. This saves staff time and ensures that the museum remains an authoritative source of information.

Metadata is information about information. For example, metadata can be used to identify a particular piece of content as being of interest to a particular audience, perhaps middle school children. This allows the museum to design a Web site where visitors can personalize the site so that the content they most wish to view is presented first.

The Web has changed a great deal since its inception and it will continue to grow and change, depending upon the community's needs and aspirations. Museums need to change as well, to adapt to the new world of instant access and wireless connectivity. Although the technology of the Web provides instant access to a museum's information, it is the content that is important, not the technology. Museums need to adopt standards and technologies that will allow them to preserve, manage and leverage that content into the future. If they do not, then the richness and educational value of our scientific heritage may be lost in an electronic sea of information.

Museum of the
History of Science, Oxford

The Museum of the History of Science is a small, university museum with an outstanding collection in a specialist area, namely early scientific instruments. It is a department of the University of Oxford, and as such is expected to contribute to the research and teaching agenda of the University. Its most prominent contribution to teaching is a Master of Science course that is conducted entirely within the Museum — taught by the curatorial staff and to a large extent shaped by the collection of instruments and the working environment of a museum. As well as its academic presence in research and teaching, the Museum also aims to be fully public — open six days a week and with a program of exhibitions, lectures, gallery talks and other events that provide a distinctive educational opportunity for visitors. We expect our position within a university to contribute to this distinctiveness, since research and scholarship can contribute to the richness of the visitor's experience. Linked to this is the fact that our collection of objects is very strong in early material, so we tend to be more object-focused than many other museums of science. Our use of the Web for education reflects these characteristics and has helped to resolve some of the tensions of being a scholarly museum with a strong public program.

The museum launched its Web site (www.mhs.ox.ac.uk) in 1995, principally as a vehicle for virtual versions of its special exhibitions (Bowen, Bennett, & Johnson, 1998a, 1998b). A series of exhibitions was begun in that year, and the first, *The Measurers: A Flemish Image of Mathematics in the Sixteenth Century*, was offered simultaneously in the gallery, on a gallery computer, and via the Internet. The Web version is still available today. This established a pattern for Web activity for several years, as each exhibition was placed on the Web, and maintained indefinitely. Each exhibition was essentially the full text of the catalog, with all the advantages of navigation and image management that an electronic edition can offer. We have no evidence that this availability was detrimental either to visitor numbers or to catalog sales, and anecdotal evidence suggested the reverse.

While it remains the museum's ambition to maintain this link between the exhibition program and the Web site, in recent years it has proved difficult to do so. Our early exhibitions helped make the case for a major grant from the UK Heritage Lottery Fund, and the Museum closed for a comprehensive project of extension and refurbishment. Equipped with new facilities, not least a dedicated gallery for special exhibitions, we have been coming to terms with a much more ambitious public program and the virtual exhibition work has not yet been re-established. This may be a local problem, rather than a common experience, but we have found that as the Museum has expanded its work — an expansion that was itself fostered and promoted by improving Web resources — it has proved difficult to maintain development on all fronts, and the Web-based work has slipped back. Clearly resource limitations are part of the problem, at a time when museum funding has not grown to meet either staff ambitions or visitor expectations. Nonetheless, a parallel gallery and Web exhibition program is too valuable to lose without a struggle and we hope to be able to revive it before long.

If the virtual exhibition program has stalled for the present, other Web work has flourished and some of this has been in line with achievements or ambitions elsewhere. Both our collections database (with images) and our library catalog are now online, as is an image library of 8,500 items — not massive but large in relation to the size of the Museum and its collection. Where our contribution may have been more distinctive has been in going beyond the presentation of our own material and towards connecting or combining distant collections.

Museums that are physically separate, even those in different countries or different continents, hold objects that are intimately related to each other in history. They may have been made by the same hand, or in the same workshop, or commissioned by the same patron, or used together in the same laboratory. They may represent closely related stages in design development. In extreme cases they may even be separated parts of what was once a single piece. Whether or not the objects have this level of intimacy, it is often enlightening to take a broad view, since a review of evidence across a number of collections may well modify or enrich one confined to a single source. In early periods of instrument making the culture of design and manufacture had aspects that were local and derived from certain traditions of production in a city or region, but there were also developments that were European in scope and depended on an economy of learning that was thoroughly international.

We have offered two, very different responses. In one case a relatively small, natural population of prestigious objects was strongly represented in a few very fine collections. These objects were closely related to each other in intellectual, economic and social respects, but the small size of the population meant that they could be considered — individually and as a group — in great detail and with a wealth of supporting material. In the second example, by contrast, we have a potentially unlimited population, and the aim here is to provide a tool for the researcher, a large database that does not itself hold detailed information on individual objects but which can be searched for unknown instances of the objects under study and for impressions of their frequency and distribution.

In the first project, four museums have combined records of their European instruments up to the year 1600 to create the *Epact* database (www.mhs.ox.ac.uk/epact). The museums are the Museo di Storia della Scienza in Florence, Italy, the Museum Boerhaave in Leiden, The Netherlands, the British Museum in London, UK, and the Museum of the History of Science in Oxford.

Here is an example of a population of objects that has a strong internal coherence and because of the richness of the four collections involved, the result also has the advantage of presenting a fair proportion of the instruments from the period now in captivity in public museums. A terminus of the year 1600 created a relatively small and manageable group of 520 instruments, and each has at least one image at three sizes. The intention was to provide as high-resolution images as was feasible, so that researchers could use the largest images to examine the instruments in detail. The user has a choice also as regards the accompanying text, for two descriptions are offered for each entry. An overview gives such systematic information as the maker, date, place, materials and dimensions, followed by some general remarks and comments on the instrument. These are intended to draw attention to points of interest and not to require any technical background. The second level of text is a detailed and technical description that seeks to reach the standards of a scholarly catalog. This is one approach to the problem of having to satisfy different audiences with different needs.

Users are offered other assistance as well, such as biographies for all the makers, information on the locations where they worked, explanations of the different types of instruments, and a glossary of technical terms. Links are provided for direct reference to this supporting information, and a range of ways of ordering the material is at the command of the user, who can also choose whether to browse by text heading or by thumbnail image. The standard facilities of a Web database have made this a catalog that is more versatile than any printed equivalent could have been, but perhaps its greatest advantage comes from the combination of collections that allows comparisons and inferences that would not be possible in a single source.

Epact was an attempt to produce the finest product that could be managed, to the best standards of scholarship and presentation. It required a great deal of effort, and twelve people from the various museums were involved in different ways. Several years of work — not, of course, full-time — and a number of meetings were needed. The result is attractive and has been well received, so that there have been suggestions that it be extended, either by involving other institutions or extending the date limit. So far, those involved have not felt inclined to reopen and extend the project they have completed, because they are well aware of what would be involved. It was a challenge to keep us all working to the same conventions and to standardize our product. But one thing that might be useful to others is that the conventions and standards we agreed on and sought to implement have been published online with the database (www.mhs.ox.ac.uk/epact/conventions.asp).

The second project, the *Online Register of Scientific Instruments* (www.isin.org), presents a complete contrast to Epact. The Register comprises a much larger population, spread over a much broader range of types and dates, has no restrictions on participating collections, and it could expand indefinitely. But the more profound differences are in ethos: where Epact aimed to be closed, complete and conforming exactly to established standards, conventions and limiting conditions, the Register is an open-ended experiment, whose future is, at least to a large extent, in the hands of its contributors and users. The conventions are minimal and its future direction and development unclear and undecided.

There would clearly be an advantage to researchers to be able to consult a single database, instead of having to keep up with the many initiatives by individual institutions. On the other hand, grand and comprehensive schemes do not work in the long term: they are unwieldy, they soon outgrow the resources and the enthusiasm of their originators, and they trespass on the legitimate interests of the keepers of collections. Collection holders rightly want to control their output. They do not want to hand over to others their responsibilities for making information available on their collections, but at the same time they want to be part of some central vehicle for making their work known and for attracting interested users. The Register seeks to answer this need: to provide the minimal facilities that will be of service to users and collection holders alike, while responsibility for information and its dissemination stays where it belongs, namely with the individuals and institutions who care for the collections.

So one principle of the Register is that collection holders contribute the information on their objects. They can include as many of their objects as they wish, and in each case they complete as many as they wish of the fields of information (there are only a few) offered by the Register. They can return to their entries whenever they like, and correct them, modify them, improve them, delete them or add to them.

The Register is like a library catalog which tells you about the existence of a book and how to find it, it does not contain detailed descriptions or histories or images: for these the user contacts the collection holder, either through a direct link to an online database where the collection holder has provided this, or by e-mail or post. At present there are 17 contributing collections, the largest being the Science Museum in London with 2,273 entries.

The disadvantage, of course, with placing all responsibility for content with contributors is that it leads to inconsistency. With Epact we had enough difficulty imposing consistency on ourselves in a relatively small group of workers dealing with a relatively coherent group of instruments. Given the scope of the Register, this would be impracticable with the slight resources at its disposal. It might not even be wholly desirable in any case, again given the ethos the Register has adopted: contributors do not relinquish responsibility and centralization is minimal.

The Register does not control or vet contributions in any way. There has been no attempt at a thesaurus of allowed terms and entries, and no particular language is required. There are no preferred forms for names of people or places. Where inconsistencies, or mistakes, appear in the indexes, it is hoped that the contributors will notice these and make correcting submissions: again it is the contributors who are responsible. In practice nothing else would be possible; the resources for a thorough vetting and correcting procedure does not exist. But in any case, there is a virtue in placing these responsibilities with contributors: the interest and responsibility is collective, and the Register will develop or not depending on the extent to which contributors and users find it valuable and want it to work. It is intended to be an opportunity, rather than a finished product. It also has to be said that the lack of success of grander projects for agreed terminology is not encouraging. The indexes generated by the Register could become vehicles for agreement and convergence, as contributors notice consensus emerging on certain terms and names, and adjust their entries accordingly.

It is in this democratic spirit of self-help, allied to the conviction that a central index of this sort could be of great use to institutions and researchers, that the Scientific Instrument Commission of the International Union of the History and Philosophy of Science has sponsored the Register. The original designer, Giles Hudson, presented the project to a Symposium of the Commission in 1998 and Jessica Ratcliff has since developed the site with the support of the Museum. The Register is a challenging and imaginative experiment, whose future rests with the community of users. For that reason, although it was launched and is managed by the Museum, it appears at a different URL (www.isin.org) from the museum's Web site (www.mhs.ox.ac.uk), where the Epact database can be found.

While the educational value of our virtual exhibitions is evident, the projects presented in more detail here, namely Epact and the Register, may seem more like tools for research, and to a certain extent that is true. Neither offers class or individual lessons tailored specifically to target groups of e-learners. Until very recently the Museum has not had the professional educational staff necessary to ensure that such material is really useful and relevant. However both projects have such potential, and Epact in particular can readily be used as an information and image resource with a great deal of supporting material for the user. Its subject matter may at first seem relatively distant from school learning, but because mathematics in the Renaissance was often closely related to practical matters, it is not difficult to apply the Epact material to social history.

Time, for example, forms the basis of many school projects, and the single most common instrument in Epact is the sundial. Many lessons can be learnt from the dials on view; the importance of time telling, for example, is seen from the number, variety and quality of the pocket dials, while the arbitrary nature of our division of the day into hours is demonstrated by the numerous alternatives in use in the past. Many of the instruments relate to warfare, while others illustrate the importance of religion or astrology in daily life. At present, students may have to do some exploration to tease out what they want to use, but the various directories and glossaries ensure that the information is there, and learning to explore has its own value.

Science Museum, London

The Science Museum (London) is part of the National Museum of Science & Industry (NMSI), which further comprises the National Railway Museum (York), and the National Museum of Photography, Film & Television (Bradford). The Science Museum has its origins with the Great Exhibition of 1851, and has resided in its present building in South Kensington since 1928. Today, the Science Museum exists to promote public understanding of the history and contemporary practice of science, medicine, technology and industry.

In regard to the online presence of the Science Museum (www.sciencemuseum.org.uk), the development of its Web site has been organic, at least for the first four years of its existence. The Web site itself was launched shortly after the Natural History Museum (NHM) (www.nhm.ac.uk) set up its own pages in 1994. It should be noted that the Natural History Museum was the first UK museum to have its own Web site presence. Both the Natural History Museum and the Science Museum benefited from the close physical co-location with Imperial College, where the relatively high-speed JANET academic network was already well established with good network connections to other UK universities and the rest of the world.

The early Science Museum site quickly developed a broad and deep hierarchy with many cross-listings. In addition to visitor information, there were approximately 12 navigation sections, a quick search and featured highlights. The sections largely represented organizational activities and to some extent, the main divisional areas of the Museum itself. These included Exhibitions, Education, Collections, Research, Commercial and Services.

Exhibitions Online

Of note at this time were the featured highlights. The Science Box series appeared under this section — a series of small, temporary exhibitions. In 1995, The Information Superhighway was one of the Science Box events that appeared on the Web and which coincided with a set of seminars and a touring exhibition component that provided outreach to the public in the area of new communication technology.

The latter half of 1998 saw the appearance of "Exhiblets," the first exclusively virtual exhibition that did not reference a physical space or exhibition in the actual museum (www.sciencemuseum.org.uk/collections/exhiblets). The name "Exhiblet" was derived from a combination of the terms "exhibition" and "Java Applet." Exhiblets represented a set of online information resources that used a specific object or collection to explain events, discoveries and personalities. Each Exhiblet is comprised of a narrative, object list and bibliography. The resources were a form of pro-active approach to enquiries in popular areas, such as the personality of Marie Curie, and to highlight collections within a historical context. The success of the medium can, perhaps, be gleaned from the Web statistics of December 1999, which shows that "Marie Curie" was a top search phrase that led visitors to the site, only exceeded by the term "science museum."

Exhiblets were a relatively low-tech addition to the Web site and were intended to be easily viewed by different browsers and printed-out as a resource for the enquiring public and school students. This complemented the availability of activity sheets that the Education Department had placed online for printing and downloading for teachers and schools.

More interactive applications were developed for specific featured objects and exhibitions in the Museum galleries. One of the earliest features to use a form of VR was the online exhibition for the Apollo 10 Command Module, which incorporated an activity entitled "Design your own rocket" (www.sciencemuseum.org.uk/on-line/apollo10).

The use of Web tools to provide interactivity and participation was encouraged under the STEM initiative. The STEM Project (Students' and Teachers' Educational Materials) was an Internet competition held by the Science Museum and sponsored by TOSHIBA and begun in 1997. STEM encouraged school visitors to create a Web site based on a particular gallery, exhibit or online exhibition. The purpose of the project was to promote the creation of a database of resources created by students and teachers for students and teachers. In this way, the project extended the interrelationship between the museum and school curriculum. Over 1,000 resources are now archived on the site (www.sciencemuseum.org.uk/learning/sheets/sheets/sheetintro.asp). Note that STEM ended in 2003.

Web Redesign

During the active period of online development and features added to the Web site, namely between 1997 and 1999, the Museum began to review its online presence and sought to provide more integration to existing content. Importantly, it looked to provide a more interactive and community-based Web experience, and to provide multiple entries to Web site content through a variety of navigation methods.

Following meetings during the summer of 1998, it was determined that there was a need to revise the existing navigation of the Web site and to provide multiple audiences with multiple information structures. One of the methods was determined to be via the offering of a subject-based navigation.

In an evaluation conducted by the Visitor Research team in 1999 (Steiner et al., 1999), recommendations indicated the need to reduce the number of highlights on the front page and the number of options for the initial navigation. Some of the section headings also came under review. For instance, whereas it was clear that the Education section contained information for the educational community, especially schools, there was also the expectation for continuing learning to be located there. Similarly the heading "Online Features" was another area of confusion for visitors who anticipated "information" about physical exhibitions. Some of the heading changes thus resulted in Education becoming "Learn & Teach" and Online Features changing to "Exhibitions Online."

These changes reveal a move toward usability in terms of language and design for a growing Web savvy audience, and possibly emphasized the variations in the needs and expectations of virtual audiences versus the physical visitors to the Museum.

Wellcome Wing

This reorganization of content and Web restructuring coincided with the development of the new, Intel sponsored, Web site for the Wellcome Wing building of the Science Museum (www.sciencemuseum.org.uk/on-line/wellcome-wing). The Wellcome Wing opened in June 2000 and focuses on contemporary science and technology, with a particular emphasis on biomedicine.

The Web site for the Wellcome Wing was the largest "microsite" at this point accessible via the museum's pages. The site was developed to support both the interactive nature of the new building and the look and feel. It contains interactives similar to those in the gallery (for example the "Pattern Wall" which can be downloaded as a Flash game or played within the Museum itself). Also of note is the VRML (Virtual Reality Modeling Language) 3-D clickable walkthrough of the Wing that gives users the chance to engage with the space itself as well as providing alternative ways into the content. The building and the Web site represented a move away from more "object-oriented" and historical rich spaces.

At the same time as the launch of the Wellcome Wing, Antenna was developed (www.sciencemuseum.org.uk/antenna), "a world first — a constantly updated exhibition devoted exclusively to science and technology news…" This exhibition brings rapidly changing exhibitions and CIPs (Computer Information Points) into both the virtual and real spaces. As time went on, the CIP development was levered into XML which means that content developed for the gallery kiosks can be re-deployed within minutes to the wider website audience. Antenna is a conscious response to the information-seeking behavior of younger audiences who are informed by and exposed to rapid news and media-led events.

InTouch (www.sciencemuseumintouch.org.uk) went a stage further in linking the real and virtual by allowing gallery visitors to create their own Web pages when they visited the gallery spaces of the Wellcome Wing. Various interactives are connected to the Web space and all are activated on-gallery using a retina scan and online using a user name sign-in system. To date, roughly 170,000 personal Web pages have been created. InTouch is also an example of *personalization* (Bowen & Filippini-Fantoni, 2004) driven design whereby the experiences offered by the interactives are user-led. The sustainability of InTouch continues to be reviewed and has provided an important foundation in this area for subsequent projects, especially in convergence (physical and virtual interdependent spaces).

Dana Centre

The Dana Centre [www.danacentre.org.uk], launched in the winter of 2003, "...marks a new direction in science communication: to challenge public perception and tackle contemporary science head on. This dynamic events space will bring the hottest themes in modern science to adults-only audiences through a program of bold and innovative events."

Similar to the Wellcome Wing, the Centre is linked to the experience of a physical building, but striking differences lie in the use of new media channels for information delivery, omission of gallery/object-oriented spaces, and its focus on adults as the key audience. The Centre itself is a highly wired building, with facilities including OB (Outside Broadcast), Web cast, Web cam, wireless, kiosk, projection, moveable stages and rigging, wired rooms, etc. The direction in which it has influenced the Web most, however, is again in the connection between the virtual and real spaces. Not only are the program and other key information displayed on the Dana Centre Web site, but also dialogue is fostered through online discussion boards (Bernier & Bowen, 2004) — and permeates the real space via the live Web kiosks and projections. There are other less obvious connections as well: the program of events, for example, is updated once online and an XML feed then powers the "what's on" projection in the building itself. There is also a projection feed of images from the NMSI Picture Library Web site [www.sciencemuseum.org.uk/piclib], chosen at random from the database.

In this community-driven space, the "4thRoom" represents another complementary innovation developed during the Web site build – a flash interactive that acts as a "skin" on the online discussion. Each person's point of view on a particular thread is displayed within this interactive as an avatar figure. Shortly the interactive will be available as a screensaver with the discussion fed remotely by the Web site. One consideration in the set up of the 4th room and debate areas of the Centre is the need for moderation and the introduction of "House Rules." The latter was undertaken specifically for the Centre and in light of its provocative content.

To support community-centric activity, and particular to a technologically aware target audience, development into the future includes an innovative 3-D chat event, use of SMS (Short Message Service) and MMS (Multimedia Messaging Service) during and after debates, "klip" technology that will provide users with live, desktop notifications of the latest discussions and events and further building of key areas of the site.

The Ingenious Project

Ingenious (www.ingenious.org.uk), funded by the NOF-digitise program (a lottery funded initiative), represents the largest scale object-based Internet project to date undertaken by the Science Museum and NMSI (to launch in the Spring of 2004). The project aim is to make publicly accessible 30,000 digitized images and accompanying records, 10,000 library records, and 10,000 object records sourced from the Science Museum (including the Science and Society Picture Library), and its sister sites: the National Railway Museum in York and the National Museum of Photography, Film & Television in Bradford (Borda & Beler, 2003).

In addition, this material is to be contextualized by several hundred pages of circa 40 topical stories aimed at life-long learners (the widest potential audience of the "microsite" projects). These topics use the primary material "to weave connections between the people, innovations and ideas that have changed our lives and the way we see the world, from the industrial revolution to the present day" (extracted from the *Vision Statement*). Through these connections, users have the further opportunity to find meaning for themselves by being presented with tools to "create" and to contribute to subject-driven debates (Borda & Bud, 2002). As an example, the Ingenious Web site allows users to custom-build their own user experience by adding resources to a "save image" clipboard, and self-market the site by passing on e-cards, personal pages and links to their friends and colleagues.

A highlight among the toolkits is the "webgallery" function that will enable users to add text and captions to images of objects, to arrange them, and to save other types of resources, as well as e-mail the end product. This creative building process provides an individualized means of making learning resources of direct significance available to the user.

On a community level (Beler et al., 2004), debates permit users to contribute and join conversations focused around contemporary issues. There are also plans to cross-link debates with the Dana Centre to build on existing tools and streamline the user experience (and moderation efforts). Due to the nature of the sharing activities and depth of authored content (especially for the topics), one issue similar to the Dana Centre is the need for libel coverage and for Ingenious to have its own set of House Rules as the Museum engages more in the area of user contributions online.

Other than the STEM project, Ingenious tools will be unique to be drawing from a comprehensive knowledge base for self-publishing and sharing, and general content syndication (Borda & Beler, 2003). Watermarking is an issue currently being discussed so that ownership and, viral marketing, can be realized. There are plans to do an extensive summative evaluation on the whole site once it becomes public. This will likely inform the development and tweaking of the tools to optimize usability first and foremost.

Science, Invention and Nature

Expanding the community-focus online, the Science Museum is leading a consortium project (again funded by the NOF-digi program) called *Science, Invention and Nature*

(SIN) (www.sinergies.org.uk). SIN is an interdisciplinary Web portal that combines information from four institutional Web sites and their respective NOF-funded projects exploring aspects of the natural and man-made worlds. The four partner Web sites linked to the portal thus far are:

- Science Museum (www.sciencemuseum.org.uk)
  – *Ingenious* (www.ingenious.org.uk)
- Natural History Museum (www.nhm.ac.uk)
  – *Nature Navigator* (www.nhm.ac.uk/naturenavigator)
- Wildscreen Trust (www.wildscreen.org.uk)
  – *Arkive* (www.arkive.org)
- Y Touring (www.ytouring.org.uk)
  – *Genetic Futures* (www.geneticfutures.com)

Internet users entering the SIN Web site can browse through different themes (also known as SINergies) and click on related links, which will also take the user to content on the four sites. The SIN Web site's main task is to collate select content from each of the sites into these broad editorially driven SINergies. One example of such a theme on the SIN portal is "Food." The SIN portal will hold an introductory text on this theme, as well as food-relevant links to the other four Web sites.

There is a possibility for the user to "drill" the other way as well. The four Web sites will each hold a SINergy button icon on different pages where theme relevant content will take the user to the SIN portal. For example, a page on "Eating disorders" on the *Y Touring* Web site will hold that SINergy button, and when a user clicks it, s/he will be taken to the SIN portal and the theme on "Food." The theme will then lead to further content on the topic "Food" through links and relevant keyword search (pre defined based on subject page on which the button resides). This concept follows the idea of a "Web-ring" in which a group of Web sites with a common theme, configured in a loop, allows the user easy access in the ring by clicking on links. Significantly, throughout this learning experience, the user is kept within a contained group of authoritative and subject-focused set of sites.

To support the concept of an "enclosure" of subject specific resources, the other principle component of the SIN Web portal is a sophisticated search engine. One of the main purposes of the portal is to provide users the possibility to search for content for relevant topics and provide them with a "complete" result across all underlying partner sites via search algorithms and metadata. The user can also search by resource "type" (e.g., image, PDF, HTML page). To achieve optimum results, the SIN portal has been designed to crawl the four Web sites (other sites may be added later) in order to create an indexed database that will be used for efficient and targeted searches. The search is a particularly key component because SIN represents a resource almost entirely comprised of *born digital* resources (i.e., Web site materials), and the site itself has minimal content — the SINergies really acting as jumping off points and a means to provide examples of the range of interdisciplinary content.

In summary, the idea behind the overall concept is to offer users something more targeted than a Google result set — here the users are provided with well-known and branded knowledge organizations from which to search across and access authoritative learning materials. The SIN site is due to launch in Spring 2004 and will then be re-evaluated.

Future Directions

Future projects are beginning to approach Web site technology, content and delivery in a more holistic way, not least the need to "join-up" both on resource and practical levels. This will mean the development of a central content management system, including industry standard storage and systems, as well as unifying many of the tasks carried out today into re-useable "toolkits." These will provide each project with an increasing array of technical solutions already in place. Effort on projects will therefore be released to focus on innovation of design and content (i.e., the user experience) rather than technology. We envisage therefore that as time goes on, the technology will become increasingly invisible to users and builders of sites within the Science Museum and the NMSI family.

Continuing efforts are being made to improve accessibility (Bowen, 2003, 2004) and interoperability across all sites, and we expect delivery to multiple platforms to begin to become an important part of what we do both online and off-line, and to extend the learning/engagement channels to multiple users and audience profiles of the Web site.

Alan Turing Home Page

The Alan Turing Home Page, now the index page to a large Web site dedicated to the life and work of the mathematician, computer scientist and codebreaker Alan Turing (1912–1954), first appeared on the World Wide Web in September 1995. At first it was hosted in my own user space on the Web server of Wadham College, Oxford University. In 1996–97 there were mirror sites in San Francisco and Chicago. But in 1997 I started paying for my own domain names and Web hosting and since then the site has been solely under its own dedicated domain name (www.turing.org.uk). It currently has between 1,000 and 2,000 visitors a day.

The background to these early developments was, of course, the very rapid expansion and growing sophistication of the Web. I was a slow starter on the Internet (for an academic), and it was not until 1995 that I appreciated how important it was going to be. Matthew Westby, one of my students at Wadham College (where I am Lecturer in mathematics) showed me HTML, and thanks to him I began life as a Web author just at the exciting point when the Web was transforming itself from a club for enthusiasts to a universal medium for global information and communication. My initial motivation for the Alan Turing Home Page was simply that on looking up "Alan Turing" on the search engines that then existed, I was aghast to discover pages of inaccurate offerings, plagiarized from obsolete texts or the result of poorly informed student projects. I had an alarming vision, which has never left me, that all my work in the 1980s, in particular my large-scale biography *Alan Turing: The Enigma* (Hodges 1983) might as well never have existed as far as Web users were concerned! More positive feelings very rapidly superseded this anxiety, because my first Web pages were noticed and appreciated at once (in particular by Jonathan Bowen, already a pioneer with his Virtual Museum of Computing), and these new contacts gave me a delight in the value of cooperative linking which, likewise, has never left me.

The Web site was started to complement that biography of Alan Turing, and its spirit has remained similar to that of the book, which I started as a multi-faceted, multi-level work back in 1977. Now as then, I work as an individual researcher and author. But there are obvious differences between writing a printed text, and authoring for the Web, which struck me right from the beginning. One is the immediacy of Web publication, both in the sense of being able to publish without the glacial timetable of book publishers, and in the sense of being unmediated by their no less icy and static conventions — to publish direct from author to reader and inviting personal response. The dynamic quality of the Web also allows correction and expansion of material to be done at any time. This means, of course, that unless fairly elaborate schemes for archiving and dating are in place, one cannot expect Web pages to play the role of printed publications in establishing definitiveness and priority. For this reason I do not write for Web pages in the same style as for a printed publication, and I am mildly surprised when other people cite ephemeral Web pages as academic references. Another difference is economic: without elaborate registration and payment schemes, impractical in my case, there is no royalty payment to the author. The economic rationale of this work has had to rest on the hope of improved book sales and a small amount from Amazon.com commissions: neither would justify the time I have spent. The shoestring finances have also dictated that the technical work must remain as it began — by doing it myself. This means that the design does not meet the standard expected by those using professional graphic design elements optimized for all platforms and screen sizes. Still less could I undertake advanced dynamic "personalization" features (Bowen & Filippini-Fantoni, 2004). However, it is tolerable.

In Web authoring I am influenced by the fact that my written style makes much use of allusion — "only connect" was one of the mottos of Hodges (1983) — and so the principle of hyperlinking seemed natural to me as soon as I saw it. In the first few days of the site, I decided on a format that has remained good ever since. There is a formal textual biography, based on my entry on Turing for the (British) *Dictionary of National Biography*. This allows for straight linear reading without the distraction of links. Parallel to it, there is a Web page "Scrapbook," visually designed to look like entries and pictures pasted into a book, which has all the links. This choice of design meant that when new ventures such as the Turing Digital Archive (www.turingarchive.org) came online, the browser could move into them seamlessly through hyperlinks. The Scrapbook is probably the most interesting part of the Web site as it is intended for hyperlinked "browsing" in a way that only the Web can offer. My policy for the Scrapbook has been from the beginning to give *annotated* hyperlinks to other sources. Web authors tend either to provide sites which are either totally self-contained, without using external links, or else to give uncritical listings of links to other pages, expressly without any responsibility or judgment of any kind regarding their content. I have never felt happy with either extreme, and try to update the links and associated commentary. The downside is, of course, that it is hard to maintain working links and compose suitable remarks. I am always aware of a large stack of tasks confronting me through the volume of relevant material on the Web. It is very different now from 1995, when there were only a few items on the Web for the Scrapbook pages to point to. All truly Web-based material is always "under construction."

After 1997 I added further sections, reflecting the new research, talks and publications that had stimulated by the existence of the Web site. In particular, a "Philosophy" section has been built round the short printed text that appeared in the *Great Philosophers* series (Hodges, 1997), and my online entry in the Stanford Encyclopedia of Philosophy (plato.stanford.edu). This more advanced material attracts about 100 users a day, many of them degree-level students from around the world. Inevitably, sometimes paragraphs from my writing must be copied, undigested, into students' work: as all lecturers know, the Web lends itself to crude cribbing. But I try to counter this tendency in the Scrapbook section by annotations that show students (and indeed anyone else) how there are many differences of view and sometimes-fierce arguments among leading current thinkers. This relates particularly to the significance of Church's Thesis, to the question of how the electronic computer came into being; and to the prospects for Artificial Intelligence. Here I encourage users to follow external hyperlinks to see these different views for themselves, and discourage mindless copying.

It also becomes apparent from enquiries that the Web site is being used by school students. Indeed, the UK National Grid for Learning references it. I am inclined to say, however, that the questions being "researched" by school students, judging from the demands for "information" that I receive, often seem to reflect the fact that they are attempting to answer questions which are far, far too difficult. There is nothing on my Web site specifically intended for school students, and on the whole my motto would be that computability is an adult subject. Of course some exceptional school students will discover my site with relish, like my book, but I make no claim that it would be of general value for education below degree level. However, this could change: the new A-level in the ethics and history and philosophy of science might make it a very worthwhile course element.

From these remarks it should be apparent that I am critical of the too-frequent assumption that Web material is, just by its nature and its availability, a learning resource of value to students of all kinds. The Web is poor in its provision of structured course learning, and is not convenient for the experience of serious reading. My site does not pretend to supply anything of what students should be getting from textbooks of computer science and computability. It can complement, but not substitute for, my own printed work. I can offer specific technical benefits — like a JavaScript Turing machine demonstration, which does much better than the static printed explanation in my book. I can give technical details that it would not be economic to print. And I can give an illustrated taste, an overall browsing experience, which can usefully excite the imagination, and be accessible to people all over the world. These seem to me the main benefits of Web material in general.

Looking into the future, I would expect to develop those features of Web material further, and especially those areas where I can give something unique that no one else would do. One of these lies in creating a "Book Update:" supplementing my printed text with new and corrected material of all kinds. If, one day, publishers let Hodges (1983) go out of print — i.e., it loses its economic viability on printed paper — then the Web site could extend into providing an e-book alternative. There is certainly room for me to enlarge and improve the role of technical design and illustration, but really it is the content that matters. Probably the area where I could best take advantage of my own particular knowledge and development, is to expand the Philosophy section into a much larger discussion of physics, computation and the human mind, for instance in relating Turing's ideas to those of Penrose (1989).

This may sound a very dry and abstruse conclusion, but actually it is a very personal one. I am in an unusual position as having come to study the development of computer science from a background of working with Penrose in relativity and quantum mechanics, and from a strong personal and political interest in Turing's individual life. Publishers, like academic departments, are very wary of cross-genre, multi-purpose work, and enforce the *either/or* mentality. The Web, with its low marginal costs and its natural interlinking, allows a *both/and* principle which suits me better, and allows me to express what I want, in the way that I choose. Although I have expressed a skeptical view about the value of Web material for traditional structured learning, readers' responses teach me that there seems to be no obvious limits to the effect that Web authoring can have on readers of all kinds and all backgrounds, and it is the response of these readers that has been the non-economic reward to this author.

Lastly, after the preceding text was completed, and just before the final editorial deadline, there was an unexpected development. I was offered, and have now accepted, a commercial arrangement with the online retailer Kelkoo (www.kelkoo.co.uk), under which I am paid for displaying links to it throughout the site. This new arrangement may, of course, be as short-lived as the famous dot-com bubble of 1999–2000, depending as it does on the marketing concepts of a multinational company, but for the time being it offers a rather better economic basis than the state of affairs that I outlined above for maintaining and augmenting the site. It also illustrates the dynamic nature of online publishing, where one has to be prepared for changes from day-to-day. You are now reading a printed text that captures the state of affairs as they stood in early 2004, but time will not stand still in cyberspace.

The business of a physical museum is a strange amalgam of detailed and obsessive long-term scholarship, the exploitation of graphic and advertising techniques, the pressure of unpredictable demands from the general public, and the necessity of negotiating financial sponsorship. Being the curator of the branch of a "virtual museum" is, albeit in microcosm, not so different. Experience shows that communication of all kinds is more like the hopeful floating of bottled messages into the oceans, than the direct linear inculcation of facts and logical deductions. Electronic publication and e-learning is no exception. One can only carry on in the hope that this new symbiosis of text and image, enhanced by the global glow of the computer screen, will spark in some readers that magic response that goes beyond "information" to real excitement and involvement.

Conclusion

In this chapter we have explored a number of case studies of science museums and their associated Web sites, as well as completely a virtual Web site concerning the history of science, in the context of e-learning. The case studies have all been pioneering in their own way and have taken very different approaches in their development and presentation of e-learning material.

Of course the Web is constantly developing with new issues and technologies. Any e-learning facility must try to keep up with these changes or it will inevitably start to look dated. For example, personalization is an increasingly important technique that allows Web sites to provide a more relevant experience based on individual requirements. Although there is considerable scope for development, major science museums such as the Science Museum in London and La Cité des Science et de l'Industrie in Paris are now exploiting this approach in their latest Web site developments (Bowen & Filippini-Fantoni, 2004).

Widening access is also an important concern, with legislation now making this a legal requirement in many countries (Bowen, 2003, 2004). Online, it is possible to make educational resources in a form suitable for many types of users, whether they are school children, the general public or advanced researchers (Bowen et al., 2001), even if they are disabled (e.g., through blindness or partial sightedness for example). The Natural History Museum of Los Angeles County was an early museum pioneer in this respect (www.ed-resources.net/universalaccess). With the ever-increasing number of technologies available on the Web, it is important that museums understand the accessibility issues when including a particular technology on their Web site (e.g., to improve the interactivity or multimedia presentation of an e-learning resource). Presenting material in multiple ways (e.g., textually, visually and aurally) is often beneficial for many end users, if suitable resources are available.

For the future, successful museums should use their Web sites to augment their educational facilities in appropriate and cost-effective ways. For example, it is an ideal medium to make material available for teachers cheaply, avoiding postage costs associated with traditional physical delivery, and for providing widely available interactive facilities while avoiding the installation of expensive hardware. It is possible to provide good facilities even when the budget is limited, as in many virtual museums. It may be feasible to leverage volunteer effort in the development of e-learning material, through liaison with universities, for example.

Overall, there is no one route to success, but the area where museums reign supreme over many other organizations is the availability of real objects and associated unique content. Using and combining these resources in novel ways can help to inspire those using museum Web sites for educational purposes in a distinctive way that is not possible for other organizations.

Acknowledgments

We are grateful to Mike Ellis, Web Manager at the Science Museum, London, for his help relating to the Science Museum section of the chapter.

References

Angus, J. (1998). Building and maintaining a large Web site. In D. Bearman & J. Trant (Eds.), *MW98: Museums and the Web conference.* CD-ROM. Toronto, Canada, April 25-27. Pittsburgh: Archives & Museum Informatics.

Angus, J. (2000). Building a Web site. *Museum International, 52*(1), 17-21.

Angus, J. (2001). Wired for accessibility. In D. Bearman & J. Trant (Eds.), *MW2001: Museums and the Web conference.* CD-ROM. Seattle, Washington, USA, March 14-17. Pittsburgh: Archives & Museum Informatics. Retrieved from *www.archimuse.com/mw2001/papers/angus/angus.html*

Beler, A., Borda, A., Bowen, J.P., & Filippini-Fantoni, S., (2004). The building of online communities: An approach for learning organizations, with a particular focus on the museum sector. In J. Hemsley, V. Cappellini, & G. Stanke (Eds.), *EVA 2004 London Conference Proceedings,* London, July 26-30, (pp. 2.10-2.15).

Bernier, R., & Bowen, J.P. (2004). Web-based discussion groups at stake: The profile of museum professionals online. *Program: Electronic Library and Information Systems, 38*(2), 120-137.

Bicknell, S., & Farmelo, G. (1993). *Museum visitors studies in the 1990s.* London: Science Museum93.

Borda, A., & Beler A. (2003). Development of a knowledge site in distributed information environments. In *Proceedings of ICHIM'03: Cultural Institutions and Digital Technology.* CD-ROM. International Cultural Heritage Informatics Meeting, Paris, September 8-12.

Borda, A., & Bud, R. (2002). Engaging with science & culture: Major missions across cyberspace to share good history. In J. Hemsley, V. Cappellini, & G. Stanke (Eds.), *EVA 2002 London Conference Proceedings,* London, July 25-26 (pp. 3.1-3.8).

Bowen, J.P. (1995, January/February). A brief history of algebra and computing: An eclectic Oxonian view. *IMA Bulletin, 31*(1/2), 6-9.

Bowen, J.P. (1996). Virtual museum of computing Web site. *IEEE Annals of the History of Computing, 18*(4), 67.

Bowen, J.P. (2002). Weaving the museum web: The Virtual Library museums pages. *Program: Electronic Library and Information Systems, 36*(4), 236-252.

Bowen, J.P. (2003). Web access to cultural heritage for the disabled. In J. Hemsley, V. Cappellini, & G. Stanke (Eds.), *EVA 2003 London Conference Proceedings,* London, July 24-25 (pp. s1.1-1.1).

Bowen, J.P. (2004, January). Cultural heritage online. *Ability, 53,* 12–14. Retrieved from *www.abilitymagazine.org.uk/features/2004/01/A53_Cover_story.pdf*

Bowen, J.P., & Filippini-Fantoni, S. (2004). Personalization and the Web from a museum perspective. In D. Bearman & J. Trant (Eds.), *MW2004: Museums and the Web conference.* CD-ROM. Arlington, Virginia, USA, March 31-April 3. Toronto: Archives & Museum Informatics. Retrieved from *www.archimuse.com/mw2004/papers/bowen/bowen.html*

Bowen, J.P., Bennett, J., & Johnson, J. (1998a). Virtual visits to virtual museums. In D. Bearman & J. Trant (Eds.), *MW98: Museums and the Web conference.* CD-ROM. Toronto, Canada, April 22-25. Pittsburgh, PA: Archives & Museum Informatics. Retrieved from *www.archimuse.com/mw98/papers/bowen/bowen_paper.html*

Bowen, J.P., Bennett, J., & Johnson, J. (1998b, January/June). Des enquêtes sur les musées en ligne: Le Virtual Library museums pages. *Publics et Musées, 13,* 115-127.

Bowen, J.P., Bridgen, R., Dyson, M., & Moran K. (2001). On-line collections access at the Museum of English Rural Life. In D. Bearman & J. Trant (Eds.), *MW2001: Museums and the Web conference.* CD-ROM. Seattle, Washington, USA, March 14-17. Pittsburgh, PA: Archives & Museum Informatics. Retrieved from *www.archimuse.com/mw2001/papers/bowen/bowen.html*

Díaz, L.A.B., & del Egido, A. (1999). Science Museums on the Internet. *Museum International, 51*(4), 35-41.

Hodges, A. (1983). *Alan Turing: The enigma.* London: Burnett and New York: Simon & Schuster.

Hodges, A. (2000). Turing: A natural philosopher. London; Phoenix (also New York: Routledge, 1999). In R. Monk & F. Raphael (Eds.), *The great philosophers.* London: Weidenfeld and Nicolson.

Maynall, W.H. (1980). *Challenge of the Chip.* London: Her Majesty's Stationery Office (HMSO).

Papert, S.A. (1993). *The children's machine: Rethinking school in the age of the computer.* New York: Basic Books.

Papert, S.A. (1996). *The connected family: Bridging the digital generation gap.* Atlanta: Longstreet Press. Retrieved from *www.connectedfamily.com*

Papert, S.A. (1999). *Mindstorms: Children, computers, and powerful ideas* (2nd ed.). New York: Basic Books.

Penrose, R. (1989). *The emperor's new mind.* New York: Oxford University Press.

Schweibenz, S., Keene, S., Worcman, K., Jaggi, K., Kraemer, H., & Karp, C. (2004). Virtual museums. *ICOM News: Newsletter of the International Council of Museums, 57*(3), 3-8.

Steiner, K., Payne, J., & Romans, J. (1999, June). *Evaluation of Web site structure.* Internal Report, Science Museum. London.

Chapter XIX

The Educational Approach of Virtual Science Centers:
Two Web Cast Studies (The Exploratorium and La Cité des Sciences et de l'Industrie)

Roxane Bernier, Université de Montréal, Canada

Abstract

This chapter looks at ways of examining informal e-learning environments to address innovative pedagogy, from two well-known institutions, where the theme of science is promoted within virtual centers, in a manner that is motivating for both online and onsite visitors. The author argues that real-time interactions such as Web casting act as a focus that enriches the people's interest and thus enhances the notion of Public Understanding of Research (PUR), while "being socialized" through the scientific community. Science centers have recently expanded their mission beyond hands-on interactive exhibits, by adopting a reflective perspective drawn from a multidisciplinary approach to technological progress; that is, covering sociological, political, historical, philosophical and even ethical issues through online conferences and live demonstrations for visitors to become involved in topical debates. This allows them to form their own viewpoints on contemporary concerns ranging from genetic engineering and sustainability to space exploration. Within the diversity of educational resources offered by virtual science centers, it is suggested that museologists should emphasize a comprehensive description of scientific-related matters, tackling subjects, people and places, rather than objects themselves in order to genuinely fulfill a social need and arouse the audience's curiosity.

Introduction

At the turn of the third millennium, the museology semantics has been deeply altered with the dissemination of content through information technologies (IT), thus encouraging Web exhibit designers to reconsider the concepts of interactivity and interaction with their visitors. Many scholars in technological education have shown that IT improves the attention span of a large audience and is stressed to be relevant, challenging, lively and straightforward learning (Sankar & Kaju, 1999). It also enables an instructor-generated collaborative process (Hiltz, 1993), which ensures useful knowledge transfer for the public.

The contribution of IT has sparked significant changes in the curators' philosophy and is considered to be at the forefront of innovation for museums (Walsh, 1992). Computer-based exhibits are reported to be of high appeal for enticing people's interest in general topics like health, history and space; it is difficult to describe a disease, feel for a historical epoch, know a recognized figure, appreciate a chemistry experiment or visualize the Big Bang through fixed images (Bernier, 2003).

Virtual museums are not an exception, because they permit an individualized exploration from home where users are likely to be best positioned to determine what information they are looking for (Bazley et al., 2002). Moreover, they offer countless narrative storytelling and multiple content presentations with respect to both the visitors' receptivity and the subject areas to accommodate varied learning styles (Bernier, 2005). Hence, the Web has rapidly imposed itself as a technological device for improving one's apprenticeship in the sciences, natural history and the arts.

The advent of the World Wide Web has indeed broadened access to museums through pedagogical use of resources and therefore managed to expand knowledge, thus enabling the development of various innovative e-learning environments, such as global digital networking (Jackson, 1997; Witcomb, 1997), real-time demonstrations (Semper, 1998; Kjeldsberg, 1999), advanced database index interfaces (Siegel & Grigoyeva, 1999; Stuer et al., 2001), interactive telepresence (Koliou et al., 2001; Barbieri & Paolini, 2001), virtual reality (Andrews et al., 2004), online conferences (Vescia, 2002; Alexander, 2004a), remote educational programs (Schaller et al., 2002; Szalay, 2003), virtual guided tours (Bernier, 2002; Korteweg & Trofanenko, 2002), and content personalization (Oberlander et al., 1997; Beler et al., 2004), as well as Web-based discussion forums (Nilsson, 1997; Bernier & Bowen, 2004). These specific types of Web presentations should help the visitors to obtain a unique "behind-the scenes" view of knowledge and reestablish their own ability to learn.

Among all types of cultural institutions found online, science centers appear particularly well suited to investigate the Web tendency; their core mission has always been to demystify major scientific accomplishments through a myriad of contexts (Oppenheimer, 1980). The purpose of science centers is to familiarize visitors with key concepts, demonstrate technical processes, clarify human perceptions, or get acquainted with social and environmental repercussions (Schiele, 1996). This can be attained through hands-on experiences based upon phenomena in nature and elementary principles of science to arouse one's curiosity (Durant, 1992). Thus, providing a taxonomy of scientific knowledge, namely to identify, describe and classify objects or events, is seen vital for supplying a comprehensive understanding of natural sciences (Davallon, 1993).

Science centers are to some extent a *social laboratory* where knowledge is sacred. This could be achieved by personalization of Web facilities, including live demonstrations with educators. The Web is said to be a terrific tool for educational innovation and has proved to be efficient with respect to online activities from a personal cognitive perspective (Brooks, Nolan, & Gallagher, 2000). Indeed, the aims of museum mediators is to invite audiences into the process of basic scientific research by showing them where specialized instrumentations are created and experiments are performed incorporating the hearts and the minds of scientists (Alexander, 2004a).

The implementation of Web casts and IT in general within discovery centers falls in line with the Public Understanding of Science (Hilgartner, 1990; Durant, 1992; Wynne, 1995; Miller, 2000; Bono, 2001) and more recently with the Public Understanding of Research (Lewenstein, 2000; Davis, 2004; Ucko, 2004). The ASTC[1], the AAAS[2], the ECSITE-UK[3], COPUS[4] and PCST[5], to name a few important organizations, provide a gateway for public engagement as well as contribute towards "scientific popularization," especially through the Web. One of the newest developments is the Internet Virtual Observatory[6], an international project started in 2001 to put the universe online and provide data on the change in nature about astronomical research (Szalay, 2003).

Nowadays, science museums and related institutions play a key role regarding the manner in which technological advancements and scientific discoveries are made accessible to the public who may be overwhelmed with detailed research results. It is understood that with the introduction of PUS and PUR, innovative programs should be developed within a multidisciplinary approach, including the humanities, to present an alternative to scientific interpretations; while researchers possess actual facts, the audience is concerned with societal dilemmas (Miller, 2000).

A basic understanding of research by the public requires three elements: "content of current research; the process of research; and potential implications or consequences [to] provide an opportunity to convey the excitement of discovery" (Ucko, 2004, p. 1). For example, the *Living Labs* at the Columbus Center in Baltimore (USA) allow visitors to screen marine chemicals for their ability to emulsify oil.

Furthermore, Orfinger (1998) recognizes that visits to virtual science museums are better than that of actual field trips, because the museum personnel can elaborate specific features in cyberspace and enhance the learning potential of exhibits with components such as Real Player clips, QuickTime VR environments, Live Picture Viewer and Shockwave Flash animations. Thus, online exhibitions are much easier to update and can be stored indefinitely. Some museologists remind us that the educational role of Web museums is as much for pre-visit as for post-visit information (Mokre, 1998; Kravchyna & Hastings, 2002) and that a bad physical experience in the real institution can be enhanced through online benefits from additional forms of content (Mintz, 1998).

As regards to IT instructional approaches, factors, such as learning objectives, characteristics of the technology used and the audience's computer literacy are reported to be of prime importance for pedagogical attainments (Brown & Atkins, 1996).

The Challenge of Virtual Science Centers

This chapter will examine how scientific centers are involved in a "cultural transition" in order to implement new vectors of activities with respect to diverse categories of visitors (Bernier, 2002). Our main goal is to explore how science online can provide meaningful learning for virtual visitors, allowing a genuine museum experience. As emphasized earlier, it will be shown that a Web-based experience is intrinsically tied to content and linked to a variety of museographic tools (e.g., QuickTime VR, Macromedia, Audio Real) as well as to specific storyboards (e.g., live demonstrations, audio interviews, video clips, online activities) according to topics debated (e.g., past, current or traditional, innovative).

More importantly, we will address the significance of information delivery models by investigating real-case scenarios of two virtual science centers having a different cultural setting in terms of pedagogical strategies (e.g., demonstrate, display, visualize) and content presentations (e.g., homepage, headings, software used). This will be done by making a comparison between La Cité des Sciences et de l'Industrie (Paris, France) and The Exploratorium (San Francisco, California), presenting some data on their public attendances and using a descriptive inventory of available online resources. We hope therefore to unveil distinctive frames of scientific knowledge and outline the features of the most attractive displays with regard to specialized audiences. However, we will concentrate our efforts on specific activities and virtual exhibitions that characterize both institutions.

The Exploratorium: A Museum of Science, Art and Human Perception

The home of the San Francisco Exploratorium occupies 110,000 square feet and was hosted as part of the Palace of Fine Arts during the World Fair held in the year 1915; it was the only building that was not torn down after the event. This science museum was founded by the physicist Frank Oppenheimer in 1969 and is the earliest science center of all. Thus, it has been regarded as a pioneer of hands-on experimentations and now for online activities. Its Executive Director, since February 1991, is Dr. Goéry Delacôte. Before that he served as a Chair of the Board of Directors and in the Scientific Council of the French National Institute of Pedagogical Research at La Cité des Sciences et de l'Industrie in the mid-1980s. Its mandate is to create a culture of learning about science and technology by "controlling and watching the behavior of laboratory apparatus and machinery" through innovative material and tools, nurturing visitors' curiosity. It focuses on cognition from physics to neurosciences, including anthropological angles and also aesthetic aspects (Delacôte, 1999).

The Exploratorium's profile can be described as follow. Its 2003-04 budget was $26,914,000 and its annual admission is estimated to be 515,000 people in 2003, with 52% being adults and 48% children; 55% of them lived in the San Francisco Bay Area and 21% from other states. Their overall collection encompasses 650 original interactive exhibits, including lectures, performances and art presentations. In addition, they contributed to formal partnerships with 15 science centers worldwide to develop didactic itineraries as well as supply touring exhibitions to 38 U.S. museum institutions (Exploratorium, 2004b).

It has extended its role in public educational programs with the creation of the Center for Media and Communication in Spring 1993, which won the 2000 ASTC (Association Scientific-Technology Centers) Award for Innovation, and the Phyllis C. Wattis Web cast studio facility, which transmits real-time demonstrations (e.g., *Live@Exploratorium*) and can accommodate nearly a hundred visitors for networked events with international researchers remotely — "whether working with NASA to broadcast a total solar eclipse from Africa, or visiting a penguin ranch near the South Pole" (Exploratorium, 2004a). As a result, Web casts benefited from a similar funding as the physical exhibitions.

The *Live@Exploratorium* concept (www.exploratorium.edu/webcasts) was initiated with the "Hubble Space Telescope Servicing Mission" launch in February 1997, from which the museum staff derived 14 presentations enabling access to the observation instrument. Their Web site received almost one thousand visitors in ten days and over twenty thousand virtual visitors for that special event. Thus, the onsite audience was allowed to interact directly with scientists while online visitors were invited to ask their questions by e-mail; for example, see "Summing It Up" on March 9, 2002 (www.exploratorium.com/origins/hubble/live/webcasts.html).

Furthermore, the Exploratorium is among the first virtual science museum sites. It has won the Webby Awards[7] for Science in 2004 and Education in 2002. It was also awarded the Best Overall Site at the 1998 Museums and the Web Conference competition[8]. Founded in 1993, it gets about 15 million annual visits and contains over 15,000 pages of educational resources; it has produced nearly 500 experiments as well as 50 live Web casts serving as a basis for training programs (e.g., *ExNet* in 1999 for exporting innovative online exhibits) in more than 100 science centers and schools globally (Exploratorium, 2004a). At least 90 percent of the nation's science museums, and 70 percent of museums worldwide, have borrowed ideas from this institution, not to mention the 11,500 copies of the *Exploratorium Newsletter* distributed bi-monthly (Exploratorium: Fact Sheet, 2004).

As for content presentation, the Exploratorium's Website (www.exploratorium.com) offers five main sections: Explore, Educate, Visit, Partner and Shop (see Figure 1); while the first two are devoted to didactic attainments, from which we will subsequently describe some of the material, the last three provide broad information on the museum. A few of the subjects found within the homepage are music, sports, biology, origins and space; "the chosen topics fall into a category of everyday science — designed to be entry points via themes the general public are interested in" (Alexander, 2004b). The homepage

Figure 1. The Exploratorium's homepage©

introduces us, for instance, to the *Cassini-Huygens Mission*, an online resource that investigates "Saturn's rings" and several moons that evolve around this planet. The *Microscope Imaging Station*, which displays highly descriptive animated visualizations of three model organisms (i.e., the frog, the sea urchin, the zebrafish) and about their embryonic developments, is also worth viewing.

Thus, Web designers have produced lively online science exhibits using a variety of interactive software technologies like Shockwave, Macromedia, Real Audio and QuickTime: see, for example, Flash (e.g., in *Science of Music*, a sound mixing experiment), QuickTime (e.g., in *Microscope Imaging Station*, a 4Mbyte clip explaining the living cells of a zebrafish) and Real Audio (e.g., in *Science of Cycling*, an historical audio description of the bicycle wheels). According to Alexander (2004b), Flash is extremely versatile for creating diversified user-friendly environments and therefore often preferred for exploiting didactic tools, whereas QuickTime is said to be inaccessible for many organizations because of firewall problems. Real Player is usable by a greater number of individuals, although the image quality is lower, with buffered downloading that may be interrupted. With respect to favorite storyboards, written and audio medium prevail, while Web casting is mainly utilized for combining audio and video in real-time broadcasts. Hence, the Web enables ways to enliven content that conventional reading media like books cannot (Alexander, 2004b).

An Innovative Approach: *Origins*

Among the topical areas of the Exploratorium's Web site, *Origins* seems to be the most appropriate and exemplary to illustrate the current trend in online activities; Web casts are considered to be at the forefront of innovation by virtual science centers. Launched in November 2000, *Origins* obtained a generous grant from the National Science Foundation in 1999 and was designed to enhance the audience's appreciation of remote scientific discoveries (i.e., through Web casts), by which to judge their uniqueness.

This heading integrates five perspectives: people, places, tools, ideas as well as a live section organized around six themes: "CERN: Matter," "Hubble: Universe," "Antarctica: Extremes," "Las Cuevas: Biodiversity," "Cold Spring: DNA," and "Arecibo: Astrobiology" (see Figure 2). These themes were selected because they unveiled a major finding in connection with the nature of matter, the universe or life itself. The staff however mentioned that it was difficult to link the subjects neatly and to show how they are connected. Consequently, they created an umbrella infrastructure, the "Pictographic Index," framing places and disciplines together (www.exploratorium.com/origins/icons.html).

The content format of *Origins* consisted of various forms of media for a specific thematic (e.g., pictures of the field, articles on people, QuickTime VR demonstrations of ongoing research as well as video interviews with scientists and audio informal observations from the Exploratorium's roving team) that did not simply describe events, but also provided an analytical view to create a genuine connection with field researchers (www.exploratorium.com/origins/antarctica/fieldnotes/index.html). "This sense of im-

Figure 2. Exploratorium: Origins©

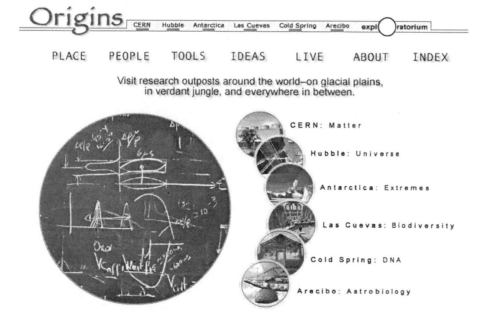

mediacy contributed to a feeling of actually being at the remote location" (Alexander, 2004a, p. 6). All the information has been organized for different degrees of learning complexity, such as a phenomenon-based program on ice, an historical viewpoint of the Human Genome Project or a prototype of a children's book about basic research, to ensure thoughtful inner reactions from heterogeneous groups of visitors. *Origins* has received over three million individuals and gets an average of 2,275 daily visits, with a vast majority of people spending nearly 5 minutes (Alexander, 2004a, p. 1).

The museum staff has devised several live Web casts related to six ongoing science research projects at recognized observatories, among which there are two from the crew located in Antarctica in December 2001 [e.g., "A Coal in the Icebox" on December 12th 2001 (www.exploratorium.com/origins/antarctica/fieldnotes/12_12ice.html)], and seven for the Cold Spring Harbor Laboratory in February 2003 that regularly included documentary pieces on the 50th anniversary of the discovery of DNA with the Nobel Laureate, Dr. James Watson (www.exploratorium.com/origins/coldspring/people/watson.html) (Alexander, 2004a). For instance, "Science in Action" provided an experiment using a scanning electron microscope to observe pollen collected in the field by the researchers (www.exploratorium.edu/origins/coldspring/people/dispatch/insight1.html).

In Winter 1998, there was "Eclipse: Stories from the Path of Totality," designed in collaboration with NASA's Education Forum[10] and Discovery Channel Online[11]; the demonstration aimed at showing a solar eclipse, only visible from the Galapagos, South America and the Caribbean (www.exploratorium.edu/eclipse/live98.html). The idea was to interest the general public (already keen to view this phenomenon) and for the Exploratorium, it was an opportunity to use the Net as a vehicle to present actual research on the interaction between the sun and the earth. The eclipse project recorded over half a million users and over ten thousand viewers to the Web cast itself; millions of people received the images via television. "Solar Eclipse" now presents a series of live Web casts (i.e., Greece in June 2004, USA in June 2002, Zambia in June 2001 and Turkey in August 1999) (www.exploratorium.edu/eclipse/index.html). "This project demonstrates the potential for opening a museum transparently to the world at large through the use of telecommunication and network while maintaining the museum's central role as an interpreter of objects and events, and as creator of social experience of discovery" (Semper, 1998).

As far as the public is concerned, the Exploratorium's staff noticed that Web casts were appealing for both onsite museum audiences and online visitors, as they differed in their needs and the ways they made use of information. "Online audiences accessed programs primarily from home and came specifically for the topic, while studio audiences viewed their experience as an extension of their general museum visit... [Although the staff intentionally attempted to appeal to novices] the questions received during the Web casts reflected a prior interest in or knowledge of the science being communicated" (Alexander, 2004a, p. 8).

The Cold Spring Harbor Web casts aimed exclusively at their online audience, targeting adults and teenagers, whereas most of the general public for "Hubble: A View to the Edge of Space" were adult males between 50 and 70 years old. This latter topic was said to be exemplary, because of the live coverage involving the first publicly available Horsehead Nebula images from the new Hubble telescope. In addition, the "Hubble Archive Web casts" are so appealing that they still attract 80 viewers weekly.

Figure 3. Exploratorium: Archived Webcasts©

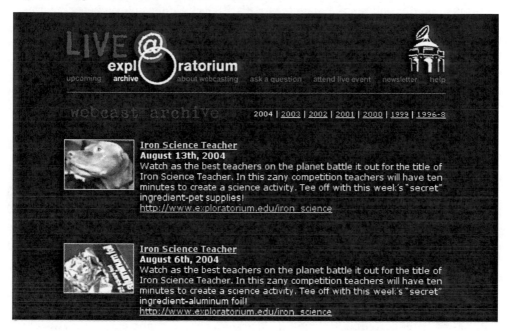

The findings of an evaluation about Web casts usage have shown that these are a resource for individuals who never attended the museum and hence created a significant demand to further engage users beyond the physical setting (Alexander, 2004a). Therefore, the principal objective of the museum personnel is to give access to more Web casts avoiding time-zone problems; they also learned that the public select them like a television listing *à la carte* and expect full-screen TV quality.

The feature, Archived Webcasts, is invaluable for virtual visitors (see Figure 3); it proffers a mosaic of selections (up to 40) dating from November 1996 to July 2004 that illustrate a solar eclipse, the making of candies and the science of sparkling (www. exploratorium.com/Webcasts/archive.html).

Lastly, the Web cast learning strategy proved to be a success for promoting live science events to a general audience as well as being useful for other scientific institutions, since it is reported that "over three million people have used our Website, double the number we initially projected [...] The importance of the *Origins* Project lies not in the fact that it brought ideas of science to the public, but rather that it brought actual scientists doing the science at the moment they were doing it" (Alexander, 2004a, p. 9).

La Cité des Sciences et de l'Industrie of Paris

The site of the Parc de la Villette served as one of the five former butchers' slaughter-houses in the year 1810 under the Emperor Napoleon I until 1986, when it was transformed as a science center during the passage of Halley's Comet. La Villette is a non-profit organization located in the 19th arrondissement of Paris; occupying 30,000 square feet, it was established in March 1986. It has welcomed 40 million visitors so far, up to 3.5 million annually, and ranks among the top Parisian cultural establishments in terms of attendances (La Cité, 2004). Its overall budget is estimated to be $100 million with three-quarter coming from government funding.

La Cité's Web site was launched in 1994 through the initiative of Joël de Rosnay, Head of Research and Development at that time; since 1999, La Cité has expanded the subjects covered by their exhibits and conferences. With over 40 online activities across 45,000 Web pages (Coiffard, 2004), it was awarded the Clic d'Or CB News award in 2003[12] for the exhibition "Challenges of the LivingWorld," that attracted tremendous public interest (www.cite sciences.fr/english/ala_cite/expo/tempo/defis/home/homef.htm). Neverthe-less, the favorite online topic is the temporary exhibition *Treasures of Titanic*, created after the success of the film and providing as informational background the following: "The Shipwrecking," "Discovery of the Wreckage," "Treasures in the Abyss," and "Scientific Findings" (www.cite-sciences.fr/francais/ala_cite/expo/tempo/titanic) (Coiffard, 2004).

This science museum is among the top six visited worldwide, as reported by Coiffard (2004). La Cité's homepage[13] contains a table of contents in three languages[14] (French, English and Spanish) (www.cite-sciences.fr/english/indexFLASH.htm): *Useful Informa-tion*, a repository of global inquiries; *Exhibitions*, an outlook on the physical displays; *Shows & Movies*, a view on external activities; *For professionals*, an insight on products and services for designing exhibits; *Cité des métiers*, a comprehensive vocational and training center; *Media Library*, an informational space on the physical library; *Lectures*, an agenda on upcoming meetings; *Education*, a school-based program section and *The*

Figure 4. La Cité's homepage©

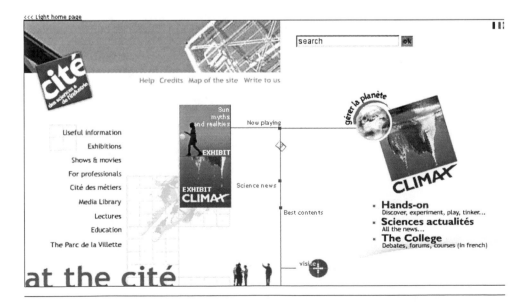

Parc de La Villette, an historical background of the location and its mission. In addition, there are timely events such as the *2nd Annual Villette Numérique Meeting* (from September 21 to October 3, 2004) inviting real visitors to develop an interest in Net art creations and other digital ventures, and the *Annual Internet Fest*, held in mid-March, presenting conferences, demonstrations and workshops related to the Web. Some of the main features are hands-on, science activities, especially designed for the Internet. However, the most trend-setting interactive resource is Le Collège, from which we will provide an in-depth portrait.

Online visitors can also find specific headings: *Now Playing*, presenting a range of temporary exhibitions (www.cite-sciences.fr/english/ala_cite/affiche/droit_fs.htm); *Science News*, giving information on topical matters (www.cite-sciences.fr/english/ala_cite/affiche/droit_fs.htm); *Best Contents*, enabling a glimpse at a selection of hands-on activities (www.cite-sciences.fr/english/ala_cite/c_coeur/droit_fs.htm); and *Visite +*, providing a detailed log of the visitor's previous visits (www.cite-sciences.fr/english/ala_cite/cite_pra/visite+/global_fs.htm) and thus offering users an access to their *Cyber Record* (via an entry ticket and a private code). One can also find a *Monthly E-Newsletter*[15], customized according to the individuals' interests and *Exhibition Presentations*, that contains most of the texts as well as bibliographic references of specific displays. Other museums such as the London Science Museum offer a similar service, for example, annotating comments and allowing personal written reflections of their visits, and even the storage of souvenir pictures from the exhibitions (Filippini-Fantoni & Bowen, 2004). Some may play back their movie clips and animations on virtual television, as at the Tech Museum in San Jose (Sato, 1999).

Figure 5. La Cité: Managing the Planet©

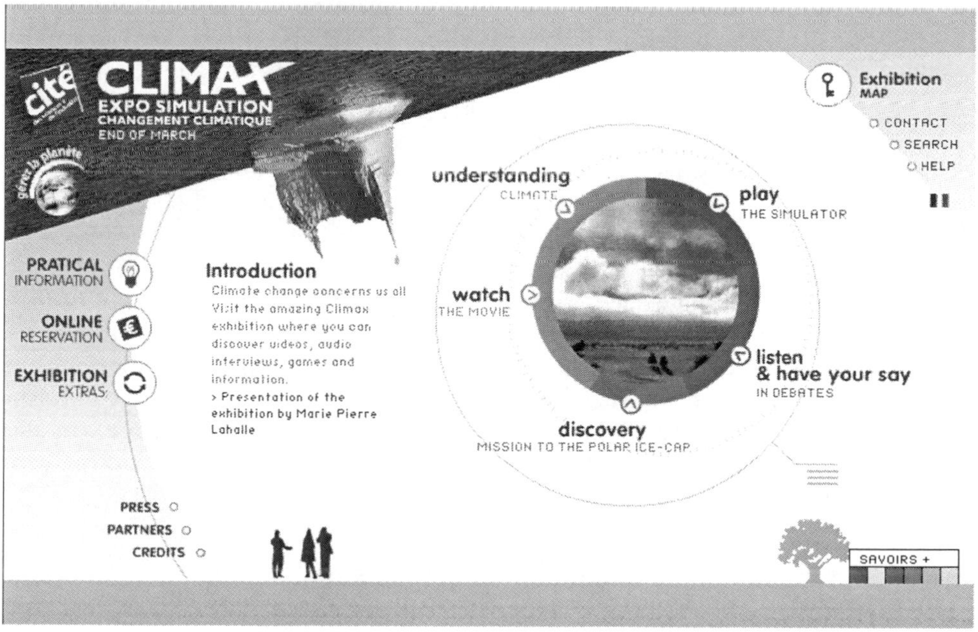

In La Cité, each department has an independent budget. Most of the chosen subject areas of their online displays are either derived from or complementary to those in the physical institution, thus providing a broader dissemination of their content, like, for example, the well-used *Managing the Planet* attraction from October 2003 to March 2006 (www.cite-sciences.fr/english/ala_cite/expo/tempo/planete/portail/glp.html). This attraction incorporates five themes[16]: "Climate," "Oil," "Sun," "Carbon"[17] and "Population," and explores the human relationship with nature and territory through hands-on activities, movies, conferences and a magazine on similar topics (see *Figure 5*); some were the results of specific features, like *Bricocité* for children to describe and understand plants, animals and objects: breeding crickets, constructing a kaleidoscope or cultivating a cactus (www.cite-sciences.fr/bricocite).

Its presentation format conforms with the exhibitions and departments found within the real museum, except for a few online resources especially designed for La Cité's Web site, like the section on "Hands-on," which gathers scientific attractions to experiment and play as well as allowing discovery in the art, physics, biology and other disciplines; these are mostly animated by Flash (www.cite-sciences.fr/english/web_cite_fs.htm). We have noted that virtual visitors are offered more brain-teasers in the French version (e.g., Environment, Logic, Chemistry, Perception, Biology) than in the English version.

There is the popular *Sciences Actualités* section that explores contemporary concerns from genetics to anthropology with the feature *Questions on Current Events* (e.g., Genes, Nanomaterials, Human Origins and the Hubble Space Telescope), although the French version focuses on events in France (e.g., the invasion of wolves in the Alps-Haute Provence and the discharge of pesticides in French rivers) (www.cite-sciences.fr/francais/ala_cite/science_actualites/sitesactu/accueil.php?langue=an).

Special Report emphasizes important present-day topics (e.g., the Greenhouse Effect, Missions on Mars, Aids Scourges), like the one on drugs entitled "Cannabis through the Scientists' Eyes," which comprises four sub-areas (plant, effects, users and regulation), including professional viewpoints (www.cite-sciences.fr/francais/ala_cite/science_actualites/sitesactu/dossier.php?langue=an&id_dossier=101).

News consists of brief worldwide headlines in the form of "pop-ups," to highlight the high abortion rate in Great Britain, Guinean children with iodine deficiency and surprising North American owls (www.cite-sciences.fr/francais/ala_cite/science_actualites/sitesactu/breves.php?langue=an). *Seen in the Press* devotes much attention to lighter events covering sociological queries to spectacular technological inventions, including Canadian squirrels stealing golf balls under the eyes of golfers, the interdiction of motorcycles in China and women's equity in Australia (www.cite-sciences.fr/francais/ala_cite/science_actualites/sitesactu/presse.php?langue=an). Additionally, *Picture of the Week* discusses, among other things, elephant excrements for manufacturing paper in Sri Lanka, the war of crickets in Maghreb and the very old Mostar bridge in Bosnia (www.cite-sciences.fr/francais/ala_cite/science_actualites/sitesactu/image_semaine.php?langue=an). There is a convenient *Index* to choose subjects by date, region or theme (see Figure 6).

Finally, educational animations like "Decipher the Scientific Actuality," inspired by newspapers' front pages, are available to further understand this particular type of

Figure 6. La Cité: Sciences Actualités©

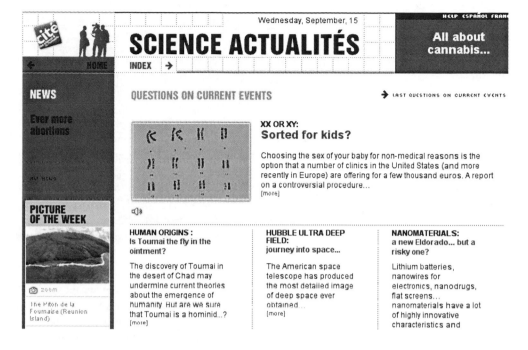

information and to develop a critical appreciation of the content provided. In the next section, we will discuss Le Collège, focusing on conferences, forums and debates, many of which relate to temporary online and physical exhibitions.

A Current Perspective: Le Collège

In an era of growing incertitude and controversial issues on science developments, progress is questioned by the general public because of their mixed feelings of fear and desires about future research outcomes. Hence, these make it difficult for people to take a position. They expect science centers to tackle relevant subjects, in fields as diverse as IT (e.g., jobs), ecology (e.g., climate), health (e.g., food), and even ethics (e.g., cloning). It is known that research can involve political decisions with nationwide implications. As indicted by Vescia (2002), people are concerned about which authorities can be trusted, who are willing to take responsibilities for endangering populations (e.g., government representatives, unions, medias, experts) as well as who will really benefit from these fast-expanding areas. In other words, people wonder what ideas are promulgated behind these developments.

The task of La Cité is not only to popularize and render a realistic picture, but also to demonstrate more commitment towards the nation. This can be achieved firstly by introducing contemporary problems and relevant subjects linked to major scientific advancements; secondly, by making specialized knowledge accessible and easily comprehensible for ordinary people through the provision of keynote speakers; and, thirdly, by letting individuals experiment with new forms of public debate via online conferences.

The Web initiative entitled Le Collège has succeeded to gather an interdisciplinary team of ten specialists, having among others an anthropologist, an economist, a sociologist, a historian, a physician and a physicist in an advisory board (i.e., Le conseil scientifique). It is divided into four sections: *Les carrefours du savoir* (*Knowledge Symposiums*), *Les samedis de l'actualité* (*Topical Saturdays*), *Colloques et événements* (*Congresses & Events*) and *Les conférences vidéo* (*Online Conferences*) and these sustain the philosophy mentioned above (www.cite-sciences.fr/francais/ala_cite/college/flash.htm); that is, to create a basis for understanding scientific achievements and create a virtual forum where people can discuss with open-minded intellectuals who relate to their concerns on scientific, political and societal issues (see *Figure 7*). Although Le Collège is only available for a French-speaking audience, it is granted a generous budget of 300,000 Euros annually. Each conference can have between 50 to 200 attendees, and the themes are mainly directed towards three topic areas: Life Sciences, Information Technologies and Environment.

With regard to the historical background of Le Collège, the first debate was held in June 2002 and revolved around the "Mad Cow disease" — considered a contentious issue on both scientific and political grounds. The conference lasted a day and introduced three themes: 1) What Happened? 2) Where are We? and 3) Management of the Crisis. The second debate, in February 2002, concerned "Climate changes" with the participation of

Figure 7. La Cité: Le Collège©

the French Center for Sustainable Development, including five discussions (e.g., future and past climates, the El Niño phenomenon, and the carbon cycle) where the invited speakers assessed on the seriousness of the outcomes, with the assistance of onsite citizens' recommendations (www.cite-sciences.fr/francais/ala_cite/conferen/climat/global_fs.htm).

Considering the outstanding popularity of the public conferences, a new auditorium which can accommodate up to three hundred persons and using the latest higher-end technologies (e.g., Web casting) was built. A specific Web site for meetings in Real Video format and related content was therefore launched in July 2002. During 2003, Le Collège received 13,700 visitors who listened to 250 eminent French researchers. Their interventions enabled La Cité to produce over 180 conferences that can be viewed online, and where a great number includes written versions of the speakers' talks, their biographies and their published books.

For its opening season in 2004-2005, Le College celebrated the international year of physics (September 25, 2004) and invited users to discover the "Concept of Emptiness" with French astrophysicists of the Commissariat à l'Énergie Atomique (CEA) and the Centre National de Recherche Scientifique (CNRS). This year's program is available at www.cite-sciences.fr/francais/ala_cite/college/04-05/conferences/programme/index.htm.

Knowledge Symposiums was established at the end of the year 2002 and numerous debates have since been organized on topics of broad interest, for example, "the impact of IT in daily life," and "the power of the sun" (related to both the onsite and online exhibition *Managing the Planet*) in April-June 2004, "the law of gender" in May-June 2004, "the global agricultural revolution" in March-June 2004, and "the Neolithic origins" in October-November 2003 (www.cite-sciences.fr/francais/ala_cite/college/04-05/conferences/programme/index.htm). *Congresses & Events* were introduced to look further into complex and critical topics in today's society and to present arguments by specialists (e.g., Public Research and Ethics in January 2004, Bamboo: Phantasmagorical Herb in February 2004, and the purpose of the Philosophy of Sciences in November 2003).

For the latest addition *Online Conferences (Now En ligne: Voir et Écouter[18])*, there was a Gorbachev conference with the man himself as guest-of-honor at Johannesburg (September 9, 2002), an inaugural banquet with the well-known French philosopher Michel Serres (September 15, 2002), an international meeting on Sustainable Development (September 21-22, 2002) and eight other debates on ethics, medicine and biotechnologies. More recently, a neuro-physiologist gave a talk on "What is Pain?" so as to give a comprehensive view of the sensory functioning of the brain. On January 8, 2004, the epistemologist Jean-Marc Lévy-Leblond presented a paper on "Quantum Revolution", explaining the meaning of antimatter. The principal objective of this feature is to present ten debates or conferences every month. Thus, a menu with a set of themes (e.g., Psychology, Energy, Astronomy, Politics, Telecommunications) is offered to help virtual visitors select their favorite topics; for instance, on Environment, see www.cite-sciences.fr/francais/ala_cite/college/archives/index_archive_glp.htm. It is possible to limit the search to video, audio or text excerpts of older conferences if desired (www.cite-sciences.fr/francais/ala_cite/college/archives/index.htm).

Finally, there is *Topical Saturdays* in which key people discuss significant subjects with the audience on how the scientific world affects our civilization, such as the future of genetics on the occasion of the 50th anniversary of the discovery of DNA. Other subjects included "Mars Online" in March 2004, "Computer Viruses" in November 2003 and "Cannabis and Teenagers" in May 2004 (www.cite-sciences.fr/francais/ala_cite/college/04-05/conferences/programme/index.htm); these provided an opportunity for the public to meet researchers, politicians and representatives of associations to debate media happenings, including important discoveries that deserve insights from a specialist viewpoint.

There are other examples of online conferences in Europe (e.g., United Kingdom, Germany, Denmark), as mentioned by Vescia (2002), but the novelty of La Cité is that they are organized in a science center and are part of a political agenda whereby the panel produces recommendations. Thus, La Cité's main goal is to become a significant informational hub on a variety of subjects involving investigations of digital technology, defense systems, civil rights, and genetics for Europeans who come from diverse cultural backgrounds. The Bionet initiative, established with the support of the European Commission and the European Collaborative for Science Industry & Technology Exhibitions (ECSITE), also features the latest discoveries and techniques pertinent to the life sciences (e.g., HIV remedies, in vitro fertilization, ageing process); it is shared by eight science centers and comes in nine languages (www.bionetonline.org).

Discussion: An Analogy Between the Exploratorium and La Cité

We have explored five fundamental elements when designing a science museum Website: (1) content presentation, (2) software technologies, (3) storyboards, (4) online activities and (5) targeted audiences.

In terms of information delivery, La Cité and the Exploratorium offer a standard homepage that consists of a table of contents on admission and memberships, temporary exhibitions, hands-on activities, online conferences, educational programs, teachers' resources and upcoming events as well as a historical background of the institution. They both contain clear headings using simple vocabulary to provide specific reading and scientific skills for visitors, although some of their features are slightly different.

With respect to their storyboards, these two science museums have managed to cross-reference with materials on the Web by using a combination of video and audio with written documents so as to let visitors choose among various learning methods for a particular topic. Delacôte (2002) believes that the crossing of the real and the virtual is necessary to widen the cultural institution's horizon using the convergence of video-conferencing and the Web, and is a key for the success of online museums.

The first element to stress about La Cité, and which is the most important distinction, is the usage of real-time demonstrations. If both institutions have features on actual space science (e.g., the Cassini-Huygens spacecraft and the Hubble telescope project). La Cité is less Web cast oriented than the Exploratorium but focuses more on online conferences, which is an approach quite akin to broadcasting. In other words, both museums emphasize live presentations (e.g., interviews, conferences, demonstrations), especially on amazing discoveries, since it is very difficult for the public to keep in touch with developments in the scholarly world. Furthermore, Web casts are known to offer a rich narrative context and work best for once in a lifetime experience for remote visitors wanting to familiarize themselves with the intricacies of an astronomical phenomenon (e.g., an eclipse) and thus raising their level of science literacy. Again, both institutions show a significant interest in informational topics. It was stressed by Vescia (2002) that people's ignorance in science lies in the fact that they are surrounded by techniques that they hardly master nor understand; therefore, they need explanations on their usage.

The second element to mention is that La Cité has several aspects in common with the Exploratorium, especially in relation to encouraging the audiences onsite to attend live broadcasts at their respective studios. La Cité favors the recording of online conferences and lectures [e.g., "D'un sexe à l'autre," launched in June 2004, for debating the ambiguity of transsexuality (www.cite-sciences.fr/francais/ala_cite/college/03-04/carrefours/seminaires/05-04-masculin-feminin/06-mercader-saladin/index.htm)], whereas the Exploratorium has a preference for experimenting with real-time demonstrations [e.g., "Eyeing the Storm", created in Spring 1997, to visualize severe storms (www.exploratorium.edu/visualize)]. Both science museums also archive their content and make use of these as the basis for online programming. Such personalized knowledge can facilitate the individual's learning process and help to organize, integrate and reuse the results of formal and informal experiences.

In addition, La Cité and the Exploratorium's online educational resources are aimed at regular visitors, newcomers, and even those who never intend to visit, and mainly concern the raising of public awareness of important phenomena in science. As supported by Semper (1998), the online usage of educational Websites should be multifaceted and "lavishly designed", with consideration for "the sweetness of physical exhibits." Starting from this last assumption, both museums' staff have created a huge amount of content to respond to the needs of virtual visitors.

The third and last element concerning La Cité is that its Web site offers more content relating to their physical facilities than the Exploratorium, and the latter does not accentuate much the European dimension in their online exhibitions, such as genetic engineering, abortion in Great Britain, massive plague of crickets in Maghreb, invasion of wolves in the French Alps as well as dangerous roads in France. Wynne (1995) demonstrated the importance of social context for lay people, because it plays a significant part in how science is interpreted; that explains why groups of citizens want to give their opinions on national issues, such as are offered in the *Congresses and Events* section of La Cité (e.g., polluted French rivers). Individuals also expect debates on technological innovations since they are part of our social reality.

Indeed, the cultural diversity of public response is of utmost importance because some topics have serious ethical implications (e.g., DNA technologies) that may concern specific countries (e.g., Austria) and less on others (e.g., UK and USA); the Germans thus established a considerable resistance to genetic testing. "In the US, for example, patents are easily obtainable for biotechnological inventions; while in the EU, they are proving much harder to secure" (Durant, 1997, p. 236). These differences cannot be explained and are not the result of ignorance, nor the product of greater or lesser levels of scientific development, but rather of particular circumstances.

Thus, it appears that both organizations have an extremely diverse museum audience and therefore provide topics accordingly, like hobbies for the Exploratorium (e.g., sports and music) and vocational occupations as well as current social issues for La Cité (e.g., jobs and cannabis). Yet, some of their content is much alike, such as special events related to the universe (e.g., solar system and glaciers) while their hands-on displays present similar subjects. By contrast, La Cité has demonstrated a more philosophical way to tackle science; that is, by anticipating the exponential growth of knowledge in techniques and social developments and providing intellectually-oriented views rather than a practical-minded perspective combined with applied science, as is the case for the Exploratorium.

On the one hand, the originality of La Cité is reliant on the fact that their online conferences are often part of a political agenda to provide views on questions of public interest, as visitors often want the right to express themselves on subject matters (e.g., Mad Cow Disease, Chernobyl, GMOs). In summary, their main priority is accessibility to research, including a democratic quest for truth and trust, since the widespread belief is that "science is for scientists." One must therefore be given the chance to express one's viewpoint in the process of building a collective consensus on ethical matters. La Cité wants to combine a humanistic viewpoint on scientific progress while being altogether highly cultural and popular. The institution has dedicated most of her partnership to large national enterprises like EDF, Cogema, Airbus and Renault (Quittet, 1992).

On the other hand, the Exploratorium aims mostly at providing virtual field trips to several observatories where significant discoveries are made in locations that most people cannot easily reach and on topics which appear rather opaque to the layperson. Hence, the biggest obstacle for the museum staff was to "bridge the setting of an active science museum dedicated to social learning[19] with an online virtual space designed for multiple-end users [and] getting information about their research (i.e., the researchers' laboratories) to the public via the Web" (Alexander, 2004a, p. 11).

Science and technology are about the future of society. That is why "the development of a society in which all individuals are or can be included in the process of reflecting on, participating in, and evaluating change are needed in public discussion [...] whether face-to-face, distance, or web-based" (UKCST, 2001, p. 3). As mentioned by the United Kingdom Council for Science and Technology (2001), the understanding of the social and ethical consequences of medical technologies is widely recognized for social and economic developments, thus relating to both environment and health. Those conferences represent indeed a great opportunity for a large proportion of individuals tackling important subject matters.

Although geographically separated and focusing on various online activities, we have shown that La Cité and the Exploratorium have much in common with respect to their content online. Vescia (2002) and Alexander (2004a) have reached the same conclusion regarding the importance of mediating of science through both renowned specialists and museum educators. Indeed, dissemination via the Web calls for a mediation in order to increase the visitors' knowledge since it has been established that the more sophisticated the content the more it needs "human mediation," as its role is precisely to provide enlightened viewpoints for facilitating the understanding of complex natural phenomenon (Bernier, 1999). This can be described as "distributed learning," as face-to-face teaching and computer-mediated communication, that enhance conventional pedagogy and therefore can overcome the barriers of meeting the needs of diverse learners, are involved (Dede et al., 2002).

We would also caution that just because something is telecast live, it does not mean that it will automatically appeal to the public. Viewers need a context in which to appreciate discoveries by reinforcing the authenticity of the physical museum with additional materials (e.g., audio comments, video-conferences, Web casts) to present a proper picture of current scientific research. Thus, the usage of sophisticated software will, in the long run, raise the overall content offering of virtual centers as well as the online visitor's knowledge to a higher level.

The most important aspect of virtual science centers is to magnify their cultural vocation into three distinct sections, artifacts, information and experience. Artifacts preserve the traditional perceptions of museums, information clearly has a place virtually, while the magic of experience can simply be viewed and explained on the Internet. The focus on science calls for an intriguing presentation, because it is inscribed within a conceptual framework for thinking about natural phenomena and more especially virtual experimentation. Needless to say, the Web is full of expectations for museum exhibit designers and virtual visitors; consequently more surveys are needed to analyze the itineraries of people accessing online science centers from their homes.

One may conclude this discussion with Freeth's argument (1998): "It's not too difficult to devise exhibits which illustrate Newtonian physics — Einstein is a different matter;" that is, with authoring software, we can provide a great deal of online educational resources on the Web to complement as well as offer an alternative to the physical museum in terms of content presentation. This helps to reproduce the enjoyable experience of the actual visit and to direct one's learning.

Limitations and Future Directions

There are five significant limitations with regard to Web casts:

1. Although Web cast training is effective, engaging, and offers a definite flexibility for providing scientific content, it is time-consuming for the museum staff.

2. Considering the educators' fees, the use of state-of-the-art visual equipment, travel expenses for the remote crew and provision of add-on written material, the costs for Web casts are relatively expensive (Alexander & Miller, 2002).

3. Unlike online activities and hands-on laboratories, the success of Web casts is harder to achieve, since they are available only in one language (Bernier & Bowen, 2004) (in our case, either French or English).

4. The lack of active participation via e-mail could be frustrating for individuals who cannot take part in the actual Web cast process, like questioning the demonstrator, which can make the experimentation less appealing for online visitors. The aim of

Web casts is to provide significant networked events so as to engage people worldwide and in real-time.

5.   Bandwidth limitations, poor audio broadcasting, screen-size, and absence of accessibility for the disabled more than affect the richness of Web casts. In addition, firewalls can block access or slow them down with intermittent audio and video transmission; however, with fast ASDL connections, the image and sound quality are less problematic.

Web casts can be seen as a boring experience, somewhat akin to satellite-based distance learning; nevertheless the museum staff have been able to do a relatively good job with their educational programs to overcome barriers of time and distance.

Overall, one must note that for Web pedagogy, Flash is more appropriate than HTML when animation and sound are central to content, because it helps to create rich multimedia interactivity (Schaller et al., 2003), while QTVR provides a means for providing a realistic 3-D walk-through, including fine control of height, size and brightness (Barnes, 2000). However, Flash is neither immersive nor standardized for all Web content and only permits Web-based audio and video applications; in addition, it is inaccessible to many disabled people and requires additional "plug-in" software (Mac Gregor, 2001). These higher-end technologies are important in cognitively authentic environments, although museum Web designers must consider both usability and user's engagement (Nielsen, 1999; Yates and Errington, 2001), as they serve a different purpose.

Despite these constraints, it is important to note that real-time displays are really about delivering information through the Web; however, one must bear in mind that online museum demonstrations should be an extension of the real institution. Thus, actual scientific discoveries should not mimic the classroom ambience to prevent the coldness of virtual assistance and avoid the pitfalls of e-learning, because it is less frequent to watch online Web cast demonstrations than to interact with those in the physical museum. To make learning friendly and efficient, one has to consider the museum setting differently from schools and relate it to personal significance and everyday experiences (Falk & Dierking, 2000).

One possible research direction would be to undertake a deep quantitative analysis on Web logs, which has become increasingly popular to measure audience satisfaction through socio-demographic data. Content appreciation such as information sought and number of pages consulted, for understanding their visiting patterns and level of curiosity for Web casts, should be studied (Ockuly, 2003; Peacock, 2002; Haley-Goldman & Wadman, 2002). Thus, the audience's profile can be invaluable for museum professionals who wish to provide a new appreciation of current research.

How well a museum Web site meets its visitors' needs leads us to another strand of analysis; that is, the consideration of online forums. They have become a significant part of the Internet for sharing ideas via text-based conversations as well as being recognized as a "social aggregate," because one can communicate electronically on specific topics (Rheingold, 1993); these users form a subculture and can set a sociability similar to the one found within physical museums because the Internet helps to expand conventional institutionalized discourse as well as encourage informal exchanges with the staff (Bernier, 1998, 2005).

For example, *NASA's Quest* helps to raise the public awareness of their goals (quest.arc.nasa.gov); flying in a shuttle, exploring distant planets, building an aircraft or getting stories about their days' work while discussing with experts, such as the Astrobiology Lecture Series (Sato, 1999) make for interesting talking points. There is also *Snacktalk* of the Exploratorium (www.exploratorium.edu/snacks/feedback.html), where users can submit comments on hands-on experiences available in many forms such as Questions, Help and Acknowledgments (Hunt, 2002). Subject-driven topics such as those provided by *Ingenious* of the London Science Museum (www.ingenious.org.uk) with access to Web forums focusing on the human genome project and other contemporary issues (Borda & Beler, 2003) also provide a type of social ontology (Beler et al., 2004).

Online communities can empower individuals in discovering objects and deepening their meanings by sharing their concerns in what is called "civic engagement" (Steinbach, 2003). However, it is important to establish dialogue under conditions that minimize polarization and crowd psychology, such as the use of skilled moderation and balanced expert information (Ucko, 2004; Bernier & Bowen, 2004). These initial forays are an indication of possible customizable resources and targeting of the visitors' interests in order to strengthen the interaction with the general public and build a global citizenship through mentoring.

In other words, museum staff should take into consideration five essential components to enhance their efficiency (KnowledgeNet, 2002): 1) Curiosity: How do we entrench the users' curiosity of exploration?; 2) Engagement: How do we get the learner fully engaged?; 3) Intensity: How do we design a memorable experience?; 4) Peer learning: How can we build communities that incorporate a social dimension?, and 5) Assessment and Feedback: How should we help individuals to acquire a greater sense of their own knowledge, and thus better organize their knowledge capital through e-learning experiences? These will help to create appealing and innovative "cyberambience" for Web museums (Barr, 1997).

Conclusion

The introduction of original live Web casts is the latest endeavor by museum educators to implement a creative museographic landscape for providing a compelling reason for people to visit online museums. On the one hand, this specific strategy from the Exploratorium was intended to refine their ways of exploring major worldwide discoveries, mostly related to applied science and technology (e.g., the Cassini-Huygen expedition, the Human Genome Project, the Hubble mission, the Antarctica journeys) as well as to offer content on objects, techniques, people and phenomena that characterize scientific developments. On the other hand, La Cité des Sciences et de l'Industrie has brought to our attention that science museums should also tackle social and ethical issues with respect to various areas of advanced technologies (e.g., health, food, information technologies, urbanization) using scholarly presentations such as online conferences and topical debates to reach out to society.

Indeed, the Exploratorium and La Cité have broadened their interpretive arena with live demonstrations and expanded their targeted audience to an international level. Judging from the analysis of log records from virtual visitors to both institutions, these Web-based resources are viewed as extremely useful for remote learning programs. These gave confidence for producing extra materials and also serve as a reliable interdisciplinary resource for schools, universities, science institutions and the general public.

However, the greatest challenge for science museums is not so much to communicate a substantial amount of information with sophisticated software, but rather that the virtual experimentation should increase interactivity. Otherwise, it could be overwhelming for non-scientists experiencing live events. In short, educators should seriously consider science centers in augmenting the interaction between people and their tools, and question their ways of designing Web resources for directly involving individuals with user-friendly environments. Five fundamental issues should be raised:

1. Are interactive technologies efficient for providing content?
2. Can museums afford them?
3. Are they appealing for online learning within a scientific format?
4. Can the audience use IT?
5. How can the engagement of visitors be increased?

Throughout this chapter, we have shown that the usage of digital broadcasting constitutes a significant rupture with the museum's conventional online features, because its "discursive interactivity" allows a personalized investigation in a domestic setting; that is, to meet the needs of educationally and culturally diverse individuals from various age groups with differing computer literacy levels and varied scientific knowledge. With the Web, the challenge of offering didactic interpretations of physical and astronomical phenomenon is greater than ever!

The emergence of Web casts as a principal means of communication has opened avenues for innovative content dissemination by virtual science museums to capture attention, stimulate imagination, sharpen reasoning, and even fulfill the vision about the future ahead of us through meaningful online activities for people. All this should help in promoting a knowledge-based society; that is, to materialize notions of science, develop the individuals' skills and demystify the visitors' understanding. The principal *raison d'être* of Web museums remains the involvement of virtual visitors.

Acknowledgments

The collaboration of both Melissa Alexander, Project Director of *Origins* at the Exploratorium and Didier Coiffard, Head of the Department Interactivité et Multimédia at La Cité provided fruitful sources of information.

Many thanks are due to Jonathan P. Bowen (Professor of Computing at London South Bank University) who helped a great deal in improving this paper as well as my mother Mrs. Stéphane Moissan and my companion Mr. Alexandre St-Pierre for providing enormous support when finalizing this chapter, by taking care of our lovely newborn Laure.

References

Alexander, M. (2004a, May). *Origins: Findings*. Final Report. San Francisco: Exploratorium.

Alexander, M. (2004b, July). E-mail correspondence with the Project Director of *Origins*.

Alexander, M., & Miller, M. (2002). Live@Exploratorium: Origins. In *Communicating the Future: Best Practices in Communication of Science and Technology to the Public Conference*. Poster Session. Gaithersburg, MD, USA, March 6-8, 2002. Retrieved August 2004, from *www.nist.gov/public_affairs/Posters/exploratorium.htm*

Andrews, S., Surendran, D., Landsberg, R., Jojola, E., Kadanoff, L., Mir, R., Podgorny, J., Rosner, D., SubbaRao, M.M., & Zhiglo, A. (2004). Cosmus: Virtual 3D cosmology in public science museums. In D. Paper, M. Roussos, & J. Anstey (Eds.), *IEEE Virtual Reality 2004 Workshop Proceedings*. Session Virtual Reality for Public Consumption.

Arnal, P. (2004, June). La Cité des Sciences mise sur l'interactivité. *Le Journal du Net*. Retrieved August 2004, from *www.journaldunet.com/0406/040608cite sciences.shtml*

Barbieri, T., & Paolini, P. (2001). Co-operation metaphors for virtual museums. In *Museums and the Web conference*. Session Engineering the Future. Seattle, WA, USA, March 14-17, 2001. Retrieved from August 2004, *www.archimuse.com/mw2001/papers/barbieri/barbieri.html*

Barnes, C. (2000). Building immersive environments using QuickTime VR: Lessons from the real world and virtual realities. In *Australasian Academic & Developers Conference*. University of Wollongong, New South Wales, Australia, April 25-28. Retrieved August 2004, from *auc.uow.edu.au/conf/conf00/papers/AUC2000_Barnes.pdf*

Barr, D. (1997). The Website: Extending the museum's mandate for lifelong learning. *Website Learning*. Retrieved August 2004, from *www.distedsys.com/documents/oma_paper/body.html*

Bazley, M., Clark, L., Bottaro, B., & Elinich, K. (2002). Think globally, act locally: The role of real teachers in community science issues. In *Museums & the Web Conference*. Session Teachers and Museums, Boston, April 17-20. Retrieved August 2004, from *www.archimuse.com/mw2002/papers/elinich/elinich.html*

Beler, A., Borda, A., Bowen, J.P., & Fantoni-Filippini, S. (2004). The building of online communities: An approach for learning organizations, with a particular focus on the museum sector. In *Electronic imaging & the visual arts conference*. London, July 26-31. Retrieved August 2004, from *arxiv.org/abs/cs.CY/0409055*

Bernier, R. (1999, January). *Initiation à l'Internet comme mode de navigation: De l'usage à l'accès au savoir*. Étude qualitative. Paris: Département Évaluation et Prospective, Cité des Sciences et de l'Industrie.

Bernier, R. (2002). The uses of virtual museums: The French viewpoint. In *Museums & the Web conference*. Session Evaluation and Experience. Boston, MA, USA, April 17-20, 2002. Retrieved August 2004, from *www.archimuse.com/mw2002/papers/bernier/bernier.html*

Bernier, R. (2003). Usability of interactive computers in exhibitions: Designing knowledgeable information for visitors. *Journal of Educational Computing Research, 28*(3), 245-272. Retrieved August 2004, from *baywood.metapress.com/link.asp?id=ea53b0arc1q33t20*

Bernier, R. (2005). Accessing heritage through the Internet: The French perception of Web museums. *Journal of Behavior and Information Technology* (forthcoming).

Bernier, R., & Bowen, J.P. (2004). Web-based discussion groups at stake: The profile of museum professionals online. *Program: Electronic Library and Information Systems, 38*(2), 120-137. Retrieved August 2004, from *dx.doi.org/10.1108/00330330410532832*

Bono, J.J. (2001). Why Metaphor? Toward a metaphorics of scientific practice. In S. Maasen & M. Winterhager (Eds.), *Science studies probing the dynamics of scientific knowledge* (pp. 215-234). Bielefeld: Transaction Publishers.

Bowen, J.P., & Filippini-Fantoni, S. (2004). Personalization and the Web from a museum perspective. In *Museums & the Web conference*, Session Personalization, Arlington, VA, USA, March 31-April 3. Retrieved August 2004, from *www.archimuse.com/mw2004/papers/bowen/bowen.html*

Brooks, D.W., Nolan, D.E., & Gallagher, S.M. (2000). *Web-teaching: A guide to designing interactive teaching for the World Wide Web* (2nd ed.). Innovations in Science and Technology Series. New York: Plenum Press.

Brown, G., & Atkins, M. (1988). *Effective teaching in higher education*. Routledge: London.

Coiffard, D. (2004, July). E-mail correspondence with the Head of the Department Interactivité et Multimédia.

Davallon, J. (1993, February). À propos de la communication et des stratégies communicationnelles dans les expositions de science. In *Conférences DEA d'information scientifique et technique*. Paris: Palais de la Découverte.

Davis, T.H. (2004). Engaging the public with science as it happens: The current science & technology center at the Museum of Science. *Boston Science Communication, 26*(1), 107-113.

Dede, C., Brown, T.B., & Whitehouse, P. (2002). Designing and studying learning experiences that use multiple interactive media to bridge distance and time. In C. Vrasidas & G.V. Glass (Eds.), *Current perspectives on applied information technologies* (Vol. 1: Distance Education). Greenwich, CT: Information Age Publishing. Retrieved August 2004, from *www.lesley.edu/faculty/tbrownlb/T502/MultipleMedia.htm*

Delacôte, G. (1999). *Towards the Year 2001.* About Us. San Francisco: Exploratorium. Retrieved August 2004, from *www.exploratorium.edu/general/directors_vision.html*

Delacôte, G. (2002). Du réel au virtuel. *Rencontres Internationales du multimedia d'apprentissage.* Session Place du secteur muséal. Quebec City, Canada, March 18-21. Retrieved August 2004, from *www.lienmultimedia.com/rima/02-03-21-07.html*

Durant, J. (1997). Deciding which stories to tell: The challenge of presenting contemporary biotechnology. In G. Farmelo & J. Carding (Eds.), *Here and now: Contemporary science and technology in museums and science centers* (pp. 235-239). London: Science Museum: London.

Durant, J. (1992). *Museums and the public understanding of science.* London: Science Museum.

Exploratorium. (2004a). *About Us.* Retrieved August 2004, from *www.exploratorium.com/about/index.html*

Exploratorium. (2004b). *Fact Sheet 2003-04.* Retrieved August 2004, from *www.exploratorium.edu/about/fact_sheet.html*

Falk, J., & Dierking, L. (2000). *Learning from museums: Visitor experiences and the making of meaning.* Walnut Creek, CA; AltaMira Press.

Freeth, M. (1998). Hands online. In *Museums & the Web conference.* Session Interactivity. Toronto, Canada, April 22-25. Retrieved August 2004, from *www.archimuse.com/mw98/papers/freeth/freeth_paper.html*

Koliou, M., Kamarinos, G., Roussou, M., Trahanias, P., Argyros, A., Tsakiris, D., Cremers, A., Schultz, D., Burgard, W., Haehnel, D., Sawaides, V., Friess, P., Konstantios, D., Katselaki, A., & Giannoulis-Giannoulopoulos, G. (2001). Enhancing museum visitor access through robotic avatars connected to the Web. In *Museums and the Web Conference.* Session Seeing Differently, Seattle, WA, USA, March 14-17. Retrieved August 2004, from *www.archimuse.com/mw2001/papers/giannoulis/giannoulis.html*

Haley-Goldman, K., & Wadman, M. (2002). There's something happening here: What it is ain't exactly clear? In *Museums and the Web Conference.* Session Evaluation and Framework. Boston, USA, April 17-20. Retrieved August 2004, from *www.archimuse.com/mw2002/papers/haleyGoldman/haleygoldman.html*

Hilgartner, S. (1990). The dominant view of popularisation: Conceptual problems, political uses. *Social Studies of Science, 20*, 52-53.

Hiltz, S.R. (1993). Correlates of learning in a virtual classroom. *International Journal of Man-Machine Studies, 39*, 71-98.

Hunt, D. (2002). *Putting knowledge to work: Bringing digital assets to the forefront and fostering virtual communities.* San Francisco: Exploratorium. Retrieved August 2004, from *www.sla.org/Documents/conf/ny/DigitalAssets.htm*

Jackson, J. (1997). The virtual visit: Towards a new concept for the electronic science center. In G. Farmelo & J. Carding (Eds.), *Here and now: Contemporary science and technology in museums and science centers* (pp. 173-179). London: Science Museum.

Kjeldsberg, P.A. (1999). Policies and practicalities in exhibiting musical instruments. In *Musical instruments or music? What is the role of a museum in a changing society? CIMCIM Conference*, Paris, June 10-14.

KnowledgeNet. (2002). Exploding the e-learning myth: Next-generation, Web-based training is here today: And it delivers the wow experience. Retrieved August 2004, from *www.knowledgenet.com/newsroom/whitepapers/elearningmyth_pf.jsp*

Korteweg L., & Trofanenko, B. (2002). Learning by design: Teachers/museums/technology. In *Museums and the Web Conference*. Session Learning in Theory and Practice. Boston, April 17-20. Retrieved August 2004, from *www.archimuse.com/mw2002/papers/korteweg/korteweg.html*

Lewenstein, B.V., & Allison-Bunnell, S.W. (2000). Creating knowledge in science museums: Serving both public and scientific communities. In B. Schiele & E. Koster (Eds.), *Science Centers for this Century* (pp. 187-208). Ste-Foy, Quebec: Éditions Multimondes.

MacGregor, C. (2001). Usable Macromedia Flash – Myth no more. *Flazoom.com 2.0*. Retrieved August 2004, from *www.flazoom.com/usability/usability_toc.shtml*

Miller, S. (2000). Public understanding of science at the crossroads. *Science Communication, Education and the History of Science*. British Society for the History of Science, London, July 12-13. Retrieved August 2004, from *www.bshs.org.uk/conf/2000sciencecomm/papers/miller.doc*

Mintz, A., & Thomas, S. (Eds.). (1998). *The Virtual and the real: Media in the museum.* Washington, DC: American Association of Museums.

Mokre, M. (1998). *New technologies and established institutions. How museums present themselves in the World Wide Web?* Internal Report. Vienna: Technisches Museum Wien.

Nielsen, J. (1999, May). Top ten mistakes in Web design: Revisited three years later. In *Alert Box: Current issues in Web usability*. Retrieved August 2004, from *www.useit.com/alertbox/990502.html*

Nilsson, T. (1997). The interface of a museum: Text, context and hypertext in a performance setting. In D. Bearman & J. Trant (Eds.), *Proceedings of Museums and Interactive Multimedia 1997* (pp. 146-153). Pittsburgh, PA: Archives & Museum Informatics.

Oberlander, J., Mellish, C., O'Donnell, M., & Knott, A. (1997). Exploring a gallery with intelligent labels. In D. Bearman & J. Trant (Eds.), *Museums and interactive multimedia 1997* (pp. 79-87). Pittsburgh, PA: Archives & Museum Informatics.

Ockuly, J. (2003). What clicks? An interim report on audience research. *Museums and the Web Conference.* Session Audience. Charlotte, NC, USA, March 19-22. Retrieved August 2004, from *www.archimuse.com/mw2003/papers/ockuly/ockuly.html*

Oppenheimer, F. (1980). Exhibit Conception and Design. In *Meeting of the International Commission on Science Museums.* Monterey, Mexico. Retrieved August 2004, from *www.exploratorium.edu/ronh/frank/ecd*

Orfinger, B. (1998). Virtual science museums as learning environments: Interactions for education. *Informal Learning Review.* Retrieved August 2004, from *www.informallearning.com/archive/1998-1112-a.htm*

O'Sullivan, D. (1990). *Transformative learning: Educational vision of the 21ˢᵗ century.* New York: Zed Books.

Peacock, T. (2002). Statistics, structures & satisfied customers: Using Web log data to improve site performance. In *Museums & the Web Conference.* Session Evaluation and Experience. Boston, April 17-20. Retrieved August 2004, from *www.archimuse.com/mw2002/papers/peacock/peacock.html*

Podgorny, J. (2004). Studying visitor engagement in virtual reality based children's science museum exhibits. Master of Arts., University of Chicago, June.

Quittet, A-M. (1992). Exposer les sciences et l'Industrie. *MScope, 3,* 54-60.

Rheingold, H. (1993). *The virtual community: Homesteading on the electronic frontier* New York: Harper Collins.

Sankar, C.S., & Raju, P.K. (1999). *Multi-media courseware to bring real-world decision making in classrooms.* Decision Sciences Institute Conference, New Orleans, LA, USA. Retrieved August 2004, from *web6.duc.auburn.edu/research/litee/media/pdfs/eval_conf_papers/dsi1999.pdf*

Sato, K. (1999). Designing science center exhibits. *Learning, Design & Technology.* School of Education, Stanford University. Retrieved August 2004, from *ldt.stanford.edu/ldt1999/Students/tita/ldt/musdesign.html*

Schaller, D.T., Allison-Bunnell, S., Borun, M., & Chambers, M.B. (2002). How do you like to learn? Comparing user preferences and visit length of educational Websites. In *Museums & the Web Conference.* Session Learning in Theory and Practice. Boston, April 17-20. Retrieved August 2004, from *www.archimuse.com/mw2002/papers/schaller/schaller.html*

Schaller, D.T., Allison-Bunnell, S., Chow, A., Marty, P., & Heo, M. (2004). To Flash or not to Flash? Usability and user engagement of HTML vs. Flash. *Educational Web Adventures* (EduWeb). Research and Evaluation Section. Retrieved August 2004, from *ww.eduweb.com/to_flash_full.html*

Schiele, B. (1996). Les musées scientifiques, tendances actuelles. Musées et Médias : pour une culture scientifique et techniques des citoyens. In *Rencontres culturelles de Genève 1996.* Muséum d'histoire naturelle de Genève, Suisse.

Semper, R. (1998). Bringing authentic museum experience to the Web. In *Museums & the Web Conference*. Session Rich Experiences. Toronto, Ontario, Canada, April 22-25. Retrieved August 2004, from *www.archimuse.com/mw98/papers/semper/semper_paper.html*

Semper, R., Wanner, N.W., Jackson, R., & Bazley, M. (2000). Who's out there? A pilot user study of educational Web resources by the Science Learning Network (SLN). In *Museums & the Web Conference*. Session Evaluation. Minneapolis, MN, USA, April 16-19, 2000. Retrieved August 2004, from *www.archimuse.com/mw2000/papers/semper/semper.html*

Siegal, P., & Grigoryeva, N. (1999). Using primary data to design Websites for public and scientific audiences. In *Museums and the Web Conference*. Session Enabling Scholarly Research on the Web. New Orleans, LA, USA, March 11-14. Retrieved August 2004, from *www.archimuse.com/mw99/papers/siegel/siegel.html*

Steinbach, L. (2003, May/June). Civic engagement in a digital age: An even greater challenge to museums. In Cultivating Community Connection. *Museum News*, pp. 27-20, 50, 52, 54 and 56.

Stocklmayer, S., & Gilbert, J. (in press). Informal chemical education In J.K. Gilbert (Ed.), *Chemical education: Towards research-based practice*. Dordrecht: Kluwer.

Stuer, P., Meersman, R., & De Bruyne, S. (2001). The HyperMuseum theme generator system: Ontology-based Internet support for the active use of digital museum data for teaching and presentation. In *Museums and the Web Conference*, Session Seeing Differently. Seattle, WA, USA, March 14-17. Retrieved August 2004, from *www.archimuse.com/mw2001/papers/stuer/stuer.html*

Stutchbury, R. (1998). Science in the Pub, Introduction. Fundation Advisory Group and Archived Programs, July. Retrieved August 2004, from *www.scienceinthepub.com/index/html*

Szalay, A. (2003). Special feature: A challenge from big science. In *Web-Wise Conference*. Session Sustaining Our Digital Resources. Washington, DC, USA, February 26-28, 2003. See also *SpaceDaily*. Retrieved September 2004, from *www.spacedaily.com/news/telescopes-01b.html*

Ucko, D.A. (2004). Production aspects of promoting public understanding of research. In D. Chittenden, G. Farmelo & B.V. Lewenstein (Eds.), *Creating Connections: Museums and the Public Understanding of Current Research*. Walnut Creek, CA: AltaMira Press Retrieved August 2004, from *www.museumsplusmore.com/pdf_files/PUR-Chap.pdf*

United Kingdom Council for Science and Technology (2001). *Imagination and understanding*. A report of the arts and humanities in relation to science and technology. July. 25 pages.

Vescia, F. (2002). Experimenting with public debate in a European science center. In *Public Understanding of Research Conference*. Minneapolis, MN, USA, September 26-29, 2002.

Walsh, K. (1992). *The representation of the past: Museums and heritage in the postmodern world*. London: Routledge.

Witcomb, A. (1997). The end of the mausoleum: Museums in the age of electronic communication. In *Museums and the Web Conference*. Session The Concept of the Museum in a World of Internetworking. Los Angeles, March 16-19. Retrieved August 2004, from *www.archimuse.com/mw97/speak/witcomb.htm*

Wynne, B. (1995). The public understanding of science. In S. Jasanoff, G. Markle, J.C. Petersen & T. Pinch (Eds.), *Handbook of science and technology studies* (pp. 380-392). Thousand Oaks, CA: Sage.

Yates, S., & Errington, S. (2001). Computer-based exhibits: A must have or a liability? In S. Errington, S.M. Stocklmayer & B. Honeyman (Eds.), *Using museums to popularise science and technology*. London: Commonwealth Secretariat.

Endnotes

[1]   Founded in 1973, the Association of Science-Technology Centers (ASTC) is an altruistic organization with nearly 500 members, including over 400 science museums in 43 countries (www.astc.org). ASTC aims at stimulating innovation through an efficient public understanding of science and technology.

[2]   Founded in 1848, the American Association for the Advancement of Science (AAAS) is the world's largest non-profit society dedicated to technological excellence across all disciplines and provides scientific information on public educational programs (www.aaas.org). They manage several Web sites: *Science Online*, *Science and Society*, *Science's Nextwave* and *EurekAlert*.

[3]   The Science and Discovery Centre Network (Ecsite-UK), formally affiliated to the European Collaborative for Science, Industry and Technology Exhibitions, was established in 2001; it represents over 80 discovery centers in the UK (www.ecsite-uk.net/index.php).

[4]   The Committee on the Public Understanding of Science (COPUS) was set up in 1987 by the Royal Society and is considered a pioneer for promoting science, engineering and technology activities on behalf of the British Association for the Advancement of Science (www.copus.org.uk).

[5]   Launched in 2001, the International Network on Public Communication of Science and Technology (PCST) encourages discussion of practices, methods, policies and conceptual frameworks related to ethical matters, social concerns as well as economical issues. The principal interested parties are practitioners, researchers and scientific communities who study PCST (www.pcstnetwork.org).

[6]   This five-year project carried out by astronomers from 17 research institutions has received a $10 million grant from the National Science Foundation.

[7]   Viewed as the Oscars of the Internet, this award was established in 1996. It contains thirty different categories and is based on six criteria: content, structure, navigation, visual design, functionality, interactivity and experience (www.webbyawards.com/main/webby_awards/index.html).

[8]   Created in 1997, the selected categories of this Web conference include Virtual Exhibitions, Educational Use, Research site, Museum Professional's site and Membership Use. (www.archimuse.com/mw98/best/index.html)

[9]   Last accessed on August 17, 2004.

[10]   NASA provided a high-bandwidth datalink connecting Aruba Island to the Internet, while the Exploratorium organized a video feed from Aruba that was used by a large group of viewers at schools and museums.

[11]   Partnering with media organizations can also be an attractive option for breaking news. This particular approach has improved assistance for museum educational programs, as they helped publicize significant events and cross-disseminate news at reduced cost (Ucko, 2004).

[12]   Introduced in 1997, this competition is held to reward the best French-based Web sites as well as to publicize innovative Web styles of use in eleven categories: the Jury's Prize, Best Media site, Best Institutional site, Best Business to Business site, Best E-Commerce site, Best Interactive Advertising site, Best Interactive Marketing site, Best Product site, Best General Audience site, Best Design site, Best Overall Interest site and Best Renowned International site. (www.clicdor.com/frameset.php)

[13]   Last accessed on August 17, 2004.

[14]   The Toubon Law, named after the French Minister of Justice (Mr. Jacques Toubon), requires that textual documents be translated into at least two other languages apart from French (e.g., English and German or English and Spanish); the law was adopted in August 1994 (Coiffard, 2004). Around 60% of La Cité's users are foreigners, so a full recasting in Spanish and English will be made available in 2005 (Arnal, 2004).

[15]   The intended visitors are the ones from La Cité; it will soon be available for all Web users (Coiffard, 2004).

[16]   Nearly 50% of online visitors show an interest on the sub-topics of temporary exhibitions (Arnal, 2004).

[17]   La Cité spends 550,000 Euros each year to improve its Web site, besides the workers' wages. The "Operation Carbon" cost about 17,000 Euros (Arnal, 2004).

[18]   Last accessed on September 21, 2004. *En ligne: Voir et Écouter* was previously entitled *Les conférences vidéo*.

[19]   Social learning, as defined by the Exploratorium, is a context that implies a public museum setting through which social exchanges are required (e.g., discussions with scientists); it also suggests multiple-end users with a wide range of needs who access *Origins* from their personal computer (Alexander, 2004b).

Further Reading

Public Understanding of Science. Quarterly journal. London: Sage. Retrieved from WWW.SAGEPUB.CO.UK/journalScope.aspx?pid=105747.

Chapter XX

Real Science:
Making Connections
to Research and
Scientific Data

Jim Spadaccini, Ideum, USA

Abstract

Almost since the inception of the World Wide Web, scientific images in a variety of fields of study have been publicly available. However, in most cases the images lacked support materials making them difficult for the public to understand. Recently science centers and other educational organizations have begun to create Web-based resources that help mediate and explain compelling scientific imagery. This chapter looks at the development of four educational Web sites that utilize actual scientific imagery. Ideum developed these sites over the last four years with the Exploratorium, NASA's Sun-Earth Connection Education Forum, and the Tech Museum of Innovation. From a developer's perspective, the creation process for each site is presented. A critical examination explains why certain decisions concerning design, site structure, technical approach, content, and presentation were made and how lessons learned from one project were applied to the next. Finally, the chapter looks at how sites that utilize "real science" can help science centers fulfill their mission of reaching the public and assisting them in better understanding scientific research and the scientific process.

Introduction

Originally developed as a tool to help scientists share information, the World Wide Web continues to be an important mode of communication for scientific inquiry. Rich science datasets in a variety of fields are publicly available, and can provide a catalyst for learning. Science centers can act as mediators, organizing information across scientific disciplines and providing tools for understanding complex scientific research. Users can gain valuable insight into the scientific process and science centers can do what they do best — make science understandable and interesting to the public.

Physical exhibits work well in showing scientific phenomena but not necessarily the scientific process (Bradburne, 1998). Providing exhibits that allow visitors to manipulate phenomena or exhibits that simply demonstrate it in compelling ways is one of the things that science centers do best (Ansel, 2003).

However, common sense dictates that due to their physical nature, most exhibits cannot be converted into effective online experiences. They tend to be pale cousins of their physical counterparts — "virtual" exhibits in the worst sense of the word. It is better to focus on creating "real" experiences for Internet visitors. In early 1996, I began working for the Exploratorium in San Francisco. Although the Web site at that time was just a few years old, this philosophy (Semper, 2001) was already in place. Over the next few years we experimented in developing various types of online resources, everything from simple experiments that online visitors could try at home to Web casting live events from remote locations.

This spirit of experimentation, coupled with the desire to create "real" experiences for Internet users, eventually led to exploring sites that utilized actual scientific imagery.

In 2000, I developed the *Solar Max* and *Auroras* sites for the Exploratorium (by that time I had formed my own firm, Ideum) that used near "real-time" data from NASA satellites and ground-based observatories.

These early experiments showed promise. In developing subsequent sites (for the Exploratorium and for other clients such as the Tech Museum and NASA) we found that creating online resources that mine data from real scientific endeavors sheds light on the scientific process. Furthermore, these types of resources provide a link, both actual and metaphoric, to the scientific community.

Additionally, developing resources that use scientific data helps ameliorate some of the problems inherent in developing traditional educational Web sites. The question of validity as to whether the user is seeing a "reliable representation of reality" (Pollock, 1999) can be addressed by the nature of these types of resources. The online visitor is seeing what scientists see. How data are presented and mediated for the visitor becomes the more important question.

During the last few years, science centers have been rethinking their mission and more actively exploring the role that science plays in society (Bahls, 2004). Science centers provide the public with tools and opportunities to better understand science and its impact on society. There is strong interest among the public too. A recent National Science Foundation survey found that approximately 9 out of 10 adults were either very or moderately interested in "new scientific discoveries and the use of new inventions and technology" (National Science Board, 2002). Furthermore, millions of people visit the Web sites of Science Centers. The Exploratorium alone boasts 15 million visitors a year (http://www.exploratorium.edu/about/about_web.html).

This chapter focuses on how science centers can in part help fulfill their mission of reaching the public through the development of Web-based resources that utilize actual scientific data. The nature of the Web allowing direct and mediated access to scientific research and data provides opportunities for science centers not found elsewhere. Science centers and other educational organizations can present data across fields of research and can make scientific data and images understandable to the public.

I'll show examples that include specific approaches for developing resources that make direct connections to scientific research and data. From a developer's perspective, I'll outline the creation process and explain how decisions concerning the content, structure, and design of these Web sites came about. I will look critically at these works, but obviously much of what I will present here is subjective. Questions concerning technology will also be addressed where pertinent. However, before we look at the example sites, it first makes sense to look at the general ways in which science centers connect the public to scientific research and communities.

Ways Science Centers Connect the Public to Scientific Research

Science centers are institutions devoted to encouraging interest in science, fostering exploration, discovery, and learning. They connect the public to scientific research and the scientific process in the following ways:

1. Transmission of scientific information: Providing public access to raw data or the results of scientific inquiry. For example, images from the Mars Pathfinder Mission or a map of the human genome.

2. Improving skills such as creativity, curiosity, collaboration, and etcetera: Creating experiences that engage the public. For example, interactive exhibits or other museum programs.

3. Experiencing and modeling the scientific process: Providing opportunities for the public to explore how and where science happens and who conducts it. For example, lab-based activities, connections to research centers, or access to scientists.

Online resources can be developed to assist science centers in making connections to scientific research and communities. In most cases, the physical environment of the museum is a richer alternative — however, there are exceptions. The nature of the Internet makes it ideal for the transmission of scientific information (this should not be confused with presenting scientific phenomenon). Web resources can reach large audiences. They are available anytime and anywhere there is a network connection.

Beyond the obvious advantages of the medium, Web-based activities can provide layers of information and different levels of complexity. Also, scientists themselves may play a role. Their participation as advisors and partners on these projects is essential and both scientists and developers benefit from this interaction (Persson, 2000). Scientists can even "appear" on the Web sites with short audio or video interviews — helping to explain what visitors are seeing.

Additionally, Web-based resources can have more longevity. Visitors can easily and repeatedly access a Web site or online resources. When exploring the more complex aspects of the scientific method, these are distinct advantages. Visitors can spend more time from home or school investigating these resources than they might in one museum visit.

Connecting to Scientific Research and Data Online

The Internet contains vast warehouses of scientific data. Everything from maps of the human genome to images of distant galaxies are available on the Web. Scientists themselves use the Internet to do research, connect with colleagues, and post research results.

While the Internet does not contain a complete set of current scientific research, and only very limited selections from past endeavors, the breadth and depth of materials available are impressive. Nevertheless, finding, organizing, and making sense of what is out there is an enormous challenge. Scientists tend to make data and research findings available for other scientists to review, not necessarily to connect with the public.

This presents an opportunity for science centers to help explain these materials and to make them more readily available to the public. Science centers can provide the necessary structure and explanations for the public to more effectively understand them and their meaning.

Of course, the particulars of the scientific data and how the resources are constructed and presented play an enormous role. All data sets are not created equal. It is easier to develop a site connecting to satellite imagery of the Earth for example; than it is to develop one explaining data sets that measure the Earth's magnetic field. Strong imagery can be more easily explained and can provide "hooks" for the public.

Also, the aesthetic quality of images can't be underestimated. Satellite images capturing features of the sun or microscopic images of cells or DNA can provide a sense of wonder. This comes not just from the beauty of the images themselves but in many cases from the knowledge that what we are seeing hasn't been seen by previous generations — that only recent technological advances make this possible.

Scientific visualizations of what can't be "seen" can also be used. However, in most cases these are developed for scientists themselves to study—not for the general public. Either way, "raw" scientific data and materials require careful explanation. Sites need to have a thoughtful balance of information, enough to adequately explain but not so much as to turn visitors away. Finally, there is a compelling aspect of connecting to "real things" that cannot be dismissed — seeing the same things that scientists do. Connecting to the same data coming from a spacecraft, a powerful microscope, or from a remote location here on Earth can inject excitement into a site.

Example 1: Solar Max

In 2000, Ideum developed *Solar Max: Your Guide to the Year of the Active Sun* (www.exploratorium.edu/solarmax), a Web site for the Exploratorium and NASA's Sun-Earth Connection Education Forum. The site is a portal created to provide visitors with information about the sun during its most active period, known as "solar maximum." What made the site unique was not that it linked to near real-time NASA satellite images of the sun, but rather that it utilizes these types of images from a variety of NASA missions, including those from non-NASA ground-based observatories. Additionally, it added explanatory materials to help users better understand what they were seeing.

At that time, each of the NASA solar missions had their own Web sites providing images of their satellite views of the Sun. However, none of them contained images from their fellow NASA missions and certainly not those from other observatories. Each mission was focused on presenting their own data and findings, not those of other missions.

In fact, the only NASA site to present images from all of the NASA solar images was the Solar Data Analysis Center (http://umbra.nascom.nasa.gov/sdac.html). However, the chosen layout format made it difficult to see these images side-by-side and make comparisons. In addition, the site did not provide links to educational resources, or news stories, nor did it contain other materials to help explain the images. This was not a flaw

Figure 1. The Solar Max portal site. Notice the near real-time images of the sun and various support materials.

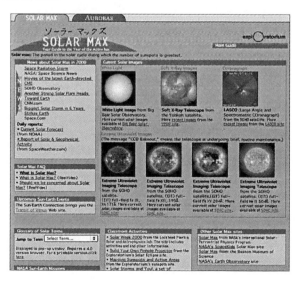

of the SDAC design but rather a question of its mission and purpose. The Exploratorium and NASA's Sun-Earth Connection Education Forum saw it as their role to provide these types of materials along with the images on the *Solar Max* site.

In addition to *Solar Max*, we developed *Auroras: Your Guide to the Northern and Southern Lights* (http://www.exploratorium.edu/auroras) which is similar in design and approach. Together these "twin portal" sites allow users to view near real-time data sets not just of the sun but also of the Earth — seeing the effects of the active solar storms. The portal design for both sites was meant to provide a compact "tool kit" not only for exploration and basic information, but also for additional opportunities to explore other sites and resources on the subject.

In addition, both sites have streaming media (Real Audio and Video) interviews with scientists. These help explain the data that users are seeing and provide insight to the larger scientific concepts presented. Featuring scientists also provides a direct connection to members of the scientific community. Importantly, including scientific researchers allows visitors to make a personal connection.

While *Solar Max* and *Auroras* do present some support materials, in retrospect they are somewhat thin on focused interpretive and interactive tools. They don't include clear, direct explanations of the individual images that appear on the site; for example, the images of the sun were captured at different wavelengths like extreme ultraviolet and soft x-ray. This presented opportunities to explain why scientists were interested in studying images of the sun produced from various points of the spectrum and to elaborate on non-visible wavelengths. Unfortunately, this potentially revealing concept was not pursued.

Other evident possibilities for exploration such as the "false colors" used with certain images were not explained. For example, some of the NASA SOHO extreme ultraviolet images are colored blue and green, to allow scientists to differentiate between the various wavelengths at which the images are captured. A visitor to the site could only wonder: why is the sun green? However, even with these shortcomings, there is some value in simply bringing these data sets together in one place. It took a science center and NASA organization that was not "mission-based" to make this happen. Additionally, the lessons learned were applied to future projects.

Example 2: Global Climate Change: Research Explorer

Following in the footsteps of the *Solar Max* and *Auroras* portal sites was the *Global Climate Change: Research Explorer* site (http://www.exploratorium.edu/climate). This project was funded by the National Science Foundation with the goal of increasing public understanding of science. Developed by the Exploratorium and Ideum in 2002, its development was shaped by many of the perceived strengths and weaknesses of the twin portal sites.

The *Global Climate Change: Research Explorer* allows the public to examine the same data that scientists and researchers analyze while studying global warming. A combination of near real-time, short-term, and long-term data helps tell the story of climate change and modern research methods.

Figure 2. The hydrosphere main page of the Global Climate Change: Research Explorer. The thumbnails of five data sets open to individual pages with more detail and explanation.

The site is broken up into five main sections, which correspond to parts of the climate system and roughly connect to areas of scientific study: atmosphere, hydrosphere, cryosphere, biosphere, and global effects (which examines models and forecasts). During the development phase, the structure of the site was quickly agreed upon but there was much discussion about what to name these categories. Some team members felt they were too complex and should be simplified.

The problem with simplifying these titles is that they loose their connection to real scientific research. Calling "atmosphere," "air" or "hydrosphere," "water" or even "cryosphere," "ice" might make the categories seem more "friendly" for visitors — but they do not accurately reflect their contents.

This issue was fairly contentious and both sides of the argument have merit. If the categories seem too complex it may turn visitors away. After all, "cryosphere" is likely not a term many people are familiar with. Yet, calling a section "ice" is not completely accurate — it doesn't begin to address the complexity of the cryosphere.

In the end, the more scientific terminology was adopted. It was decided that since the *Global Climate Change* site's mission was to explain climate by exploring these complex systems — to begin that process by calling them something else would not be the best approach. Furthermore, it was argued that the roots of the more technical terms were fairly well known "hydro," "cryo," and "bio." Finally the argument was made that if we didn't introduce these terms to the public, who would?

Global Climate Change presents a cross-section of scientific disciplines including physics, chemistry, biology, geology, meteorology, and oceanography. The data comes from a variety of research institutions and is not displayed together anywhere else. Presenting the data and results of scientific study together provides a foundation to tell a larger story.

A glossary of terms and answers to common questions, first introduced in *Solar Max* to further interpret and explain the content, were refined and expanded the *Global Climate Change* site. Some improvements can be attributed to the fact that the *Global Climate Change* site contains a large number of data sets and these tools were applied to more pages, but there were also fundamental changes in the overall approach. A glossary of terms was improved, so it appears within the context of specific data and topics. Each page of data has its own set of terms whose definitions are visible without having to launch a separate page. In *Solar Max* the user must select a pop-up field to view glossary terms, which then appear in a pop-up window.

Answers to common questions were expanded and added to all of the data. Additionally, the nature of the questions was expanded to get a sense of how scientists conduct research, test theories, and come to conclusions. While many of the questions focus squarely on the science, "How much carbon dioxide do forests take out of the atmosphere?," other questions explore scientific methodology, "How can we tell the difference between short-term fluctuations in the weather and long-term climate change?" This was a departure from *Solar Max*, where the questions were simply about the science, "What is solar max?" and "Should we be concerned about solar max?"

Figure 3. Atmospheric Carbon Dioxide Records from Mauna Loa (1958-2000) from the Atmosphere section in the Global Climate Change site.

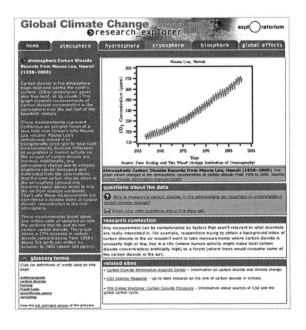

In addition to providing an answer to a question associated with each data set, a mechanism for receiving questions from the public was developed. While limited resources prohibited individual responses to these questions, this information provided insight into how users were interacting with each data set and what questions the data presented.

Like *Solar Max*, the *Global Climate Change* site is "portal-like." The layout is very compact and there are links to related sites from each data set page. The links are presented within the context of the specific data set, providing more information on the subject if visitors are interested.

While there were certainly many similarities in approach, the biggest difference in developing *Global Climate Change* lies in the amount of basic support material. A text narrative provides an introduction to each section. Each data set contains a detailed description, caption, and "research connection."

These three areas explain why this data is important in studying climate change, how the data was collected, how it is studied, and how scientists use the data to test theories and develop conclusions. Each diverse data set provides an opportunity to explore the science and research methods. For example, information on how sea surface temperatures are measured with microwave energy, how ice core samples are used by scientists, or how coral bleaching occurs are explained within the context of the data sets shown.

While the support materials provide comprehensive explanations for users interested in the data, the argument could be made that these materials themselves are too dense. Still, the portal nature of the site allows data and support materials to be presented together, helping users to make important connections. Also, basic navigation is simple and the structure of the site is clearly visible at all times.

However, pages often seem crowded and are likely overwhelming for some visitors. Additionally, the site relies too heavily on text, with only one short video interview answering one of the "questions about the data" on the entire site. This interview came from another existing Exploratorium project. Unfortunately, there wasn't enough funding to interview scientists specifically for this project. While scientists didn't appear on the site they did play a role as advisors. They reviewed the text that appears on the site and pointed team members to additional data sets and related Web sites.

Example 3: The Sun-Earth Viewer

The *Sun-Earth Viewer* (http://ds9.ssl.berkeley.edu/viewer/flash/flash.html) is yet another resource exploring NASA's near real-time solar data sets. It was first developed in 2003 and later revised in 2004 by Ideum for NASA's Sun-Earth Connection Education Forum (SECEF). It should be noted that SECEF is not a science center. However, their focus and audiences are much the same. They reach school children of all grades, families, and curious adults. Additionally, they work collaboratively with a number of science centers and other educational institutions.

The *Sun-Earth Viewer* is built entirely in Flash — an interactive Web-based multimedia tool developed by Macromedia. Beyond the basic connection to scientific data, the *"Viewer"* has some similarities to the *Solar Max* and the *Global Climate Change* sites.

First the data itself comes from a variety of sources: NASA satellites, a NOAA (National Oceanic and Atmosphere Administration) satellite, as well as ground-based observatories. Additionally, the *Viewer* is portal-like in structure, very dense but easy for users to compare images and view support materials. Also, like the other sites, it relies on links to additional sites and sources for deeper explanation.

There are some significant differences. While *Solar Max* and *Global Climate Change* simply presented the data along with varying degrees of explanation, the *Sun-Earth Viewer* allows users to zoom and pan images of the Sun, viewing features such as sunspots and solar flares. Additionally, a scale tool showing the size of the Earth is provided for perspective on some of the solar images.

While *Solar Max* and *Global Climate Change* simply linked to the data, the *Viewer* integrates it, allowing users to actively explore — viewing and comparing details from various images. Resources that allow users to explore and make connections on their own seem to be more effective and follow the widely accepted practices of museum educators (Hein, 1991).

In addition to providing an improved method to view solar images, the *Viewer* contains a wide and varied assortment of support materials. Detailed illustrations, computer visualizations, and satellite video footage that examine a variety of Sun-Earth phenomenon are presented. These materials all include associated text explanations.

Finally, over 40 minutes of short video clips with scientists and other experts examine common questions about the Earth and its relationship with the active sun. Although the site contains much less text than the *Global Climate Change* site, the use of dynamic videos helps answer the most common questions about the Sun-Earth connection.

Figure 4. The Sun-Earth Viewer showing a close-up image of the Sun. Notice the scale tool on the bottom right. The four tabs on the top provide sections with additional support materials for the visitor.

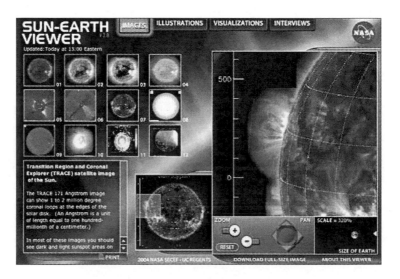

Progress in Web technology and improvements in Flash in particular, played a very important role in the development of the *Sun-Earth Viewer*. This type of interaction with near real-time data was simply not possible back in 2000, when *Solar Max* was first developed. Flash as a development tool wasn't yet sophisticated enough to link to near real-time data.

Other new features in Flash such as the ability to embed video and allowing for external text (or XML), make it a better solution for developers. It can also reach more visitors. The *Sun-Earth Viewer* only requires the Flash plug-in. Macromedia's Flash has developed into a standard over the last few years and is installed on the vast majority of visitors' systems. *Solar Max* and *Global Change* required users to have the RealVideo Player (http://www.macromedia.com/software/player_census/flashplayer/).

While the technological approach was different, many of the lessons learned from *Solar Max* and *Global Climate Change* have been applied to the *Viewer*. Also, since there have been two versions of the *Viewer*, lessons learned in the first version were applied to "2.0." The site retains the portal-like structure and explanations are provided within the context of the data presented. Multimedia, in the form of interviews and visualizations, helps explain important scientific concepts and phenomena, so the site does not depend entirely on long text passages. Scientists participated both as advisors and as participants.

Still, in retrospect there are areas for improvement. There is no glossary of terms, though many of the complex descriptions would benefit from this feature. In addition, while the *Viewer* technology works well for near real-time solar images, the window may be too small for viewing illustrations. With some of the illustrations it is difficult to read the labels when the entire illustration is visible, and when you zoom in it can be hard to see how the whole illustration works.

Figure 5. The Sun-Earth Viewer with an illustration of the Electromagnetic Spectrum

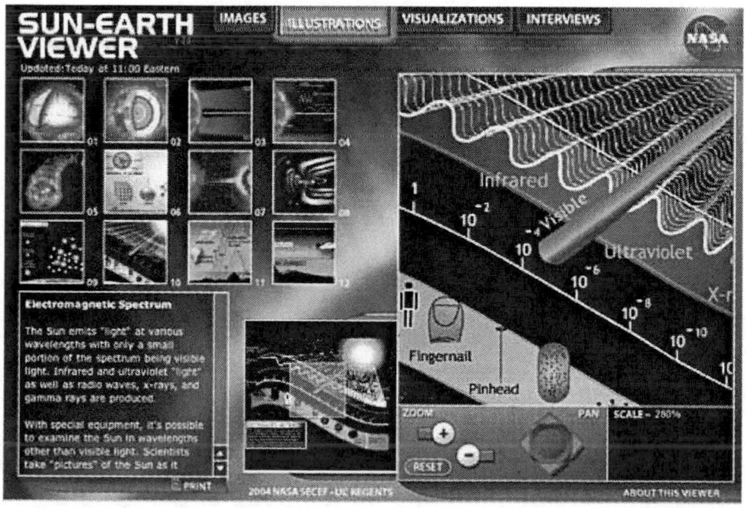

Looking back, a different layout for the illustrations and a different sized window would probably be the best approach. Still, even this solution would pose more problems. If the viewer window were larger then something else (on an already crowded screen) would have to be made smaller, be on scroll mode, or otherwise be hidden. Increasing the size of the entire window is not an option. The window at 760x540 pixels is already the maximum size for an 800x600 pixel computer monitor (with room for browser controls, tool bars, etc.). If the window size were larger some users would have to scroll to see portions of the viewer.

Another issue worth examining is the site structure, which is based on the type of media rather than the topic. A structure that reflects the Sun-Earth system would be ideal. Perhaps something like: the sun, the solar wind, and the Earth's magnetosphere; or the solar interior, solar surface, solar wind, Earth's magnetosphere, auroras, and etcetera. These terms were explored but the structure that appears was chosen for a number of reasons.

First, the site's focus is on the Sun-Earth connection: looking at how events on the sun can affect the Earth. While the sun and the Earth's magnetosphere could be looked at as separate systems, the goal of the site was to show where these systems connect. After all, the mission of the Sun-Earth Connection Forum is "to increase science literacy and focus attention on the active sun and its effects on Earth" (The Sun-Earth Connection Education Forum, 2001).

Also, there were practical reasons, namely that the type and availability of media and resources were somewhat uneven. As stated earlier, data sets are not created equal. There are more solar images than images of the Earth's magnetosphere, for example. This is a reality in developing these types of sites. It is not that data on the magnetosphere is sparse; it is just not available in a format that allows for it to be easily presented to the public. When building Web sites that utilize real scientific data, the structure of the site needs to reflect the educational goals of the site as well as the available materials.

Figure 6. The Sun-Earth Viewer showing the SOHO EIT images during CCD bakeout

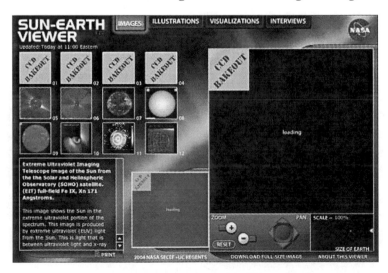

A concern in developing sites that utilize near real-time data is that sometimes for a variety of reasons the data becomes unavailable. For example, the *Sun-Earth Viewer* is very dependent on images from the Extreme Ultraviolet Imaging Telescope (EIT) aboard NASA's SOHO (Solar and Heliospheric Observatory). The first four images come from this device. Every three months the CCD (Charge Coupled Device), which takes EIT images on SOHO, goes through a process called "bakeout" for 7 to 10 days. This temporary shutdown is a necessary maintenance procedure (http://umbra.nascom.nasa.gov/eit/CCD_bakeout.html).

A simple method for swapping out images hasn't yet been developed. Only associated text files can readily be updated since they are external from the Flash application. This issue is one we hope to address later this year. When developing the second version of the *Viewer*, images from NASA's TRACE (Transition Region and Coronal Explorer) satellite were added. TRACE full-disk images are only available when SOHO EIT images are not. This provides some coverage of the sun for our visitors. Of course this was not done for our benefit, but rather for scientists studying the sun.

Example 4: Zooming Into DNA

Zooming Into DNA (http://www.thetech.org/exhibits/online/ugenetics/zoomIn/) was developed in 2004 for the Tech Museum of Innovation. Created on a small budget, this was one of four features developed by Ideum with the Tech Museum for an online exhibition on Genetics. *Zooming Into DNA* shows microscopic images of cells, chromosomes, and DNA. All of the images are "real" scientific images.

The original idea for this exhibit came from an online version (http://www.thetech.org/exhibits/online/genome/intro4.html) created by the Tech Museum back in 1996. The Tech Museum felt that the concept was strong and there was room for improvement. In

Figure 7. Zooming into DNA, showing Fibroblast Skin Cells. Notice the "About This Image" window is open

researching this project we also came across a similar DNA Zoom interactive. It was a more polished Flash version (http://www.biotechnology.gov.au/biotechnologyOnline/ interactives/DNA_zoom_interactive.htm) developed years later by Biotechnology Australia.

Using both of these exhibits as starting points we first wondered whether it was worth pursuing a new version. There was a strong desire not to duplicate what was already out there. During our initial research we also came across some very interesting images of cells, chromosomes, and DNA, unlike the two other "zoom-ins," which relied either entirely or in part on illustrations. It was decided to create a version that utilized actual scientific images.

Using real images at the core of the site provided some unique benefits. First, the images themselves are very compelling. Like the images of the sun, they have a great aesthetic quality. They can be a catalyst for learning, sparking a visitor's interest. Additionally, the fact that they are real presents some interesting questions: What am I seeing? What scale is this? How was this image taken?

Questions about the images provide a path to explore the scientific process. Certainly the tools — whether they are satellites or microscopes — play an important role in the modern scientific endeavor. Although, interestingly, this also fits into the Tech Museum's stated mission to engage people "…in exploring and experiencing technologies affecting their lives…" (http://www.thetech.org/about/mission.cfm). Exploring the technologies used to create the images was also an end in itself.

Along with the ability to learn "About this Image," a scale tool was added to *Zooming Into DNA*. This time a series of images capturing bacteria on the "head of a pin" was used. A short "movie" allows the user to play the sequence starting with a pin being held between two fingers and zooming all the way to the point where bacteria can be seen on the pinhead.

One possible area for confusion is that the "zoom in" is not perfectly linear. Rather, the "zoom in" follows a conceptual path, as if one were to physically travel from outside to inside cells, finally arriving inside the nucleus. For example, "nuclear pores" at 300,000X appear before the "human karyotype" at 5,000X. This was done because nuclear pores are a point of entry from a cell's cytoplasm to inside its nucleus where DNA resides. They allow communication between a cell's nucleus and its cytoplasm.

The human karyotype image connects more easily with the chromosomes and DNA that appear at the end of the zoom in. The vast differences in scale are due to both the imaging technique and the fact that the karyotype is a diagnostic test — so we're not seeing chromosomes in their normal state. At first glance some users may view zoom-in progression as an error. This was a trade-off. Either retain the purity and simplicity of zooming in a linear fashion or restructure the site to better match the content. We chose the latter.

Interestingly, if we decided to do an illustrated version it would have likely been linear and the subject of "nuclear pores" would probably not have been included. We came across the image by chance while researching *Zooming Into DNA*, found the image fascinating, and decided that it should be included.

Finally, there are areas that could have been improved. Like the *Viewer,* some of the terminology is rather advanced and unfortunately there is no glossary tool. Also, the budget did not allow for multimedia interviews with scientists (although scientists did play a role in advising). Therefore the explanations are mostly text-based. Perhaps like the *Viewer,* there may be a 2.0 version sometime in the future?

Evaluation

There's certainly anecdotal evidence that these resources have been both popular with the public and seem to provide educational value. Additionally, *Global Climate Change* and *The Sun-Earth Viewer* were honored (out of 1,700 entries) with a Pirelli International Multimedia Award in 2004 and 2005. Still there haven't been any formal evaluative studies done. There is one voluntary user survey currently underway on the *Global Climate Change* site, but unfortunately the results are not complete at the time of this writing.

There is no way beyond interpreting crude Web statistics to tell how users are interacting with these Web sites. Through these data we know how many people are visiting and how long they stay. But we have no sense how they are using them or what they are taking from these sites. Formal evaluation could reveal how effective they are educationally and could also examine questions of usability.

Beyond evaluation for these specific projects, there is little in the way of evaluation available for science-based educational resources in general. A comprehensive visitor study was conducted by the Exploratorium and Science Museum, UK in 2000 (Semper, 2000). However, it focused on different types of educational resources, none that utilized actual scientific data. Furthermore, the Web has changed dramatically in four short years. The sites themselves are dated, much of the technological aspects for the study are also out of date. Still some important information about how these sites are used can be found.

While the need for the more formal and specific evaluation is clear, developers can learn much from what others have done. One of the most engaging sites that utilizes and explains a wide variety of data is NASA's Earth Observatory site (http:// earthobservatory.nasa.gov/). This site clearly organizes the images and, like the examples in this chapter, it has supplemental materials and other tools for exploring the data.

Developers can also learn from their own past endeavors. Critically looking at your own work, examining anecdotal feedback from visitors or partners, and viewing Web statistics can help create a cycle where future sites improve. While each project utilizing real scientific data has different content, and available resources can change, ideas and lessons learned from one project can be applied to others.

Usability

In developing effective resources that use scientific data, usability is an important aspect. Due to the compact nature of these resources, questions of usability are magnified. Usability for the Web has been defined in many ways. In the simplest sense, a Web site should function correctly, be efficient to use, easy to learn, easy to remember, error tolerant, and subjectively pleasing (Brinck, 2002).

It's a tall order, complicated by the fact that you want to be able to provide interpretive tools and support materials in close proximity to the images that users are seeing. The more elaborate the tools and supporting text or media — the more complicated the design process becomes.

Ideally, users would have the ability to view an image (and possibly compare it with others) while being able to review support materials with the image still visible. This is not always possible, but it is certainly preferable.

Layering materials so they only appear as needed can help address the lack of space. However, this adds complexity to the site. Users will have to find these elements, and then decide whether to use them. Finally they need to be able to easily open and close these windows. Additional elements can add "an additional burden on the user" (Nielsen, 1993).

Creating scrolling text fields or pop-up windows is another way to keep these sites and pages compact. But again, users need to scroll, and open and close the pop-up windows.

Close coordination between content developers and designers can help create sites with a high degree of usability. While design considerations shouldn't always shape content development, flexibility is important; again, even more so when screen space is so restricted. With the projects we've been involved in, we've either developed the content internally or worked closely with museum writers as well as outside scientists and other experts.

Conclusion

In the introduction I mentioned that we can "mine" scientific data, and I think the analogy is a good one. There is an opportunistic quality in creating these types of sites. Like miners looking for valuable resources, as developers we look for compelling or interesting images. Sometimes finding a great image sparks your interest in the topic and forces you to rethink the structure or content of your site. Developing a process with time for discovery, research, and "inventory" at the beginning of each project is extremely important.

The aesthetic quality of the images can provide "hooks" for visitors. Stunning visual images can help inspire visitors to dig deeper to learn more. Still, the choice of images needs to be balanced with the educational goals of the site. The beauty of the images can only be weighed as one factor. How the image fits into the site as a whole, what it shows, how it compares to the other images — all of these factors need to be addressed.

Additionally, near real-time data has an exciting quality. However, in many cases science requires time; therefore viewing changes in data over long periods of time is required to observe results. Where this is the case, this message needs to be made clear to visitors through materials that accompany real-time data, or by using them in conjunction with short-term or long-term data sets.

Providing support materials that help explain what visitors are seeing is paramount. Presenting raw data alone serves only the technical purpose of "mirroring." It's important to have explanations that are specific to data shown and that hopefully anticipate the types of questions visitors may have. A glossary of terms and links to other related sites can also help users expand their understanding of the materials. These materials are particularly important for those who are new to the topics presented.

Illustrations, audio, or video can make a site more engaging and compensate for different learning styles. Audio or video interviews with scientists who work directly with the data can help the visitors better understand the data sets and the scientific process, as well as making personal connections with the people who work in science.

Creating tools that allow users to manipulate data lets visitors make their own discoveries. Additions such as the ability to zoom and view detail can make a site more engaging. Also, we've seen how tools can help explain scale and provide perspective for satellite-based or microscope images.

Still more can be done. Authoring software such as Flash has only just recently become sophisticated enough to allow developers to create new ways to view data. In future projects we hope to explore more direct ways of comparing data (including real-time data) such as "onion skinning" which would allow users to overlap images and change their transparency. We'd also like to develop tools that would allow users to examine real-time data over time — to be able to view changes in the data.

Additionally, for static images more can be done with labeling of key features. (For example, semi-transparent layers that could show the key parts of a cell could be turned on or off by the user.) By directly showing the parts, users would gain insight into the workings of subjects they are viewing. There could be a series of these labels, each examining different aspects of what the image or data show. There are many possibilities.

Beyond better tools for examining and manipulating data, improvements in the support materials themselves could make for better resources. Improved layering of information for different levels of expertise would broaden the appeal of resources, allowing them to reach novices and experts alike.

Another point worth making is that these types of resources don't have to live on the Internet exclusively. The *Sun-Earth Viewer* is on the museum floor at Chabot Space and Science Center in Oakland, California. Having resources that can be used on the Web and the museum floor is appealing to museums and other organizations that have tight budgets for exhibit development.

Electronic exhibits that aren't available on the Internet can also connect to these types of data. In 2004, we helped develop an electronic exhibit on Fuel Cells. One of the sections explaining the benefits of clean-powered Fuel Cells connects to near real-time maps demonstrating air quality in Los Angeles. Seeing "today's smog levels" not only helps explain the topic, it helps connect the user with the sense of immediacy and relevancy. This exhibit is currently on display at the California Science Center in Los Angeles.

Regardless of how resources that utilize data sets are built or where they ultimately reside (on the Internet or as exhibits) there are, of course, limitations to keep in mind. First and foremost, not every area of scientific inquiry can be explored in the way outlined in this chapter. The Internet is a visual medium, and strong visual data is needed. In some areas of research, the data simply doesn't exist.

Also, many data sets are not visually appealing or visual at all. Scientific data is created for scientists, not the general public. Most are difficult to explain and require a great deal of mediation. Finally, we've seen how real-time data can become unavailable, leaving sites without key images.

Yet even with these limitations, educational resources that utilize scientific data do provide opportunities for science centers and other educational organizations to reach the public in interesting ways. They take advantage of qualities of the Internet, instead of attempting to work around its limitations. If science centers want to develop effective Web-based educational resources, new models and methods need to be explored.

As demonstrated, sites that utilize real scientific data introduce users to various topics in the world of science. They act as catalysts, peaking interest in science through strong and compelling imagery. In addition, thoughtfully designed tools for interpretation and reference deepen a visitor's understanding of scientific concepts.

While there are many different approaches and more experimentation and evaluation needed, sites that utilize scientific data can provide an engaging educational experience for Web visitors. They can help foster creativity, curiosity, and discovery. Furthermore, sites that utilize scientific data can directly explain and model the scientific process, providing opportunities for visitors to explore "real science." All of these qualities help science centers fulfill their mission to reach the public and to help them better understand science and the role that it plays in society.

Acknowledgments

Many thanks to Andrea Bandelli and Michelle Williamson for their valuable assistance in formulating some of the ideas and examples presented in this chapter.

References

Ansel, J. (2003). Real, simple and new. *The Informal Learning Review,* No. 63.

Auroras: Your Guide to the Northern and Southern Lights. (n.d.). Retrieved March 18, 2004, from *http://www.exploratorium.edu/auroras*

Bahls, C., Park, P., Heyman, K., & Hunter, P. (2004). Science museums exhibit renewed vigor. *The Scientist, 18*(6).

Bradburne, J.M. (1998). Dinosaurs and white elephants: The science center in the 21st century. *Public Understanding of Science*, 7(3), 237-254.

Brinck, T., Gergle, D., & Wood, S. (2002). *Designing Web sites that work: Usability and Web*. San Diego, CA: Academic Press.

DNA Zoom Interactive. (n.d.). Retrieved March 20, 2004, from *http:// www.biotechnology.gov.au/biotechnologyOnline/interactives/ DNA_zoom_interactive.htm*

Exploratorium: About Our Web Site. (n.d.). Retrieved March 25, 2004, from *http:// www.exploratorium.edu/about/about_web.html*

Global Climate Change: Research Explorer. (n.d.). Retrieved March 18, 2004, from *http://www.exploratorium.edu/climate*

A Handful of DNA. (n.d.). Retrieved March 20, 2004, from *http://www.thetech.org/ exhibits/online/genome/intro4.html*

Hein, G. (1991). The museum and the needs of people. Retrieved March 11, 2004, from *http://www.exploratorium.edu/IFI/resources/constructivistlearning.html*

Macromedia Flash Player Statistics. (n.d.). Retrieved March 18, 2004, from *http:// www.macromedia.com/software/player_census/flashplayer/*

NASA Earth Observatory. (n.d.). Retrieved March 18, 2004, from *http:// earthobservatory.nasa.gov/*

National Science Board. (2002). *Science and engineering indicators – 2002.* (NSB-02-1). Arlington, VA: National Science Foundation.

Nielsen, J. (1993). *Usability engineering.* San Diego, CA: Academic Press.

Persson, P.E. (2000, January/February). The changing science center: Sustaining our mission into the 21st century. *ASTC Dimensions.*

Pollock, W. (1999, September/October). Science centers on the Web. *ASTC Dimensions.*

Semper, R. (2001, November/December). Nodes and connections: Science museums in the network age. *ASTC Dimensions.*

Semper, R., Wanner, N., Jackson, R., & Bazley, M. (2000). Who's out there? A pilot user study of educational Web resources by the Science Learning Network. In D. Bearman & J. Trant (Eds.), *Museums and the Web: Selected paper from an international conference, 2000.*

Solar Data Analysis Center. (n.d.). Retrieved March 18, 2004, from *http:// umbra.nascom.nasa.gov/sdac.html*

Solar Max: Your Guide to the Year of the Active Sun. (n.d.). Retrieved March 18, 2004, from *http://www.exploratorium.edu/solarmax*

The Sun-Earth Connection Education Forum. (2001). *The Sun-Earth Connection Education Forum* [Brochure]. Berkeley, CA: SECEF.

The Sun-Earth Viewer, 2.0. (n.d.). Retrieved March 18, 2004, from *http:// ds9.ssl.berkeley.edu/viewer/flash/flash.html*

The Tech Museum of Innovation: About Us: Our Mission. (n.d.). Retrieved March 20, 2004, from *http://www.thetech.org/about/mission.cfm*

What's a CCD bakeout, anyway? (2004). Retrieved March 11, 2004, from *http:// umbra.nascom.nasa.gov/eit/CCD_bakeout.html*

Zooming Into DNA. (n.d.). Retrieved March 20, 2004, from *http://www.thetech.org/ exhibits/online/ugenetics/zoomIn/index.html*

About the Authors

Leo Tan Wee Hin has a PhD in marine biology. He holds the concurrent appointments of director of the National Institute of Education, professor of Biological Sciences in Nanyang Technological University, and president of the Singapore National Academy of Science. Prior to this, he was director of the Singapore Science Centre. His research interests are in the fields of marine biology, science education, museum science, telecommunications, and transportation. He has published numerous research papers in international refereed journals.

R. Subramaniam has a PhD in physical chemistry. He is an assistant professor at the National Institute of Education in Nanyang Technological University and honorary secretary of the Singapore National Academy of Science. Prior to this, he was acting head of physical sciences at the Singapore Science Centre. His research interests are in the fields of physical chemistry, science education, theoretical cosmophysics, museum science, telecommunications, and transportation He has published several research papers in international refereed journals.

*****

Steven Allison-Bunnell is senior writer and producer with Educational Web Adventures, LLP (USA). Along with developing online learning interactives for museums and other informal education organizations, he has contributed to Eduweb's research applying learning theory and evaluation methodologies to online informal learning materials. He has developed non-fiction online content since 1995, when he was the founding nature editor of the Discovery Channel Online. He holds a PhD in science and technology studies from Cornell University, where he studied the history of natural history museum exhibits, and a BA in biology from the Robert D. Clark Honors College of the University of Oregon.

Jim Angus joined the National Institutes of Health (USA) in 2001 as Web project manager. He has a diverse background with undergraduate degrees in biology and geology, and graduate training in immunology and molecular biology. His prior employment was at the Natural History Museum of Los Angeles County where he established and managed the Museum's Molecular Systematics Laboratory. In 1992 he became interested in information technology as applied to scientific research and assisted in the development of the Museum's IT infrastructure. In 1994 he produced one of the first museum Web sites, which was subsequently named "best educational use" at the 1997 *Best of the Web* awards sponsored by Archives and Museum Informatics. He currently serves on the board of directors of several professional organizations including the *Museum Education Roundtable* and frequently speaks on Web design and accessibility of Web sites.

Patricia Barbanell holds a doctorate from Columbia University and has extensive experience in both K-12 education and museum education. She has worked for more than 20 years developing integrated programs that serve both museums and schools. Her specialties are integrated arts, multicultural programming and technology integration. She has presented scores of presentations at professional conferences and has published several papers in professional journals. She is past president of NYS Art Teachers and NY Council of Educational Associations and helped to write the NYS Learning Standards for the Arts.

Saulo Faria Almeida Barretto (saulo@umc.br) teaches mechanical engineering courses at Universidade de Mogi das Cruzes, São Paulo, Brazil. He received his BS in civil engineering from Universidade Federal de Sergipe (1986) and his MS in finite element methods from Universidade de São Paulo (1990). He received his doctoral degree in boundary element methods from Universidade de São Paulo (1995). In 1999 he moved his research interests to work with learning Web environments and since then he has been conducting research projects funded by Brazilian funding agencies (FAPESP and CNPq) and the European Community.

Alpay Beler is an information systems architect and digital technologies/new media strategist. He has more than 10 years experience of development and management experience in the educational and heritage sectors, with particular knowledge of digitization technologies, e-commerce, R&D, cross platform networks and Web product design. Alpay has recently completed a major Web infrastructure at The Science Museum (London), managing the development of a £1.2m government-funded Web site that draws on the resources of three national museums and which offers personalized activities and community building tools. He was also a key team member of a pilot project for educational content delivery to mobile phones, consulted on the JPEG2000 standard for the British Standards Institution, and currently is responsible for the implementation of an online student information system at Birkbeck College.

Jim Bennett is the director of the Museum of the History of Science, University of Oxford (UK), which has a strong commitment to making its exhibitions and collections available on the Web. He also directs the master's course offered by the Museum. Previously Jim was at the Whipple Museum of the History of Science, University of Cambridge. He is attached to Linacre College and the Faculty of Modern History at the University of Oxford. His research work is on the history of practical mathematics from the sixteenth to the eighteenth century, the history of scientific instruments and the history of astronomy.

Roxane Bernier has been a researcher at the Department of Sociology of Université de Montréal (Canada) since 1999 and investigates visitors' perception with regards to the usability of information technologies in cultural institutions. She was affiliated three years with the Centre d'Étude et de Recherche sur les Expositions et les Musées at Université Jean Monnet (France) in the mid-1990s and since 2003 has been a fellow member at the Centre Interuniversitaire de Recherches sur les Sciences et la Technologie in Quebec. Dr. Bernier has worked as a consultant for the Direction des musées de France, the Cité des Sciences et de l'Industrie in Paris, the Quebec Museum of Civilization in Quebec city, the Canadian Museum of Civilization in Gatineau, and more recently for the Canadian Heritage Information Network in doing audience analysis. She coordinated a special issue on IT for the European museology journal *Publics et Musées* and contributed to peer-reviewed journals like the *Educational Computing Research*, *Behavior and Information Technology* and *Program*. Additionally, Bernier was invited to write articles for the cyberart magazines *Parachute* and *Archée* as well as participated in international conferences including CATaC 2000, AoIR 2001, Museums and the Web 2002 & 2003 and the Visitors Studies Association Conference 2004.

Bronwyn Bevan is director of the Center for Informal Learning and Schools, which is based at the Exploratorium in San Francisco. Her work focuses on partnerships between cultural institutions, science agencies, and schools. Bevan has also conducted numerous research and evaluation studies examining arts integration into the K-12 curriculum, and the role of arts-based cultural institutions in supporting instructional change

Ann Borda has held strategic and operational roles in academic and cultural organizations and recently held the position of head of multimedia collections at The Science Museum, London. Among the projects in which Ann has been involved are Fathom.com, an e-learning collaboration led by Columbia University, and a large-scale Web initiative to bring cultural collections online across several national organizations in the UK. Ann received her PhD at the University of London in Information Science (1999) and has published in the areas of e-learning, informatics, metadata applications and content delivery. Ann is the chair of the CIDOC multimedia group and assistant editor of the journal *Multimedia and Information Technology* (MmIT) (www.mmit.org.uk). Currently she is a visiting research fellow at the Institute for Computing Research, London South Bank University.

Jonathan P. Bowen (www.jpbowen.com) is professor of computing at London South Bank University, where he is deputy director of the Institute for Computing Research. Previously he was at the University of Reading, the Oxford University Computing Laboratory and Imperial College. He has been involved with the field of computing in both industry and academia since 1977. As well as computer science, his interests also extend to online museums. Bowen established the Virtual Library museums pages (VLmp) in 1994, a Web-based directory of museum Web sites worldwide that has since been adopted by the International Council of Museums (ICOM). He was honorary chair at the first *Museums and the Web* conference in 1997 and has given presentations at each conference since then. He guest edited two special issues of the *Museums International* journal concerning online museums. In 2002, Bowen founded Museophile Limited (www.museophile.com), a spinout company from London South Bank University with the aim to help museums online, especially in the areas of accessibility, discussion forums and collaborative e-commerce. Bowen is a fellow of the Royal Society for the Arts and holds an MA in engineering science from Oxford University.

Horace Dediu is a software developer and programmer who founded the consulting firm *Handheld Media* (USA) in 2001. Horace has published articles in the areas of information retrieval theory and electronic document formats. He teaches corporate seminars on product usability and mobile information access. Horace researches user interface design and human factors in computing, and helped shape the *Open Ebook* file format standards. His ability to explain the technology and its potential for broadening information access services, have made him a popular speaker at conferences such as *Seybold* and *Library*. In addition, Horace counts as clients companies such as Nokia, JP Morgan Chase and Verizon.

Suzanne Dewald has been the development officer of the Schenectady City School District (USA) for more than a decade and has led the authoring of dozens of successful grants that have helped to move the Schenectady City Schools into a leadership role in integrated technologies. She has 20 years of experience authoring funded programs and directing projects.

Nicoletta Di Blas graduated in classics and obtained a PhD in linguistic sciences from the Catholic University of Milan, Italy. She currently teaches theory of communication at the Polytechnic of Milan. Her research interests focus on linguistic themes, on usability and advanced educational applications, particularly in the cultural heritage field.

Lynn D. Dierking is associate director at the Institute for Learning Innovation (USA) and completed her PhD at the University of Florida. She is internationally recognized for her research on the behavior and learning of children, families and adults in free-choice learning settings and has published and spoken extensively in these areas. Her research priorities include the long-term impact of free-choice learning experiences on individuals and families and the development and evaluation of community-based programs. Dr.

Dierking has worked in a variety of settings: the Smithsonian Office of Educational Research, University of Maryland's College of Education and at the Smithsonian's National Museum of American History.

Denise P. Domizi is a doctoral student in instructional technology at The University of Georgia (UGA) (USA). She holds a BA in psychology (1991) and an MEd in instructional technology (2003), also from UGA. Her research interests include: informal learning environments, designing and implementing technology-infused learning opportunities in science centers and museums, and learning communities. Ms. Domizi is currently employed as a research assistant in the Learning and Performance Support Laboratory at UGA.

Michael Douma is the executive director of IDEA (the Institute for Dynamic Educational Advancement) (USA) where he designs and implements information systems that use technology to facilitate the learning process. His research and teaching interests center on interactive teaching methods, user interface design, and accessibility. Michael has presented papers on the interface between learning and technology in *Archimuse* *"Museums and the Web," American Chemical Society*, the *British Museum*, and *Centre de la Recherche Scientifique*. Through IDEA, Michael recently collaborated on educational initiatives with Brandeis University, Native American Cancer Research, the National Gallery of Art, and the Prado Museum (Madrid).

John Falco has served as superintendent of the 9,000 student urban school system of Schenectady, NY for three years and served for six years as deputy superintendent. In addition to Project VIEW, he has led development of several transformational projects in the Schenectady Schools' Capital Region Science Education Partnership (CRSEP) (USA), a multi-district Local Systemic Change Initiative funded by the National Science Foundation. Dr. Falco has been named NYSCATE Superintendent of the Year for outstanding leadership in integrating technology. Dr. Falco holds a doctorate in educational administration from Seton Hall University, where his research centered on improving reading skills for struggling emergent readers.

Silvia Filippini-Fantoni is a research student at the Sorbonne University in Paris, where she is working on her PhD based on *Personalisation through IT in museums*. She graduated in contemporary history from the University of Milan and has experience in working as a researcher at the European Centre for Digital Communication (Heerlen, The Netherlands), the McLuhan Institute (Maastricht, The Netherlands) and the Louvre Museum (Paris), where she focused on developing personalization applications for the new Web site. The results of work carried out so far have been presented at international conferences and seminars (EVA, ICHIM, Museums and the Web), where positive contributions toward the hypothesis have already been received and have led to interesting collaborations with Dédale on a European Union study about *Cultural Institutions as New Learning Environments*, La Cité des Sciences et de l'Industrie in Paris and the Canadian Heritage Information Network (CHIN).

Stefan Göbel received a diploma in computer science at the Technical University of Darmstadt, Germany in 1997. He worked as a researcher in the GIS Department at the Fraunhofer-Institute for Computer Graphics. The topic of his graduation deals with graphic-interactive user guidance to geospatial data archives. Since July 2002 he has been head of the Digital Storytelling Department at ZGDV Darmstadt e.V. Here, he is the project leader of numerous activities and projects within the research field of interactive digital storytelling and edutainment applications. Since 2003, he has chaired the international conference "Technologies in Interactive Digital Storytelling and Entertainment" (TIDSE) and the Forum for Knowledge Media Design (KMD-Forum) as speaker.

Kathryn Haley Goldman, MA, is a senior researcher at the Institute for Learning Innovation (USA). She has a bachelor's degree in anthropology from Bryn Mawr College and has extensive training in educational measurement, statistics and evaluation in learning and technology. She has worked on evaluation projects with the National Aquarium in Baltimore, Disney's Animal Kingdom, and the Cleveland Museum of Art. Previously Haley Goldman worked in several departments of the U.S. Holocaust Memorial Museum, primarily concentrating on audience research. Her research priorities include the long-term impact of museum visits and investigation of free-choice learning in new media environments.

Michael J. Hannafin is the Charles H. Wheatley-Georgia research alliance eminent scholar in technology-enhanced learning, professor of instructional technology, and director of the Learning and Performance Support Laboratory (LPSL) at the University of Georgia (USA) — positions he has held since 1995. After earning his doctorate in educational technology from Arizona State University in 1981, he held academic positions at the University of Colorado, Penn State University and Florida State University, and directed centers at both Penn State and Florida State. As director of the LPSL, he provides leadership for and supports the efforts of several university research scientists, technical support staff, and graduate students as they identify, pursue, and implement R&D initiatives related to learning and performing with technology. He has served as principal investigator (PI) or co-PI on research funded by the U.S. National Science Foundation, Department of Education, and the Department of Defense as well as through several private foundations. Dr. Hannafin has published more than 100 journal articles, textbooks, chapters and reviews, and has earned international awards for both journal articles and textbooks. He has given numerous invited addresses throughout the world related to research focusing on developing and testing frameworks for the design of student-centered learning environments. His current work includes a focus on evidence-based inquiry — the use of direct evidence featuring real-time video capture, coding, and analysis of performance data in assessing the effectiveness of learning environments.

Susan Hazan is currently curator of new media and head of the Internet Office at the Israel Museum, Jerusalem (since 1992), identifying, and implementing electronic architectures for the gallery, and outreach programs. Selected projects include QuickTime, VRML environments, video conferencing, an interactive, and online school curriculum, (museum@school), as well as the comprehensive institutional Web site in English,

Hebrew, Russian and Arabic. Her master's and PhD research at Goldsmiths College, University of London in Media and Communications focuses on electronic architectures in the contemporary museum. Hazan has published numerous publications on new media in education, art, and museums and regularly presents at international conferences.

Janette R. Hill, PhD, is an associate professor of instructional technology in the College of Education, University of Georgia (UGA) (USA). Dr. Hill holds a BA in communications from the University of North Florida in Jacksonville (1988), and an MSLS in library and information science from The Florida State University (1990), where she also completed a PhD in instructional systems design in 1995. Her research areas include community building in virtual environments, resource based learning, and the use of information technologies for purposes of learning. Dr. Hill teaches undergraduate and graduate level courses in the instructional technology program at UGA.

Andrew Hodges was born in London in 1949 and studied mathematics at the University of Cambridge. His postgraduate and postdoctoral work was with Sir Roger Penrose in the development of twistor theory, applicable to problems in fundamental physics. He is now attached to the Mathematical Institute at the University of Oxford (UK) and is a lecturer at Wadham College, Oxford. His interest in the computing pioneer Alan Turing developed partly through mathematics. His *magnum opus* is the definitive biography entitled *Alan Turing: The Enigma*.

Anja Hoffmann holds a diploma degree in media system design from the University of Applied Sciences in Darmstadt, Germany. As a member of the Digital Storytelling Group at the Computer Graphics Centre, she has conducted research in this area since 1998. Her interests concentrate on interactive digital storytelling and interaction design for augmented and mixed reality applications. She has devised concepts and interfaces for digital systems, which allow users to create and experience interactive narratives. Her understanding of human cognition, technology and design led her to focus on user-centred knowledge media design.

Ido Iurgel has a master's degree in philosophy, social psychology and linguistics from the Ruhr-University of Bochum, and a master's ("Diplom") in computer sciences from the Technische Universität Darmstadt. His special interest is in the integration of humanistic sciences and computer science. From 1995 to 1996, he worked in Brazil, his native country, in the industrial application of computer graphics. In 1996, he became a member of the research group "Phenomenology and Contemporary French Philosophy," led by Professor B. Waldenfels at the University of Bochum. From 1996 to 1999, he was granted a postgraduate research scholarship from the German Research Council ("Deutsche Forschungsgemeinschaft") and worked on philosophy of emotion and metaphor. From 1999 on, he was a research assistant at the Department of Digital Storytelling at the ZGDV in Darmstadt, where he has been developing graphics and AI of virtual characters. He finalized his degree in computer sciences with a master about virtual personalities in a mixed virtual/real human dialogue group. Since 2002, he has been a full time scientist in the department, working within several projects that employ virtual humans for education and entertainment.

Billie J. Jones is assistant director for writing support at Shippensburg University (USA), in Shippensburg, Pennsylvania. She completed her doctoral work at Bowling Green State University in 1998, and has been teaching writing in Pennsylvania since then. Her training in rhetorical analysis has prepared her to look at subjects as diverse as trauma narratives, memorials and museums, including their Web presences. In all of this work, she treats these diverse subjects as rhetorical texts that make meaning through their words and design.

Andreas Lorenz completed his master's degree in computer science at the University of Kaiserslautern (Germany) in 2001, and joined the research group "Information in Context" at the Fraunhofer-Institute for Applied Information Technology in Sankt Augustin (Germany) in spring 2002. He is a research associate and commenced his PhD in the research field of multi-agent systems. His further research interests include user-adaptive systems, mobile and nomadic systems, evolutionary algorithms, and software engineering. He was responsible for software-design and implementation in the LISTEN-project.

Daniel Tan Teck Meng graduated with a BSc (Honors) from the National University of Singapore, majoring in zoology. He is currently a senior manager (Exhibition Group) at the Singapore Science Centre. Prior to this, he was chief operations officer of ChainFusion Ltd., a software development house specializing in the development of enterprise level Web applications for major multinational corporations. Earlier, he had spect five years as a senior scientific officer at the Singapore Science Centre, developing life sciences programs and organizing various scientific exhibitions and national scientific competitions.

Dianna Newman is associate professor at the University at Albany (SUNY) and director of The Evaluation Consortium at Albany. She has served on the board of directors for the American Evaluation Association; assisted in writing the Guiding Principles for Evaluators which serve as the professional guidelines for practice, and is currently on the national Joint Committee for Standards in Evaluation. She has served as evaluator for several federal and state funded technology-based curriculum integration grants and is currently developing an innovative model of evaluation that will document systems change resulting from technology-based curriculum integration in K-12 and higher education settings.

Joan C. Nordbotten is an associate professor in information and media science at the University of Bergen, Norway. Her main interests include multimedia database management, HCI (human computer interaction), and Web exhibit construction. She teaches and has published numerous papers on these topics. She has also functioned as a judge in several Web exhibit competitions. She currently heads the "Virtual Exhibits on Demand" project, documented at http://nordbotten.ifi.uib.no/VirtualMuseum/VMwebSite/ VEDweb-site.htm, in collaboration with Bergen Museum. The project has funding from the Norwegian Research Council to develop tools for integration and presentation of multimedia data from multiple museum sources.

Teresa Numerico (PhD, history of science) teaches humanities computing courses at the Philosophy Faculty, Computing for Museums in the Science Museums Management master's courses of the University of Bologna (Italy) and new media theory and techniques at the University of Salerno. She co-edited (with A. Vespignani) *Informatica per le Scienze Umanistiche* (Computer Science for Humanities Studies, Mulino, 2003), has published various papers on the history and philosophy of computer science and is about to publish a book on *Alan Turing and Machine Intelligence*. She has also worked as a business development and marketing manager for different media companies. Currently she is a visiting researcher at London South Bank University, having been awarded a Leverhulme Fellowship.

Paolo Paolini is a full professor at the Polytechnic of Milan, Italy, and a lecturer at the University of Italian Switzerland. He has a master's degree and a PhD in computer science from the University of California in Los Angeles (UCLA). He has been active in the following research fields: database modeling and systems, document modeling, hypertext and multimedia models, multimedia authoring systems, design methods for the Web and multi-channel applications. He is the scientific coordinator of NET-LAB, a network of laboratories dedicated to researching advanced communication by means of new technologies. NET-LAB includes HOC-LAB (Polytechnic of Milan, Milan and Como campuses), TEC-LAB (University of Italian Switzerland) and SET-LAB (University of Lecce, Italy).

Renata Piazzalunga is currently president of the Information Technology Research Institute, Brazil. She received her BS in architecture and urbanization from Universidade de São Paulo (USP) in 1991 and her MSc in urban design, also from USP in 1998. She received her doctoral degree in cyberspace architecture at USP. She is currently engaged in researching how the information society can influence the way of creating spaces and representations in architecture. She is also engaged in researching interactive computer-based learning environments involving cognition systems development for the Web.

Caterina Poggi is a PhD student in information engineering at the Polytechnic of Milan, Italy. She graduated in communication science at the University of Italian Switzerland in Lugano. Her research interests focus on multimedia applications for edutainment, specifically 3-D virtual worlds for education, virtual

Hannu Salmi has been working since 1984 at Heureka, Finnish Science Centre, currently as the head of research and development. He completed his PhD at the University of Helsinki, Faculty of Education (1993). The main theme of his research has been informal learning, motivation and science education. Recently he has been in charge of several European projects related to public understanding of science and research of scientific literacy. These projects focus on developing new types of ICT-based educational solutions and open learning environments

David T. Schaller is principal and founding partner of Educational Web Adventures, LLP (USA), based in St. Paul, MN. He is responsible for the overall creative direction of the company and the perpetual quest for the sweet spot where learning theory, Web technology, and fun meet. Dave has over 15 years of experience in natural history and social science interpretation, working in print, exhibit, and Web media. In recent years he has led Eduweb's research and evaluation efforts, publishing several papers and frequently presenting at museum conferences. Dave holds an MA in geography and museum studies from the University of Minnesota and a BA in humanities from Macalester College.

Oliver Schneider holds a diploma degree in television technology from the University of Applied Sciences, Wiesbaden, Germany. His special research interest are narrative and authoring environments. From 1991 to 2000 he worked as a sound engineer and from 1996 to 2000 he worked also as a camera assistant. Additionally he had been a trainer for camera assistants from 1997 to 2000. From 1999 to 2000 he was responsible for a film compositing software as an application engineer. Since 2000, he is a full time scientist at the Department of Digital Storytelling at the ZGDV Darmstadt e.V. (Germany) and from 2002 he has held a supplementary teaching position for sound engineering at the University of Applied Sciences in Ulm/Germany.

Jim Spadaccini, Founder of Ideum (USA) (www.ideum.com), has managed and directed a wide range of media exhibits and Web sites for a diverse set of clients. Former director of Interactive Media at the Exploratorium in San Francisco, Jim was a major contributor, lead designer, and manager of the three-time Webby award-winning Exploratorium Web site (Best Science Site 1997-1999). He has received a Smithsonian Computerworld Award (1999) and an Association of Science and Technology Centers Award for Innovation (2000). Jim taught for San Francisco State's Multimedia Studies Program from 1995 to 2003 and more recently has been teaching for the Cultural Resource Management Program at University of Victoria, British Columbia.

Marcus Specht is a post doctoral researcher at the Fraunhofer-Institute for Applied Information Technology, Germany. He is head of the Mobile Knowledge Group and coordinator of the European project RAFT on field trip support for mobile collaboration. He received an MS in psychology and a PhD in adaptive learning technology from the University of Trier (Germany). He has rich experience in intelligent tutoring systems and the integration of ITS and Web-based tutoring adaptive hypermedia, ITS (ELM-ART, InterBook, AST) and ubiquitous e-learning. His main research interests are adaptive learning and training systems, knowledge management, contextualized computing, and intelligent interfaces. He coordinated the technical development of a highly scalable e-learning platform for design and architecture in the EU-founded WINDS project and was project leader of the LISTEN project on audio augmented environments.

Ramesh Srinivasan is a doctoral candidate at the Harvard Graduate School of Design (USA). His research is focused on the interactions between digital media, communities, education and learning, and the built environment. He has published his research in several conference proceedings and journals including the *Journal of Knowledge Management*, *Journal of Digital Libraries*, and *Journal of Urban Technology* (upcoming). Srinivasan holds an MS degree from the MIT Media Lab and a BS from Stanford University. He is the designer and creator of several interactive media projects including PhotoGlas, a Web-based news photomontage system; Village Voice, a digital community system that employed dynamic collage and community-driven architectures to connect Somali refugees within the Boston area to re-connect and preserve their histories and cultures; Public Body, a physical installation created for the pedestrians within a tunnel in downtown Boston, and Tribal Peace, an intertribal media system designed to share knowledge, and cultural narratives across a dispersed set of Native American reservations.

Andreas Zimmermann is a research associate and PhD candidate at the Fraunhofer-Institute for Applied Information Technology in Sankt Augustin (Germany). He received an MS in computer science from the University of Kaiserslautern (Germany) in 2000. After one year of business work he joined the research group "Information in Context" to acquire his doctoral degree in artificial intelligence. His further research interests include areas like user modeling, personalization, contextualization and nomadic systems. He was responsible for user modeling and implementation of context-awareness in the EU-founded LISTEN-project.

Index

Symbols

3D graphics 309
3D modeling 142

A

accessibility 151, 275
accessible museum Web sites 368
accountability 97
active learning 128, 236
active zones 233
actual visit 282
adaptability 273
adaptive annotation 54
adaptive
    content 274
    hiding 54
    navigation support 54
    ordering 54
    presentation 54
adaptivity 273
AHA Project 277
analogy 173
animation 128
antenna 382
art 94

artificial intelligence 206
artistic interventions 97
Association of Science-Technology
    Centers 369
asynchronous 95
audio streaming 141
augmentation layer 54
augmented reality 205, 257
authenticity 179
authoring process 193
avatar 312

B

bar-coded ticket 283
Berkeley Museum of Paleontology 373
Best of the Web competition 367
blogs 100, 155
born digital resources 385
brochureware 369
broker 55

C

captions 133
case studies 371
casual information seekers 228
Cats! exhibit 373

Center for Informal Learning and Schools
    69
Cité des Sciences et de l'Industrie 283
cognition 195
cognitive and affective domains 176
collaborations 99
collaborative filtering 274
color coding 137
community 155, 384
community knowledge 95
computer graphics 157
computer information points 382
constructivist 128, 195, 295
    learning 164
consumer knowledge 55
content 147
    databases 228
    personalization 394
    repository 375
    -based filtering 274
    -management system 52
context parameters 57
contextual model of learning 28
cookies 274
creative play 168
cultural institutions 69
culture 94
customization 273
cyborg 96

D

Dana Centre 383
data modeling 139
databases 152
decision trees 151
designing Interfaces 260
digital simulations 172
distributed authorship 100
DNA 435
domain
    knowledge 55
    model 55
dramaturgy 193
DVD 97
dynamic exhibits 236

E

e-community 276
e-learning
    center 229
    environments 394
    site 229
e-mail 155
education 367
educational programs 394
edutainment 191
effectiveness 128
Electronic Guidebook Research Project 284
electronic interactives 164
ELENA project 278
emotion 195
empowerment 97
entertainment 194
Epact database 377
ethical implications 409
evidence 156
exhiblets 381
experimentation 94
expert users 168
expertise knowledge 56
Exploratorium 69, 94, 284, 396

F

feminist critique 96
formal e-learning 277
forums 155
frame of mind 262
Franklin Institute 368
free-choice learning 28, 164

G

game 237
global climate change 86, 428
goal based scenario 183
Gopher 372
graphical user interface 128

H

Haraway, D. 96
home page 373

HTML (hypertext markup language) 374
Hubble Space Telescope 397
human
    cognition 173
    element 275
    -computer interaction 144
hyperbolic magnification 144
hyperlinked story 230
hypertext 98

I

icons 145
immersion 193
implementation 128
In Touch project 282
individual interest 277
individualized exploration 394
informal
    e-learning 278
    learning environments 111
information
    architecture 164, 217
    brokering 55
    browsing 230
    dissemination 230
    filtering 52
    gathering 230
    items 56
    presentation 128, 246
    retrieval 243
    seekers 229
Ingenious project 280
innovative
    interface 97
    systems 95
inquiry process 172
integrated databases 244
integration schema 245
intelligent agents 225
intent participation 70
interaction 216
interactive
    exhibit 236, 397
    experience 192
    learning 228
    museum exhibits 164

    site 229
    software technologies 398
    storytelling 193
    videoconferencing 293
interactives 97, 382
interactivity 128, 158
interface 144, 222, 268
International Council of Museums 370
Internet users 229
InTouch 383

J

JANET academic network 380
Jodi Mattes Access Award 368

K

K-12 293
knowledge 94, 199
    delivery 147
    -based society 414

L

La Cité des Science et de l'Industrie 369,
    396
learning 29
    communities 155, 295
    environment 192
    methods 191
    styles 394
    -by-doing 236
lifelong learners 281
live demonstrations 395
Live@Exploratorium 80, 397
localization 152
location model 54
log analysis 348
London Science Museum 282

M

magic lens 130
magnifying glass 233
marketing 194
mediation approach 369
meta-narratives 95
metadata 375

microsite 384
mixed reality 206
MMS 383
modernity 94
multi-tiered exhibits 231
multidisciplinary approach 395
multimedia 164, 199
museum
 displays 367
 exhibit 231
 learning collaborative 72
 mediators 395
Museum of Science & Industry 369
Museum of the History of Science 376
museum-based learning 70
MyExploratorium 284

N

narrative 178
National Grid for Learning 388
National Museum of Ethnology 286
National Museum of Science & Industry
 380
National Science Foundation 372
Natural History Museum 380
Natural History Museum of Los Angeles
 County 371
navigating 143
navigation bar 231
networked events 397
NMSI Picture Library 383
NOF-digitise program 384

O

on-demand exhibits 239
online
 activities 396
 audience 400
 collaborations 95
 communities 276
 conferences 394
 interactives 164
Online Register of Scientific Instruments
 378
ontology 220
Oppenheimer, F. 94

organization 147
origins 80

P

PDA (personal digital assistant) 283
pedagogic 94
pedagogical attainments 396
personal agenda 285
personalization 53, 151, 273, 383
physical
 phenomena 165
 principle 171
point to point videoconferencing 305
popular culture 97
post-visit experience 282
pre-visit phase 285
previous knowledge 276
problem solving 230
public awareness 413
public understanding of research 395

Q

qualitative 133
quantitative 134

R

ratings 155
real world learning 149
real-case scenarios 396
real-life 140
real-time demonstrations 394
resource-based learning 111
rhetoric 3
role-play game 171
rule-based filtering 274

S

SAGRES system 278
San Francisco Exploratorium 368
Science and Society Picture Library 384
science centers 111, 348, 424
Science Learning Network 368
science museum 381
science museum education 368
Science Museum of Minnesota 368

science museum Web sites 367
scientific
      discourse 95
      discoveries 395
      imagery 424
      method 180
screen design 150
searching 150
semantic matching 241
simulated realities 257
simulating 101
simulation 139, 172, 238
single-tiered exhibits 231
site map 231, 373
situated knowledge 96
situational interest 277
SMS 383
social
      aspects 207
      information filtering agents 217
      issues 410
      policy 175
      relevance 175
Solar Max 427
space model 54
spherical magnification 143
Spyglass 130
STEM Project 381
stereotypes 63
storyboards 396
storytelling 192
student-centered learning 147
subsequent experience 277
Sun-Earth Viewer 431
surface phenomenon 172
symbolic intervention 97
synchronous 95
systematized knowledge 94

T

teacher's guides 149
technological innovations 409
text databases 152
text Only 369
text-mining 348
textbook 231

thumbnail index 244
toolglass 130
toolkits 386
topical debates 414
training 200
TryScience 368
Turing, A. 386
Turing machine demonstration 388

U

underlying principle 172
understanding 177
University of Oxford 376
usability 164, 412
user
      friendly 144
      interface 144
      model 55
      profile 273

V

video games 164
virtual
      characters 199
      classroom 309
      communities 155
      devices 140
      dimension 253
      exhibit 229
      experiments 238
      galleries 371
      library museums pages 370
      museum 98, 266, 389
      Museum of Computing 370
      projects 238
      Science Center 28
      science centers 2, 29
      science museums 7
      tour 233
      visitors 396
      worlds 195
Visite Plus service 283
visualise 95
voting 155
VRML 98, 382